CSR, Sustainability, Ethics & Governance

Series Editors

Samuel O. Idowu, London Metropolitan University, Calcutta House, London, UK

René Schmidpeter, Cologne Business School, Cologne, Germany

In recent years the discussion concerning the relation between business and society has made immense strides. This has in turn led to a broad academic and practical discussion on innovative management concepts, such as Corporate Social Responsibility, Corporate Governance and Sustainability Management. This series offers a comprehensive overview of the latest theoretical and empirical research and provides sound concepts for sustainable business strategies. In order to do so, it combines the insights of leading researchers and thinkers in the fields of management theory and the social sciences – and from all over the world, thus contributing to the interdisciplinary and intercultural discussion on the role of business in society. The underlying intention of this series is to help solve the world's most challenging problems by developing new management concepts that create value for business and society alike. In order to support those managers, researchers and students who are pursuing sustainable business approaches for our common future, the series offers them access to cutting-edge management approaches.

CSR, Sustainability, Ethics & Governance is accepted by the Norwegian Register for Scientific Journals, Series and Publishers, maintained and operated by the Norwegian Social Science Data Services (NSD)

Samuel O. Idowu • Stephen Vertigans
Editors

Sustainability in Global Companies

Theory and Practice

Springer

Editors
Samuel O. Idowu
London Metropolitan University
London, UK

Stephen Vertigans
Robert Gordon University
Aberdeen, UK

ISSN 2196-7075 ISSN 2196-7083 (electronic)
CSR, Sustainability, Ethics & Governance
ISBN 978-3-031-77973-2 ISBN 978-3-031-77971-8 (eBook)
https://doi.org/10.1007/978-3-031-77971-8

This Springer imprint is published by the registered company Springer Nature Switzerland AG
The registered company address is: Gewerbestrasse 11, 6330 Cham, Switzerland

Foreword

As more and more industries are being held accountable for finding a balance between economic growth and environmental stewardship, globally, the discourse on sustainability has grown exponentially in both dimension and depth over the past few years. Our global challenge is the dual one of human development and environmental sustainability for our love of freedom, our regard for our precious natural heritage, and our concern for patterns of historical injustice that persist. In that case, this is a much-needed book and it sets the parameters of what can be examined in how sustainability is being embedded in the everyday business of companies across a range of sectors in the global economy.

Sectors such as aviation, banking, energy, financial services, food, manufacturing, technology, transportation, and tourism, as discussed in this book, are central to the sustainability debate so there is clearly scope for greater research engagement with the topic. These are high resource-consuming sectors that at the same time have scope for bringing in change through new practices. This book's collection of chapters explores how companies in these industry sectors are addressing the challenges of sustainability, situated within the lens of Sustainable Development Goals (SDGs) of the United Nations with a horizon of 2030.

The contributors to this volume offer insights into the way companies in various parts of the world conceptualize and operationalize sustainability. This book seeks to develop a guide for an understanding of international business and sustainability by theory and empirical evidence. This inclusive volume provides readers with fresh perspectives on the diverse set of activities involved in addressing sustainability across broad facility types, from manufacturing to commercial buildings, to products and services.

This book will be a precious asset for students, academics, corporate leaders, policymakers, and anyone interested in sustainable development, combining both practical case studies and theoretical expositions in the chapters.

Current questions that this book answers include where top businesses are heading in sustainability, the sectors best placed to create big breakthroughs, and the major supply chain shifts that can reduce carbon footprints. Moreover, the book discusses how corporations promote sustainable consumption, the impact of corporate social responsibility initiatives, and the competitive advantages for global corporations resulting from sustainable management.

This is complemented by the book's discussions of top-down/quick-impact measures used in developing and developed countries, and by the counterpoint perspective of effectiveness at the international, national, and local levels (and in the implementation of these measures) from the point of view of the unsustainable system of Western corporations. It also aims to investigate the strategies and concepts behind the theoretical foundation of sustainability and helps the reader to understand the theories behind the problems and solutions by focusing on company models.

I am confident that this book will not only enhance comprehension but also inspire actions toward achieving a sustainable present and future for future generations.

Embark on this expedition of exploration and education—together we will uncover the various and innovative implementations of sustainability taking place worldwide. The editors of this book, Samuel O. Idowu and Stephen Vertigans, deserve recognition for their significant contribution to the ongoing discourse on sustainability and its increasing importance for the future of global business. Through their efforts, they have shared the valuable insights of the contributors in this book, adding to the collective wisdom on the subject.

Faculty of Economic Sciences,
Lucian Blaga University of Sibiu,
Sibiu, Romania
June 2024

Cristina-Roxana Tănăsescu

Preface

This book focuses on how sustainability is practiced in companies in different sectors in the global economy—examples include aviation, banking, energy, financial services, food, manufacturing, technology, transportation, tourism, and many more. These corporate sectors are at the forefront in the debate about business and sustainability both in their usage of finite resources and in their potential to deliver meaningful solutions. Consequently, we asked contributors to focus on how sustainability is understood and practiced in different companies around the globe in terms of the UN Sustainable Development Goals 2030. Chapters incorporated both theoretical and applied examples that can help improve levels of insights into international business approaches to sustainability providing opportunities to learn and disseminate what is delivering short-, medium-, and long-term solutions and problems. These insights were able to incorporate the spectrum of activities required in order to achieve sustainability through industrial processes, supply chains, and consumer behavior.

The multidisciplinary nature of the fundamental issues meant we were able to generate interest from academics and practitioners from many disciplines who have knowledge and/or experiences of sustainability and international business theories, practices, and processes. Contributors were able to either explore practices or enhance understanding through theoretical exposition. The intended readership is expected to be students and academics from across related disciplines, corporations, policymakers, and members of societies who study or have an interest in sustainable development. It is hoped that the end product of this project meets these requirements satisfactorily and the information needs of our global readers in every corner of our great planet. After all, it's all about our common future and how sustainability would assure the future and this generation and those that would exist on planet Earth when they are no longer here.

London, UK
Aberdeen, UK
June 2024

Samuel O. Idowu
Stephen Vertigans

Sustainability in Global Companies: An Introduction

International attention and critical focus on business sustainable narratives, policies, strategies, intentions, actions, and impacts have grown considerably in recent years. Although concerns remain about the motivations behind why some businesses undertake and promote their sustainable practices, ultimately they are being judged on impacts and not public relations. In essence, if companies are not implementing strategies that are improving their sustainability, then climate change targets are not going to be met, irrespective of government and civil society interventions. Across growing numbers of public and private organizations, strategic frameworks are being devised, websites declare commitment to sustainable practices that often revolve around supply chains and technology, and governance and regulatory arrangements are self-assessed to be exemplary while stakeholder engagement is reported to be at the cornerstone of policy development and implementation. Although there has been an increase in reporting and regulatory bodies that can provide credence to annual sustainable statements, much of the public pronouncements lack independent monitoring and evaluation. This collection of chapters is designed to help unpick actions from the pervasive strategic rhetoric that surrounds sustainability and underpinning concepts such as balanced scorecard, corporate social responsibility (CSR), environmental, social, and governance (ESG) criteria, and triple bottom line (TBL). This collection aims to help the reader be better informed about the range of policies and practices being adopted around the world and the extent to which they are contributing to a reduction in their environmental footprint, improving social well-being and securing economic viability.

The international focus within the book is emphasized by sections on activities and policies in African, Asian, European, and South American companies with studies of various locations, including Bolivia, Bulgaria, England, Ghana, Greece, India, Latvia, Nigeria, Poland, Portugal, Romania, Turkey, and Uzbekistan. This broad range provides the collection with a spread of insights, experiences, and businesses that help challenge the Western dominance in the sustainable development literature. In so doing, both similarities and differences within and between different parts of the world are noticeable. Overall, there is a sense of progress across the chapters. However, as the following introduction to the chapters indicates, this optimism is guarded, recognizing the low starting point and the urgent requirement for progress to be accelerated.

How and what people learn about sustainability is increasingly recognized to be an area requiring greater insights. This area is recognized in Chap. "Educating for Change and Collaboration: Sharing University Students' Stakeholder Expectations About Improving the Impact of Retailing Companies' Sustainable and Socially Responsible Practices" by Vesala-Varttala, Nunes, and Garrido's study of university students in Portugal, representing 20 nationalities, and their analysis of CSR practices in retailing. Through learning about student expectations surrounding stakeholders and development, the authors are able to highlight the role universities can play in educating and empowering students that will enable impactful change within companies.

The roles of family businesses can often be overlooked with sustainable practices. In Chap. "The Importance of the Family Businesses for Fundamental Sustainable Improvements: Cases from Bulgaria, Romania and Uzbekistan," Pavlov, Puiu, and Alieva explain in their comparative study of family companies in Bulgaria, Romania, and Uzbekistan, such intergenerational organizations can both play a key role within regional sustainable development and extend their reach into international markets. And, as with the preceding chapter, Pavlov et al. also include students in the collection of data, contributing to the knowledge base and providing opportunities to inspire future generations of researchers and advocates for sustainable practices.

Across concerns about climate change and constraints on sustainable development, the aviation industry is at the forefront. Panagiotopoulos takes on these issues in Chap. "Sustainability in Aviation: CSR & Air Transportation" within his exploration of international initiatives and a case study of the UAE-based Emirates Airlines. Alongside the better-known negative impacts of aviation, the chapter examines attempts to improve energy management, air emissions, and noise control. Through critical analysis, the chapter concludes with practices that indicate where further improvements can be made in the global aviation industry.

Implementation, disclosure, and compliance of Baltic listed and private companies' ESF practices are the focal point of Zumente, Bistrova, and Lāce's study in Chap. "Frontier Markets Compliance with Global ESG Practice: Sample of Baltic Countries." While acknowledging that scope for improvements remains, the authors point to a positive shift in corporate trends. Through examining the extent to which ESG practices have been adopted, the importance of stakeholders and the ways in which internal and external pressures, such as from the financial market, encourage adoption.

Moving south of the Baltic countries, Mazur-Wierzbicka and Cierniak-Emerych examine, in Chap. "Practical Implementation of the Idea of Sustainable Development: The Perspective of Polish Companies," how sustainable development is understood and practiced in Poland. They identify the significance of legislation to protect the environment and promote sustainable development. Similar to Zumente et al.'s analysis of the scope for greater improvements in the Baltic states, Mazur-Wierzbicka and Cierniak-Emerych also believe that progress is being made in

Poland, notably around CSR programs, environmental investments, and increasing social and ecological awareness within Polish businesses.

Food and agricultural sustainability is integral to efforts for sustainable development which Abreu, Oliveira, and Tomé discuss in Chap. "Agriculture Sustainability: Strategies of the Olive Oil in Portugal" with relation to olive oil production in Portugal. By focusing on the implementation of sustainable practices within such a traditional industry, they are able to highlight how olive oil producers can both reduce their environmental footprint and improve the quality of their produce.

One of the long-running debates within sustainable development discourse surrounds the extent to which organizations can be trusted to self-regulate or whether national and international monitoring and punishment can be meaningfully implemented. The EU taxonomy for sustainable economic activities and the transition to a greener economy is a recent attempt by the European Union toward regulation that Frostenson examines in Chap. "The EU Taxonomy for Sustainable Activities: What It Really Implies for Companies." This intentionally transparent self-assessment tool has resulted in multi-layered direct and indirect consequences. At the micro-level, while recognizing benefits, Frostenson concludes that based on knowledge available and the timespan, the anticipated outcome of the chain of action from transparency through to transformation remains to be proven.

International organizations such as the EU understandably are anticipated to have an important role in the direction of sustainability within Europe. By comparison, more localized developments are relatively neglected. Yet, as Lincoln and Croad detailed fieldwork discovered, in Chap. "Impact of Developing Iconic Cultural Centres on Sustainability and Socio-Economic Development: A Case Study of the Turner Contemporary Margate," local cultural initiatives such as the Turner Contemporary also have important roles to play in encouraging wider engagement and broader inclusion in the drive to promote well-being and social and economic benefits. This initiative was to help transform impressions of the town but Lincoln and Croad conclude that without broader and more extensive policies, the growing appeal of Margate would exclude those with least resources to reside and participate, thereby threatening the sustainability of economic regeneration and achievement of localized SDGs.

Recognition of the importance for how CSR strategies and practices are communicated has grown. Organizations have recognized both the need to engage and inform stakeholders and transmit integral messages while seeking to avoid overpromotion and accusations of "greenwash" that accompanied a number of CSR campaigns that were associated with disproportionate corporate promotion rather than meaningful outcomes. Arguably part of the challenge facing organizations has been that the demographics of professionals responsible for communication strategies have informed communications through medium with which they are most comfortable. An obvious consequence is that they are not communicating with people who do not share the same preferences or demographic backgrounds such as age. Hence, communication strategies led by older professionals tend to miss younger generations. The emergence of a gamification approach as outlined by Sağlam and Ertürk in Chap. "Gamification Approach in CSR Communication: Case of Jotun Türkiye" provides innovative ways to connect to hitherto hard to reach groups. Through semi-structured interviews, the authors outline how gamification can improve CSR stakeholder engagement and improve the global brand as a platform for green creativity that is compatible with organizational sustainability goals.

Sustainability in African contexts attracts attention in part because produce, goods, and services are closely interwoven within international supply chains and financial markets. The cocoa and chocolate industry in Ghana is, as Sarpong discusses in Chap. "Sustainability Issues in the Cocoa and Chocolate Industry: Building a Promising Future for a Beleaguered Industry," at the forefront of challenges between supply and demand with the former affected by deteriorating weather conditions and illegal mining and the latter associated with the expanding consumer demand for chocolate in international markets. Facing challenges surrounding sustainability, Sarpong's fieldwork with groups of farmers identifies the significance of national, regional, and international initiatives albeit that hitherto have often made slow progress. Without more concerted and more specific commitments from international companies and governments. Sarpong argues the viability and sustainability of cocoa farmers will continue to be vulnerable to weather and market fluctuations.

International businesses also feature in Chap. "A Multi-Sector Assessment of Sustainability and Socially Responsible Practices of International Businesses in Nigeria" and Lincoln and Diamond's multi-sector assessment of practices in Nigeria. The chapter considers international business practices and determines which are making the biggest contributions to environmental sustainability and CSR practices. This analysis acknowledges that although there are international contributions toward sustainable development, the rigor and extent of these applications were weaker than approaches in their "home" countries. Moreover, generic applications failed to take into account local diversities and requirements, which causes problems for affected communities at odds with the international businesses' declared outcomes.

The growing presence of Nigeria within the African economy is heavily informed by the oil and gas industry which remains one of the most controversial globally and more regionally because of activities within the Niger Delta. In Chap. "Conceptions of Sustainability and their Impact on Practices in Oil and Gas Corporations Operating in the Niger Delta," Richard-Osu and Buckler focus on perceptions of sustainability and impact in this controversial region. Through interviewing HR and sustainability professionals, they explain how the lack of common understanding about the meaning of sustainability creates problems for identification, implementation, and ultimately achievements toward SDGs. Identifying shared understandings and shifting organizational culture to allow for SDGs to be better integrated throughout hierarchies within the energy sector would contribute toward the better implementation of sustainable business practices.

Notoriety concerning transnational actions in the Niger Delta and the economic dominance of the oil and gas sector in Nigeria has tended to deflect attention from other significant sectors of the national economy. Okafor, Asogwa, Agbata, and Okaro broaden the focus, in Chap. "Sustainability Reporting Practices in Nigeria: A

Study of Firms Quoted on the Nigerian Exchange Group," to look at the limited level of sustainability reporting and disclosures among companies quoted on the Nigerian Exchange. Noting some variations in the levels of reporting, with environmental disclosures recorded least across sectors, the authors raised concerns about the poor levels of reporting. They conclude with optimism by highlighting how the adoption of IFRS S1 and S2 is expected to lead to improvements in sustainability disclosures across the majority of Nigerian companies.

Returning the focus to the significance of education and relocating to India, in Chap. "Inclusion of Sustainability into Business Education: Understanding the Student's Awareness, Knowledge, Attitude and Beliefs—A Study Based on India," Ghosh investigates university student perceptions about the inclusion of sustainability into business education. Analysis of the data collected from students through an online questionnaire led the author to discover the support and commitment for sustainability and the achievement of SDGs. Despite this commitment, Ghosh identified a distinction between reported attitudes and barriers to the inclusion of sustainability in the curricula which included student and staff disinterest in further teaching. Moreover, a "commitment gap" was found between students' open support and their unwillingness to undertake sustainable practices which raises wider concerns about the extent to which business leaders of the future will deliver the actions required in order to meet forthcoming sustainability crises.

The book concludes in Chap. "Enhancing Sustainable Supply Chains in Bolivia: Aligning CSR Practices with the SDGS" with emphasis on the importance of supply chains within sustainability more broadly and within Bolivia in particular. Herbas-Torrico, Arandia-Tavera, Zurita-Lara, and Leoni-Peinado examine shifts in sustainable supply chains. Progress was reported against a number of SDGs with progress restricted by limited investment and initiatives. Achieving sustainability continues to face challenges with complexities to be overcome. Participants recognized how social and political insecurities impact on supply chains and the delivery of SDGs. This recognition is a fitting way to conclude this multi-layered collection of chapters, namely that the environments in which businesses operate are influenced by much more than economic factors. Consequently, business approaches to sustainability and the achievement of SDGs must both align their related approaches within sustainable supply chains and broader developments to factor in the wider consequences of their actions and the regional, national, and international social and environmental factors which influence the capacity and capabilities for sustainable development.

London Metropolitan University
London, UK
Samuel O. Idowu

Robert Gordon University
Aberdeen, UK
Stephen Vertigans

Acknowledgments

We wish to express our gratitude to all those fantastic contributors who have helped us in meeting the objectives of this great book. It is evident that without their hard work, it would have been impossible for us to have this end product. The lead editor expresses his gratitude to all these great individuals who have contributed to the success of the book, and it is something he cherishes immensely. The two editors are grateful to those of you who have stood by them with their impressive chapters in the book. Many of them despite their busy schedules felt obliged to help them in putting together this very fine informative addition to the literature on how sustainability is being embedded in global companies. We are also grateful to Professor Cristina-Roxana Tănăsescu for her great Foreword to the book. And my good friends and brothers—Michael Soda, Samson Owokoniran, Lawrence Ojelade, Yinka Odunlami, Alfread K Idowu, Layo Akinfala, Emmanuel Ayo Osho, Samson Nejo, Kehinde Jolaoso, and Olusheyi Bada. I am also grateful to my friend and colleague Dammer Sahi. Thank you all.

We would like to express our gratitude to our publishing team at Springer headed by the Executive Director—Books, Christian Rauscher, Barbara Bethke, Dr Prashanth Mahagaonkar—Executive Editor, Dr Bijita Majumdar, Anette Weiss, Rocio Torresgrosa, and other members of the publishing team who have supported this project and all my other projects.

Finally, we apologize for any errors or omissions that may appear anywhere in the book; please be assured that no harm was intended to anybody.

Contents

About the Editors

Samuel O. Idowu, PhD, is Professor of Accounting and Corporate Social Responsibility at the Guildhall School of Business and Law, London Metropolitan University, where he is currently the course leader for the MSc Corporate Social Responsibility and Sustainability and Advanced Diploma in Professional Development (ADPD) Corporate Social Responsibility and Sustainability. Samuel is also Professor of CSR and Sustainability at Nanjing University of Finance and Economics, China. He is a fellow member of the Chartered Governance Institute, a fellow of the Royal Society of Arts, a Liveryman of the Worshipful Company of Chartered Governance Institute, and a named freeman of the City of London. He is the CEO and President of the Global Corporate Governance Institute an international network of CSR scholars. Samuel has published over 100 articles in both professional and academic journals and contributed chapters in several edited books and is the Editor-in-Chief of three major global reference books by Springer—the *Encyclopedia of Corporate Social Responsibility* (ECSR), the *Dictionary of Corporate Social Responsibility* (DCSR), and the *Encyclopaedia of Sustainable Management* (ESM), and he is Series Editor for Springer's *CSR, Sustainability, Ethics and Governance* books. Samuel is the Editor-in-Chief of the *International Journal of Corporate Social Responsibility* and the *American Journal of Economics and Business Administration.* Samuel has been in academia for more than 30 years winning one of the Highly Commended Awards of Emerald Literati Network Awards for Excellence in 2008 and 2014. In 2010, one of his edited books was placed in 18th position out of 40 top Sustainability books by Cambridge University Programme for Sustainability Leadership, and in 2016, one of his books won the outstanding Business Reference Book of the year of the American Library Association. In 2018, he won a CSR Leadership Award in Cologne, Germany, and in 2019, he won the 101 Most Impactful CSR Leaders Award in Mumbai, India. Samuel is on the Editorial Advisory Boards of the *International Journal of Business Administration.* He has been researching in the field of CSR since 1983 and has attended and presented papers at several national and international conferences and workshops on CSR. Samuel has made a number of keynote speeches at international conferences and workshops and has written the *foreword* to a number of leading books in the field of CSR and sustainable development. And he has examined a few PhD theses in the UK, Australia, South Africa, the Netherlands, and New Zealand.

Stephen Vertigans is a sociologist and Emeritus Professor at Robert Gordon University, Aberdeen, UK. His research interests include corporate social responsibility, political violence, and community development. At present, he is researching into geothermal systems in East African low-income housing. In his wide-ranging publications, Stephen applies a sociological approach to advance knowledge and understanding of social, political, economic, and cultural processes that shape individual and community behaviors and activities. Positioning CSR within these wider processes can provide better insights into the successes and failures of CSR policies. Recently, he founded idid (International Direct Impact Donations www.ididonations.com) to help improve income streams and sustainability for Kenyan community-based organizations.

Contributors

Rute Abreu Instituto Politécnico da Guarda, CICF-IPCA, CISeD-IPV, CITUR, Guarda, Portugal

Amaka E. Agbata Department of Accountancy, Nnamdi Azikiwe University, Awka, Anambra State, Nigeria

Deniza Alieva Management Development Institute of Singapore in Tashkent, Tashkent, Uzbekistan

Carlos Alejandro Arandia-Tavera Alicorp Bolivia, Santa Cruz de la Sierra, Bolivia

Ilja Arefjevs BA School of Business and Finance, Riga, Latvia

Chinedu U. Asogwa Department of Accountancy, Nnamdi Azikiwe University, Awka, Anambra State, Nigeria

Jūlija Bistrova Faculty of Engineering Economics and Management, Riga Technical University, Riga, Latvia

Sarah Buckler Robert Gordon University, Aberdeen, UK

Anna Cierniak-Emerych Wrocław University of Economics and Business, Wrocław, Poland

Jane Croad Robert Kennedy College, Zurich, Switzerland

Brendhain Diamond University of Liverpool, Liverpool, UK

Egemen Ertürk Maritime Faculty, Dokuz Eylul University, İzmir, Turkey

Magnus Frostenson School of Business, Society and Engineering, Mälardalen University, Västerås, Sweden

Susana Garrido University of Coimbra, Coimbra, Portugal

Sumona Ghosh Post Graduate and Research Department of Commerce, St. Xavier's College, (Autonomous) Kolkata, Kolkata, India

Boris Christian Herbas-Torrico Tecnologico de Monterrey, Guadalajara, Mexico

Panagiotopoulos Φ. Ioannis Department of Business Administration, Mediterranean College - Derby University, Athens, Greece

Nataļja Lāce Faculty of Engineering Economics and Management, Riga Technical University, Riga, Latvia

Pedro Alejandro Leoni-Peinado Universidad de Cadiz, Cadiz, Spain

Adebimpe Adesua Lincoln University of Liverpool, Liverpool, UK

Ewa Mazur-Wierzbicka University of Szczecin, Szczecin, Poland

Carmina Nunes University of Coimbra, Coimbra, Portugal

Gloria O. Okafor Department of Accountancy, Nnamdi Azikiwe University, Awka, Anambra State, Nigeria

Sunday C. Okaro Department of Accountancy, Nnamdi Azikiwe University, Awka, Anambra State, Nigeria

Ermelinda Oliveira Instituto Politécnico da Guarda, Guarda, Portugal

Daniel Pavlov Entrepreneurship Center of the University of Ruse "Angel Kanchev", Ruse, Bulgaria

Silvia Puiu Faculty of Economics and Business Administration, University of Craiova, Craiova, Romania

Oluchukwu Jane Richard-Osu Robert Gordon University, Aberdeen, UK

Bayram Bilge Sağlam Maritime Faculty, Dokuz Eylul University, İzmir, Turkey

Sam Sarpong University of Central Lancashire, Preston, UK

Francisco Tomé Instituto Politécnico da Guarda, Guarda, Portugal

Tanja Vesala-Varttala Haaga-Helia University of Applied Sciences, Helsinki, Finland

Ilze Zumente Faculty of Engineering Economics and Management, Riga Technical University, Riga, Latvia

Pamela Mirtha Zurita-Lara YPFB Refinación S.A., Santa Cruz de la Sierra, Bolivia

Part I
Sustainability in European Companies

Finland with Portugal
Bulgaria with Romania and Uzbekistan
Greece
Latvia
Poland
Portugal
Sweden/Norway
United Kingdom
Turkey

Educating for Change and Collaboration: Sharing University Students' Stakeholder Expectations About Improving the Impact of Retailing Companies' Sustainable and Socially Responsible Practices

Tanja Vesala-Varttala, Carmina Nunes, and Susana Garrido

1 Introduction

With the rise of purpose-driven organisations, universities are also feeling the pressure to take on a larger role in making an impact in society and the world, looking to transform their teaching and learning to "secure a livable future for humanity and the natural world" (Sterling, 2021). To address and explore this growing need, The Journal of University Teaching and Learning practice (JUTLP) recently launched a call for proposals for a special issue on "Purpose-Driven Learning in Universities: Its Role and Social Impact". So far, research has focused on new types of purpose-driven higher education worldviews, learning frameworks, pedagogies, and case studies fostering sustainability transitions and developing learner competencies to drive large-scale social changes (e.g. Andrew et al., 2023; Brundiers et al., 2021; Haski-Leventhal, 2020; López-López et al., 2021; Lozano & Barreiro-Gen, 2021). Increasing attention has also been given to the importance of collaboration between universities, industry, and other stakeholders, focusing on both successes and practical challenges of establishing and maintaining long-term partnerships (e.g. Rybnicek & Königsgruber, 2019).

Although many inspiring case studies of universities' purpose-driven efforts to create societal impact do exist, higher education is still subject to criticism of not being responsive and adaptive enough (Stewart et al., 2022). In response to this criticism, an increasing body of case studies and theoretical research has emerged

T. Vesala-Varttala (✉)
Haaga-Helia University of Applied Sciences, Helsinki, Finland
e-mail: tanja.vesala-varttala@haaga-helia.fi

C. Nunes · S. Garrido
University of Coimbra, Coimbra, Portugal
e-mail: carmina.nunes@fe.uc.pt; garrido.susana@fe.uc.pt

S. O. Idowu, S. Vertigans (eds.), *Sustainability in Global Companies*, CSR, Sustainability, Ethics & Governance, https://doi.org/10.1007/978-3-031-77971-8_1

discussing the growing role of universities in diverse "social innovation" ecosystems (Păunescu et al., 2022). As development steps in purpose-driven university research, learning, and teaching, the JUTLP editors suggest that universities increase their efforts to engage students in "changing their world" (Graham & Moir, 2022). They also point out that there is a need for universities to find ways to advance towards a more systemic approach requiring "critical reflexivity" (Cunliffe, 2016) and ability to re-evaluate our assumptions, interactions, and our way of being in the world in the light of moral and ethical considerations (Cunliffe, 2016; Sterling, 2021).

To foster the role of universities in driving future-oriented changes and improvements in sustainable development, university educators in business and management face the need to adopt more participatory pedagogical approaches that take learners out of classrooms to engage with businesses and nonprofit organisations and their stakeholder groups. The better higher education students and world-of-work organisations learn to analyse, share, and critically discuss their values, opinions, and expectations, the easier it is for them to form employment relationships, collaborate, and commit to shared goals and value creation. For the present study, we created an international and collaborative learning project which encouraged students to critically explore their stakeholder expectations towards retailing companies with an aim to propose future-oriented development ideas to improve the impact of retailers' CSR and sustainability practices. As the retail sector is the largest private employer in Europe and serves some 450 million EU consumers daily (European Commission, 2024), it is already actively present in university students' everyday lives. Because of retailing companies' regular contacts with high numbers of employees and consumers and through their extensive local and global supply and value chains, the sector is well placed to form impactful partnerships and drive impactful changes towards a more sustainable future.

The main objective of this chapter is to analyse and share the stakeholder perceptions, expectations, and development ideas of international university students regarding the current sustainability and social responsibility activities of a selection of retailing companies. Our study focuses on students' stakeholder views and ideas because university students are an important stakeholder group for retailing companies now and in the future. In addition to being customers and current and future employees, they also represent potential future managers, entrepreneurs, and decision-makers in the field. This study is an addition to the body of scholarly work exploring university students' perceptions, competences, attitudes, values, and expectations regarding sustainable development and CSR, with emphasis on the role and power of stakeholders in improving the impact of companies' sustainability and CSR activities.

2 CSR, Shared Value Creation, and Stakeholder Theory

In the current business environment, Corporate Social Responsibility (CSR), Triple Bottom Line (TBL), and ISO 26000 are fundamental conceptual frameworks guiding organisations towards more sustainable and responsible operations. For the purposes of the present study, we provide a concise exploration of the interrelations of these frameworks, focusing on core subjects and dimensions of social responsibility and their alignment with Sustainable Development Goals (SDGs). Over the last two decades, CSR has changed the way the world does business (Hou et al., 2016). CSR consists of three key concepts: "corporate", which covers the business; "social", which refers to society; and "responsibility", which implies an ethical relationship between corporations and communities around them. In its broadest sense, CSR encompasses all the constituent elements and stakeholders of an organisation's operations, along with the society in which it operates (Carroll, 1979).

The strategic importance of CSR has grown steadily in recent years, as more businesses recognise the inevitability of their commitment to positively impacting society and protecting the environment while pursuing profitability. CSR obligates companies to operate ethically and improve stakeholders' quality of life, contributing to the well-being of employees, communities, and society at large (Carroll, 1991). The concept encompasses economic, legal, ethical, and philanthropic or discretionary responsibilities, urging companies to look beyond profit-making and consider broader economic, societal, and environmental impacts.

CSR connects the benefits of organisations with the needs of their stakeholders. Stakeholder theory (Freeman et al., 2004) starts with the premise that business practices are intertwined with values. The theory urges businesses to clarify and crystallise to themselves and their stakeholder groups the collective understanding of their purpose and the value they generate to key stakeholders. Furthermore, managers are encouraged to clearly articulate their preferred approach to conducting business, particularly the relationships they aim to establish with stakeholders to fulfil their objectives. All stakeholders should comprehend the complex process of value creation and exchange (Freeman et al., 2004). Stakeholder theory proposes that treating all stakeholders well creates synergy (Parmar et al., 2010). In other words, how a company treats its customers influences the attitudes and behaviour of its employees, and how a company behaves towards the communities in which it operates influences the attitudes and behaviour of its suppliers and customers (Cording et al., 2014).

Falck and Heblich (2007) explain how by strategically committing to CSR, companies can do well while doing good: they can both generate profit and make the world a better place. In a similar vein, Porter and Kramer (2006) introduce the seminal concept of shared value, advocating for integrating CSR into business strategy to reach competitive advantage by creating economic value while addressing societal needs. In fact, some businesses have embraced creating shared value (CSV) as an alternative to traditional CSR, focusing on generating economic value that simultaneously addresses societal needs and challenges. Shared value creation encompasses corporate strategies and initiatives that enhance competitiveness while fostering social and economic progress within the communities where the company operates. Shared value creation can be achieved through three main avenues: redefining productivity along the value chain, innovating products and markets, and fostering local cluster developments (Porter & Kramer, 2011). What is needed is a proposition in line of which the entire organisation delivers, experiences, and captures value while managing costs and risks to attain superior outcomes for stakeholders. The principles of sustainable value creation diverge from conventional CSR approaches in their holistic focus on long-term value creation and broad stakeholder collaboration (Chandler, 2020).

Orlitzky et al. (2003) examine the relationship between CSR and monetary performance, finding a shy but positive correlation between the two. Margolis et al. (2007) also conducted a meta-analysis on the association between CSR and financial performance, suggesting its contingency on various factors and advocating for meticulous further research. However, according to Murray et al. (2010), the sustainability challenges companies encounter are so severe and multifaceted that CSR-related collaboration between governmental bodies, NGOs, businesses, and civil society is inevitably needed. Thus, fostering collaboration among a broad group of stakeholders emerges as a logical strategy to advance towards a more responsible and environmentally sustainable global economy.

A contemporary perspective articulated by Haski-Leventhal (2022) underscores the importance of strategic CSR, which can be construed as a comprehensive and forward-looking approach to corporate responsibilities. This approach integrates stakeholders and ethical considerations into business practices while leveraging the firm's resources and brand to address societal and environmental concerns. A holistic approach to CSR implies integrating every facet of the company's operations and guiding decision-making with ethical and responsible criteria. Strategic CSR advocates shifting from short-term gains to long-term sustainability and solutions. Embracing broad responsibilities implies that companies acknowledge a duty beyond profit maximisation. Stakeholder integration emphasises the importance of considering all individuals who engage with or are impacted by the company's activities. Ethical behaviour underscores the necessity for moral courage in making decisions, particularly in challenging and controversial circumstances. Finally, leveraging the company's resources and brand to address societal and environmental challenges entails skillful use of competitive advantages and strategic positioning to define the company's role in society and the community.

The stakeholder theory has become progressively relevant as organisations recognise the importance of engaging with and considering the interests of all those affected by their operations. This broader perspective goes beyond the traditional focus on shareholders and includes employees, customers, suppliers, the local community, and other relevant stakeholder groups. Stakeholder theory, as proposed by Freeman (1984), emphasises the interrelation between businesses and their stakeholders, including employees, customers, suppliers, communities, governments, and the environment. Clarkson (1995) underlines the need to be specific about a company's relations with different stakeholder groups when assessing corporate social performance and presents a stakeholder framework that distinguishes between internal stakeholders (e.g., employees, shareholders) and external stakeholders (e.g. customers, suppliers, community).

The stakeholder theory framework helps organisations identify and prioritise stakeholder groups based on their importance. Mitchell et al. (1997) propose a stakeholder model that examines stakeholders' power, legitimacy, and urgency in determining their salience to the organisation. This model guides organisations in prioritising their stakeholder engagement efforts based on their level of influence and importance. From a management angle, Donaldson and Preston (1995) provide a comprehensive overview of stakeholder theory, discussing its implications for corporate governance and management practices. They highlight the importance of considering stakeholders' interests in organisational decision-making. In exploring the consequences of stakeholder management for organisational survival, reputation, and success, Freeman et al. (2007) argue that companies that effectively engage with stakeholders are better positioned to achieve long-term sustainability and competitive advantage. Effective stakeholder engagement is essential for building trust, managing risks, and enhancing organisational legitimacy. Companies must acknowledge and address stakeholders' interests and expectations to achieve long-term sustainability.

3 Triple Bottom Line (TBL)

The TBL model has become an influential approach to sustainable business all over the world (Svensson & Wagner, 2015), as it takes a holistic view on sustainability transition by focusing on the interrelated impacts of environmental quality, social well-being, and wider economic benefits. The TBL framework, introduced by Elkington (1997), expands the scope of corporate performance evaluation beyond financial metrics to include social and environmental dimensions. It advocates for businesses to measure success based on three pillars: People, Planet, and Profit. By considering their operations' social and environmental implications, organisations can pursue sustainable development while maintaining financial viability.

Hart and Milstein (2003) developed a framework for creating sustainable value based on the TBL approach, discussing the challenges and opportunities in integrating non-financial metrics into organisational performance evaluation. Similarly, Gray (2006) examined how social, environmental, and sustainability reporting contribute to creating value from a Triple-Bottom-Line (TBL) perspective, and Schaltegger et al. (2017) explored different business cases for engaging with sustainability, including the TBL approach. These studies emphasise the importance of aligning business strategies with social and environmental objectives to achieve long-term success and discuss how companies can derive competitive advantage by integrating sustainability into their core strategies and operations.

The TBL framework has also been criticised for being challenging to measure and quantify, making it hard for companies to fully assess their impact on social and environmental issues. Despite criticisms, proponents argue that conceptualising and organising their operations in terms of CSR and the TBL is crucial for businesses to thrive in the twenty-first century (Pereira & Martins, 2021). It is, however, important to critically evaluate the potential drawbacks and limitations of the application of CSR and TBL approaches to assessing business performance and competitiveness (Du et al., 2010; Henriques & Richardson, 2004). To develop the approach to sustainable performance, Lozano (2015) takes a holistic perspective on corporate sustainability drivers, including the TBL framework, and examines the factors influencing organisations' adoption of sustainability practices and their impact on business performance and societal outcomes. Such critical research is needed because balancing economic, social, and environmental objectives within CSR and TBL frameworks may lead to trade-offs and conflicting priorities. For example, pursuing environmental sustainability goals might entail higher costs or compromise short-term financial performance (Elkington, 1997). Furthermore, despite their holistic approach, CSR and TBL frameworks may still face criticism for their focus on short-term financial returns. This emphasis on financial metrics could overshadow long-term sustainability considerations and broader societal impacts (Hart & Milstein, 2003). Moreover, implementing robust CSR and TBL measurement systems requires significant resources, including time, expertise, and financial investment. Small and medium-sized enterprises (SMEs) and resource-constrained organisations may struggle to allocate resources for comprehensive sustainability reporting (Burritt & Schaltegger, 2010).

4 ISO 26000

ISO 26000 is an international standard for corporate social responsibility that helps organisations address their social responsibilities and guides them to implement CSR practices effectively. Focusing on the core subject of organisational governance, the Clause 6 of the ISO 26000 standard outlines six additional and interrelated core subjects of social responsibility: human rights, labour practices, the environment, fair operating practices, consumer issues, and community involvement and development (Fig. 1).

These core subjects provide a comprehensive framework for organisations to assess social responsibility across different aspects of their operations. Altogether, the ISO 26000 standard offers more than 450 recommendations related to its main principles and core subjects.

The principles and recommendations of ISO 26000 help organisations contribute to SDGs. For example, Subclause 6.4.4.2 of ISO 26000 under the core subject "Labour practices" states that to pursue the goal of SDG1 (No poverty), "[a]n organization should pay wages at least adequate for the needs of workers and their families. In doing so, it should take into account the general level of wages in the country, the cost of living, social security benefits and the relative living standards of other social groups". As another example, subclause 6.7.3.2 of ISO 26000 under the core

Fig. 1 Holistic approach to organisational governance (adapted from ISO 26000, 2018)

subject "Consumer issues" contributes towards SDG 12 (Responsible consumption and production) and states the following: "[w]hen communicating with consumers, an organization should provide complete, accurate, and understandable information that can be compared in official or commonly used languages at the point of sale and according to applicable regulations" (ISO 26000, 2018).

Marrewijk (2003) discusses the conceptual foundations of CSR and corporate sustainability, highlighting key themes such as stakeholder engagement, ethical behaviour, and environmental stewardship. These concepts align closely with the core subjects outlined in ISO 26000, emphasising the importance of holistic approaches to social responsibility. Matten and Moon (2008) propose a conceptual framework for understanding implicit and explicit CSR dimensions, encompassing visible actions and underlying values and principles. This framework helps identify critical issues such as ethical conduct, transparency, and accountability, which are central to ISO 26000 core subjects. Bansal and DesJardine (2014) argue that business sustainability is increasingly becoming a strategic imperative for organisations, driven by environmental concerns, stakeholder pressures, and competitive dynamics. ISO 26000 provides a framework for integrating sustainability principles into organisational behaviour, influencing decision-making processes and stakeholder relations. Despite implementation challenges, ISO 26000 can potentially drive positive changes in organisational behaviour, fostering transparency, accountability, and stakeholder engagement and facilitating the accomplishment of SDGs.

To sum up, CSR, stakeholder theory, TBL, and ISO 26000 align with the broader objectives of sustainable development as outlined by the SDGs. These frameworks promote responsible business practices contributing to poverty alleviation, social equity, environmental conservation, and economic prosperity. By addressing core subjects such as human rights, labour practices, environmental sustainability, fair operating practices, consumer issues, and community development, organisations can actively contribute to achieving SDGs. By integrating these frameworks into their operations, organisations can create shared value for stakeholders while advancing the broader goals of sustainable development.

Against this theoretical background, we set out to explore university students' CSR and sustainability expectations with the help of these research questions:

RQ_1: What kind of perceptions and expectations do international higher education students have about retailing companies' CSR and sustainability activities in view of the three dimensions of the triple-bottom-line (TBL) framework?

RQ_2: What kind of development ideas do students propose in view of the six core social responsibility subjects of ISO 26000?

5 Methods and Data

This research applied a qualitative approach utilising content and thematic analysis techniques to investigate how international higher education students perceive Corporate Social Responsibility (CSR) and sustainability practices of their chosen retailing companies and how they would develop those activities. The higher education students worked in multicultural and multigender groups to produce 36 project reports (R1–36) with CSR development suggestions, focusing on the activities of 17 retailing companies altogether. Some companies had just one student group and some had 2–3 student groups analysing their activities, CSR reports, websites, and other communication materials. We used a convenience sample of students from the Faculty of Economics at Coimbra University in Portugal. The selection was made by one of the researchers who teaches CSR, ethics, and sustainability to large numbers of international students. Convenience sampling is an efficient way to achieve a general overview of a phenomenon of interest, especially when it would otherwise be challenging to access informants and convince them to participate in long-term qualitative reflection and assessment (Peterson & Merunka, 2014).

The sample consisted of 166 young higher education students (of 20–25 years) who enrolled in the CSR and Business Ethics course at Coimbra University during the academic year of 2022–2023. These students represented 20 nationalities: Portuguese, Italian, Polish, Uruguayan, Bulgarian, Spanish, Czech, French, Slovakian, German, Swedish, Belgian, Norwegian, Ukrainian, Turkish, Arabian,

Slovakian, Finnish, Dutch, and Greek. The course attracted a diverse body of students, as Coimbra University hosts many Erasmus exchange students annually. The collaborative learning project was integrated with the Erasmus+ research and development project "LEARN&CHANGE—Collaborative Digital Storytelling for Sustainable Change", coordinated by the Haaga-Helia University of Applied Sciences, Finland.

University students and their stakeholder expectations towards retailers were chosen as the focus of the study because research into CSR often tends to take a corporate perspective instead of adopting a stakeholder perspective (Gupta & Pirsch, 2008; Ramasamy & Yeung, 2009). When implementing their CSR strategies, retailers must know their stakeholders' needs and understand how, for example, consumers or employees perceive and respond to their CSR activities. Consumer surveys show evidence of growing expectations towards retailers: they are expected to behave responsibly in their business practices and monitor the sustainability of their suppliers' activities. To improve the impact of CSR in the retail industry, companies should adopt a holistic approach, ranging from immediate business practices and employee concerns to supply chain practices and growing consumer pressures (Rahdari et al., 2020).

The teacher in charge of the CSR and Business Ethics course gave a group assignment to students, asking them to critically analyse websites and CSR reports of retailing companies of their choice. Additional data from news reports and stakeholder surveys or interviews could also be used. As case companies, most of the student groups chose companies representing the fast-moving consumer good industry: Lidl, Intermarché, Minipreço, Carrefour, Pingo Doce, Auchan, Continente, Mercadona, Aldi, and a Finnish retail chain S Group. Other students chose to analyse data from companies representing the fashion industry: 8000Kicks, H&M, Mango, and Zara. The rest of the groups analysed data from IKEA, McDonald's, and Litocar. For the present study, the companies were anonymised, and they are referred to as Company 1–17.

6 Content and Thematic Analysis

Based on their field research, students wrote team reports about the companies' CSR and sustainability practices, reflecting on their stakeholder expectations and proposing development ideas in the form of future-oriented action plans. The researchers of the present study analysed the students' reports (R1–36) to categorise their expectations and development ideas. Figure 2 illustrates the research process.

Qualitative content and thematic analysis methods were applied to the data set. Content analysis has a well-established history, and many methodological sources delineate its key tenets. Holsti (1968) regards content analysis as a technique for making inferences by systematically identifying specified characteristics of messages, and Krippendorff (1980) defined it as a research technique for making replicable and valid inferences from data that can then be discussed and interpreted in

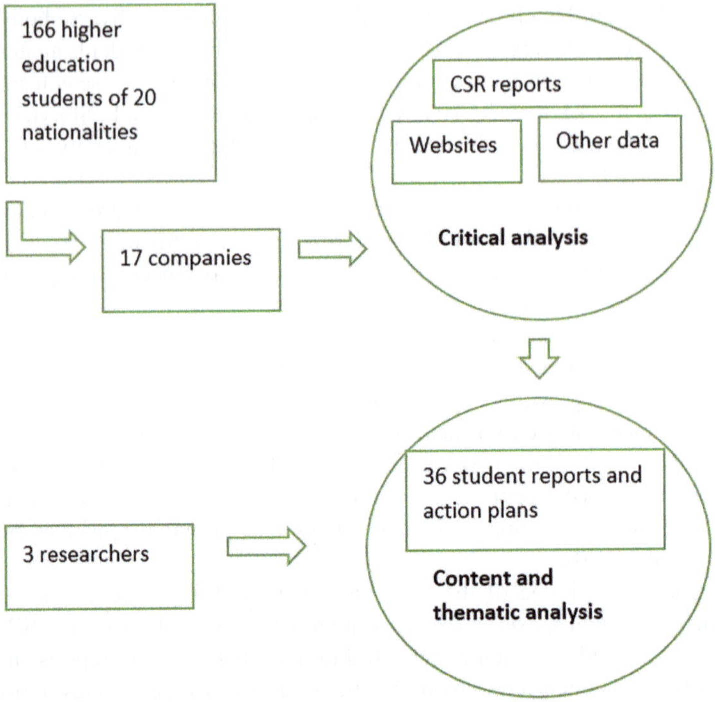

Fig. 2 Student sample, data gathering, and data analysis

connection with their context. Different process descriptions of the implementation of content analysis exist. We apply the steps suggested by Stempel (1989):

1. Formulation of the research question or objectives: By clearly stating the research questions or objectives, researchers ensure that the analysis focuses on relevant content.
2. Selection of communication content and data sample: Identifying relevant data to answer the research questions and determining the appropriate communication content.
3. Developing content categories: Content analysis is a research method that involves categorising data to answer specific research questions. The category system developed is the cornerstone of content analysis. Berelson (1952), for example, believed that the success of content analysis depends on the clarity and appropriateness of its categories. According to Chadwick et al. (1984), the categories should be mutually exclusive and a piece of content should belong to only one category. In practice, this may be difficult to achieve, but even with partly overlapping categories, or content with connections to several categories, it should be possible to decide to which category a piece of content primarily belongs. As stakeholder expectations about sustainability and CSR form a complex phenomenon with several dimensions that are holistically interrelated and

intertwined, it was sometimes challenging to apply single categorisations to given pieces of student reflections. In such cases, the primary categories were negotiated among the three researchers.

4. Confirming units of analysis: The choice of units of analysis depends on the data and research goals. To get an overview of the balance of student perceptions, we first organised their CSR expectations into three main categories according to the TBL. After that, we categorised students' CSR development ideas in more detail by classifying them according to ISO 26000 core subjects. The added value of this two-part analysis was that we gained further insight into the quality of students' development ideas. Due to the partly overlapping and intertwined nature of our data (the same piece of content spreading over several units of analysis), we grouped the ideas into thematic types and quantified them under their primary content category. In so doing, we were able to divide the data into specific subcategories, to demonstrate the breadth and variety of the development ideas, and to outline which categories and themes appeared as specific areas of interest among our sample of young international university students.
5. Preparing a coding schedule and checking inter-coder reliability: In content analysis, it is essential to provide training to the coders to ensure reliable coding. It is recommended to use multiple coders even in a small-scale study. The coders can check the inter-coder reliability between themselves, which helps enhance the quality and reliability of the analysis (Chadwick et al., 1984). In the present study, the researchers discussed and adjusted their coding decisions iteratively during the entire data analysis process.
6. Analysing the data: The research problem and units of analysis guide the data analysis, helping to highlight patterns and relationships to be explored and interpreted. In qualitative research, content analysis can intertwine with characteristics of thematic analysis, which allows for a meaningful amalgamation and thematisation of content categories (see, e.g., Braun & Clarke, 2006). Data-driven content or thematic analysis aims to convert recorded "raw" phenomena into data categories and themes, which can then be treated scientifically so that a body of knowledge may be built up. Theory-driven content or thematic analysis, in turn, starts with a set of theoretical categories and applies them to the data to arrive at a systematic understanding of leading trends, recurring themes, and key messages. In practice, the analysis process is often iterative and can include both data-driven exploration and theory-driven categorisation elements (see, e.g., Elo & Kyngäs, 2008). In line with the steps suggested by Stempel (1989), the analysis process is defined in Table 1.

7 Data Analysis and Key Results

To provide answers to our research questions, 36 student reports were analysed in two rounds and with the help of two main categories and nine units of analysis (Table 1). RQ1 was answered by categorising the data according to the dimensions

Table 1 Content and thematic analysis process

Steps	Characterisation
1. Formulation of the research questions	• RQ_1: What kind of perceptions and expectations do international higher education students have about retailing companies' CSR and sustainability activities in view of the dimensions of the triple-bottom-line (TBL) framework? • RQ_2: What kind of development ideas do students propose in view of the six core social responsibility subjects of ISO 26000?
2. Selection of communication content and data sample	• 166 higher education students' project reports and action plans (*n* = 36) with development ideas created based on their critical analysis of CSR reports, websites, and other data focused on given retailing companies' CSR and sustainability activities.
3. The first round of data analysis to test and develop the main content categories	The main categories used were: • Category 1—Triple-bottom-line dimensions • Category 2—Six core subjects of ISO 26000
4. Confirmation of the units of analysis	Based on the first round of data analysis, the following units of analysis were confirmed: Category 1—Triple-bottom-line dimensions: Unit 1—Environmental (ENV) Unit 2—Social (SOC) Unit 3—Economic (EC) Category 2—The six core subjects of ISO 26000: Unit 1—Human rights (HR) Unit 2—Labour practices (LP) Unit 3—Environment (Env) Unit 4—Fair operating practices (FOP) Unit 5—Consumer issues (CI) Unit 6—Community involvement and development (CID)
5.-6. Scheduling the second round of data analysis; checking inter-coder reliability; and summarising and discussing key results	The coding was performed in March–April 2024 by the three researchers who independently coded the units and continuously checked the inter-coder reliability of all categorisations. In the process of results interpretation, the content items were organised into thematic subcategories and tabled for presentation and discussion.

of TBL, further dividing each dimension into thematic subcategories of students' CSR development expectations (Table 2). RQ2 was answered by categorising the data according to the six ISO 26000 core subjects and grouping the students' ideas into development themes under each core subject (Table 3). The development ideas and their subgroups were quantified to illustrate their diversity and focus in relation to each core subject and development theme.

Table 2 Students' expectations categorised and thematised according to TBL

1. Environmental dimension	2. Social dimension	3. Economic dimension
1.1 Fostering circular economy	**2.1** Respect for basic human rights	**3.1** Offering jobs to the disadvantaged
1.2 Commitment to waste reduction	**2.2** Non-discrimination	**3.2** Fair pay and benefits
1.3 Sustainable sourcing	**2.3** Diversity, equity, and inclusion in the workplace	**3.3** Fair incentives to boost productivity and retention
1.4 Greener energy use	**2.4** Well-being at work	**3.4** Workforce planning to optimise workloads
1.5 Environmental and animal protection	**2.5** Opportunities for skills and career development	**3.5** Investing in employee healthcare and childcare
1.6 Fostering sustainable and plant-based food consumption	**2.6**. Fair benefits and performance evaluation	**3.6** Investing in skills and career development
1.7 Tracking environmental impacts in the value chain	**2.7** Open salary policies	**3.7** Innovations in technology, methods, tools, processes, and business models to boost sustainability
1.8 Transparent and unbiased communication with clear targets and KPIs	**2.8** Active stakeholder engagement and interaction	**3.8** Supporting local production and distribution
1.9 Inspections, audits, certifications, and labels to monitor sustainability	**2.9** Ethical code of conduct for stakeholders	**3.9** Investing in sustainable product lines and production methods
1.10 Environmental community initiatives and strategic partnerships to widen their impacts	**2.10** Consumer education, guidance, and support	**3.10** Using company brand and resources to set benchmarks and best practices for responsible business
	2.11 Sustainable and fair marketing	**3.11**. Building long-term partnerships with stakeholders to create shared value and broad economic, societal, and environmental benefits
	2.12 Professional stakeholder communications	
	2.13 Purpose-driven and regular stakeholder events	
	2.14 Promotion of sustainable and healthy lifestyles	
	2.15 Fostering community involvement through volunteering and donations	

7.1 Data Analysis According to TBL

Table 2 demonstrates students' CSR and sustainability expectations towards the 17 retailing companies, categorised according to the TBL framework. During our data analysis, we identified 36 thematic subcategories of environmental, social, and economic expectations.

In terms of **the environmental (ENV) dimension**, students highlighted the importance of committing to circular economy in its various forms. The importance of transparent and reliable information and progress reports related to environmental objectives and targets was highlighted by many student teams. Students also expected companies to commit to responsible sourcing, waste and pollution management, green energy, animal welfare, and biodiversity preservation. Strategic partnerships with environmental organisations were seen to play an important role in generating more comprehensive environmental benefits and impacts. The

Table 3 Students' development ideas categorised and thematised according to ISO 26000

Core subject	Thematic subcategories arising from the data
1 Human rights (11 development ideas)	1.1 Fundamental rights (6) 1.2 Inclusion and non-discrimination (5)
2 Labour practices (35 development ideas)	2.1 Inclusive and fair working conditions (17) 2.2 Transparent and fair benefits (13) 2.3 Human development and training (5)
3 The environment (44 development ideas)	3.1 Sustainable resource use and waste management (27) 3.2 Sustainable energy and transport (7) 3.3 Protection of the environment and biodiversity (5) 3.4 Tracking progress of environmental impact (5)
4 Fair operating practices (34 development ideas)	4.1 Promoting responsibility in the value chain (19) 4.2 Active stakeholder interaction (10) 4.3 Responsible business innovation (5)
5 Consumer issues (30 development ideas)	5.1 Guiding and educating consumers for sustainable consumption (18) 5.2 Consumer support, services, and interaction (8) 5.3 Fair marketing (4)
6 Community involvement and development (40 development ideas)	6.1 Community development (14) 6.2 Charity partnerships, volunteering, and donations (12) 6.3 Promotion of public health (9) 6.4 Education and skills development (3) 6.5 Technology development and access (2)

following quotations demonstrate students' expectations about the environmental dimension of retailing companies' activities:

> There is an evident problem concerning the lack of data and transparency. We are still unable to find information on the environmental objectives achieved. They are also missing regular reporting of production volume, even though there is evidence of overproduction and waste. Data collection processes should be improved to enable evaluation of the working circumstances, environmentalist use of materials and reasonable amount of production (Company 16, R8)

> The company should improve their solutions of electricity storage and use of renewable energy. (Company 4, R14)

> The company should adopt strategies to reduce plastics and packaging. (Company 8, R31).

> They are producing a lot of environmental waste, which is meeting a lot of critique from their customers. All of their packaging should be biodegradable. (Company 3, R15)

> Important actions to take are reducing water waste by adopting more efficient processes and machines. (Company 15, R13)

> As part of their recycling plans, the company could conduct a waste audit to identify the types of recyclable waste generated by their clients, inform them of the results, and show them sustainable alternatives to what could happen in the future if they continue wasting. (Company 2, R22)

> Concerning environmental campaigns, we suggest the creation of a refill station for water bottles in their stores and, also, the implementation of reverse vending machines that would allow the exchange of plastic bottles for discount vouchers. (Company 11, R7)

The social (SOC) dimension of the TBL framework was reflected in students' reports in the form of expectations concerning human rights, non-discrimination, and diversity, equity, and inclusion in both societal and work contexts. Companies were also expected to contribute actively to employee well-being and skills and career development. Active and dialogical stakeholder interaction both in the workplace and among all stakeholders was also seen as crucial to successful and impactful CSR. The need for fair and trustworthy marketing and communication practices was highlighted throughout students' reports. According to students' CSR expectations, partnering with charities and other relevant organisations to invite and channel volunteering and donations benefiting local communities was an important part of retailing companies' societal role. The following key ideas from students' reports exemplify their expectations:

> We believe the company can do more by offering equal opportunities to every worker ensuring the possibility to progress in their career, promoting a responsible and diverse recruitment, making sure to have a team including minorities and people in disadvantage, guaranteeing an inclusive working environment. (Company 5, R6)

> The company could engage with the communities in which it operates by supporting local charities and events. (Company 7, R35)

> We recommend community philanthropy, which valorizes diversity, unity, and sustainable and value-driven growth. (Company 12, R5)

> Engaging in local activism and donating to institutions whose work is associated with children with eating disorders and other diseases related to poor nutrition is also a proposal to consider. (Company 8, R17)

The economic (EC) dimension, in terms of investments and long-term economic impacts, was brought up frequently in student reports, as students were well aware that developing sustainability and CSR requires tangible economic resources and investments in courageous and innovative solutions. However, students felt that those investments would translate to profits and broader economic impacts in the form of increased employee commitment, loyalty, and productivity as well as improved stakeholder collaboration and shared value creation throughout the value chain. All in all, students expected retailing companies to use their brand, resources, and extensive stakeholder networks to set good examples and benefit society and the planet at large. Regarding the economic dimension, here are some illustrative examples:

> A big company with significant profits and ample funds should also shoulder social responsibility. Such a company should see that improving the lives of its workers and society at large is not only the right thing to do but also a means to generate growth and a way to stand out. (Company 5, R6)

> As a huge multinational company with significant presence in many countries, the company has a social responsibility to ensure that its employees are paid fairly and that its operations have a positive impact on the communities in which they operate. (Company 3, R30)

> When done right, CSR can be a huge uplift to the brand's recognition, leading to broader awareness and potentially increasing sales as well. (Company 9, R1)

We propose work-life balance policies such as allowing flexible working hours, offering the option of teleworking, giving the birthday off, or even offering childcare services. These measures make the employee feel more valued and therefore become more productive. (Company 13, R10)

Investments in innovative management measures are needed to keep track of stock and reduce food waste: inventory management systems such as smart shelves and other ways to optimise processes to monitor the expiry date of products. (Company 10, R27)

7.2 *Data Analysis According to ISO 26000*

Through the content and thematic analysis of students' CSR development ideas according to the core subjects of ISO 26000, it became evident that their development focus was most often directed towards the environment, followed by community involvement/development and labour practices. All in all, however, the data included diverse development ideas covering all six core subjects. The ideas were coded according to the six units of analysis. Altogether 194 development ideas were identified in the data. These were quantified according to type and thematised to form 20 subcategories (Table 3).

Students brought up issues relating to **human rights (HR)** in different contexts throughout the data, for instance in connection with discussing their expectations of employees' equal treatment and non-discrimination in the workplace (labour practices), the rights of consumers belonging to vulnerable groups (consumer issues), and the innovative management of food waste and the distribution of excess food to benefit poor families and others in need (community involvement and development, environment, and fair operating practices).

Beyond such specific contexts, students also discussed fundamental human rights expectations and development ideas that they felt fell within the sphere of retailing companies' CSR responsibilities, such as monitoring the abiding by human rights laws and regulations throughout the supply and value chain and using company resources to improve living conditions in general. The human rights reflections of our data were grouped under two main themes: fundamental human rights and social inclusion and non-discrimination. The following quotations illustrate students' expectations regarding retailing companies' activities related to human rights:

The living and working conditions and non-discrimination of employees are extremely important for consumers. Consequently, we think the company should be more concerned about inclusion and accessibility. The recruitment of disabled people should be prioritised as a way of giving the same work opportunities to the whole community and promote inclusion. (Company 11, R7)

If the company wants to take the rights of workers seriously, it has to take initiative. This should include a proper examination of their suppliers, setting benchmarks for living wages and supporting worker unions. Information is missing about the living wages, gender breakdown of employees, number of migrant workers or contract workers, and more. The data about employees can be evaluated to find out whether the work opportunities are the same for all and to check that not too many employees work for a minimum wage and that the company does not exploit immigrants. (Company 16, R8)

Among our sample of university students, **labour practices (LP)** were regarded as one of the key areas in need of attention and further development. One reason for this may be that many young university students work in retailing companies to finance their studies or at least know fellow students who do, which makes them interested in labour practices in the industry. Students' expectations and development ideas were most often focused on guaranteeing inclusive and fair working conditions, such as involving all employees in dialogue and decision-making; creating and maintaining a good work atmosphere and culture of mutual respect; and being open and transparent about data and progress indicators affecting the business and everyone's work.

In addition, students expected retailers to offer a diverse set of benefits adding value to employees and delivered on the basis of reward systems that are transparent and equal for all. The benefits mentioned included those related to well-being and health, work–life balance, and salary incentives based on high-quality service. As yet another important development theme, students proposed skills and career development and training opportunities for both staff and managers, including CSR trainings, accessibility trainings to help customers with special needs, and management trainings to increase the sense of community and shared commitment in the workplace. The following quotations illustrate the wide range of students' expectations:

> We want to create a workplace where everyone is treated fairly and equally, so we propose that the company conducts a salary audit to identify any discrepancies in pay based on gender, race, or other issues. The company should also encourage open communication and feedback from employees on how the company could improve its diversity and inclusion efforts. (Company 2, R22)

> The company could arrange 'Staff–Manager Roundtable' sessions to discuss current challenges and results of employee surveys. (Company 2, R20)

> Introducing the concept of 'Employee Voice,' involving employees in decision-making. (Company 10, R18)

> They could provide employees with training and courses to advance their careers and gain knowledge of the latest technologies and skills in the industry. (Company 11, R3)

> The company should put in place some ethical guidelines to assist each employee in considering the proper attitude to take, particularly when confronted with delicate situations that they may have to handle. (Company 10, R2)

The ISO 26000 core subject area that attracted the most diverse set of development ideas from students was **the environment (Env)**. Higher education students are clearly concerned about environmental matters and strongly believe that retailing companies should constantly work to increase the positive environmental impacts of their business activities through developing strategic CSR and sustainability partnerships and initiatives. The area inviting the most development ideas was sustainable sourcing, resource use, and waste management, with students proposing many kinds of practical ways to transfer to more sustainable materials and to reduce and manage waste.

Several ideas were also proposed related to increasing the use of sustainable energy and transport solutions and protecting and respecting the environment, animal welfare, and biodiversity. Finally, students emphasised the need for companies to determine concrete targets and KPIs for their environmental impacts and to openly keep publishing unbiased reports of their environmental progress. The following quotations illustrate students' development ideas:

> We propose using LED lights since they are smoother and energy efficient. Our primary suggestion is to bring in natural light, either by moving the staff break room to an area with windows or renovating the current breakroom. (Company 5, R19)

> Establish a responsible 'way to work' benefit by investing in yearly bus cards or bike acquisitions for employees. (Company 2, R20)

> Creating a repository of used plastic packaging inside the shopping center for customers to place their packaging there to be later recycled and transformed into new ones. (Company 5, R6)

> The use of disposable packaging can be reduced by introducing reusable containers for bulk products. (Company 6, R25)

> The next step could be to get the Oeko-Tex® Made in Green label, which also certifies that the product has been manufactured in environmentally friendly facilities under safe and socially responsible working conditions. (Company 1, R23)

> The company should assess the success of its sustainability efforts by establishing specific goals, keeping track of performance metrics, and carrying out routine assessments and audits. (Company 10, R26)

> The company could also develop a trade campaign, for example, for every kilo of plastic, glass, and metal deposited in the store's bins the client would receive a discount coupon (Company 2, R22)

Students' expectations and ideas concerning **fair operating practices (FOP)** were heavily focused on promoting and monitoring responsibility in the value chain, including negotiating fair deals with local farmers and suppliers, improving the transparency of supply chain sustainability, and creating criteria, standards, and partnerships with external certification bodies to extend the reach and impact of sustainable development activities throughout the value chain and beyond.

Students proposed ideas for active and dialogical stakeholder interaction and engagement, stressing the need to inquire about, monitor, and seek to fulfil stakeholders' expectations and needs as an integral part of strategic decision-making and business development and regeneration. Concrete ideas about innovative business solutions and investments contributing to sustainable development were also suggested in students' reports and action plans. The following quotations illustrate the students' ideas and proposals:

> We propose searching for one or more local farmers in every company's geographic area; working out a good communication concept to explain the main focus of the collaboration project to the farmers; negotiating the contract details with the farmers who want to participate; and including fair trade salary and high quality and organic product standards in the contract. (Company 6, R16)

> The company has already implemented various eco-friendly practices and some CSR strategies and analysis. However, this will not be enough unless they develop better communication channels/messages and transparency policies to make consumers and stakeholders feel involved, listened to and valued. (Company 1, R4)

> The company should pay more attention to the conditions in the countries they source from. Therefore, inspections should be increased and take place more regularly to ensure that the suppliers maintain sustainable working practices. (Company 17, R11)

> The company must implement a functioning communication network between multiple stakeholders; negotiate with them in regard to critical issues and seek voluntary agreement; anticipate their concerns and try to influence the stakeholder environment; and allocate resources in a manner consistent with stakeholder concerns. Start by strengthening ties between the company and the general public, by investing in initiatives that promote sustainability and social inclusion. (Company 7, R9)

In terms of **consumer issues (CI),** students expected retailing companies to be active in educating consumers to make more sustainable purchasing choices and guiding them towards more sustainable lifestyles. Students presented a great variety of ideas about how to inform and educate customers in practice, including creative ways to actively promote healthier and plant-based food and more sustainable products. Ideas were also proposed for making it impossible or hard for consumers to purchase products that harm the environment (e.g. not offering products containing plastics or increasing their price). Students also suggested making recycling and reuse solutions as well as trading in second-hand products easily available to consumers.

Students expected retailing companies to organise campaigns, workshops, and events to involve consumers in creating a more sustainable society and to help them find long-term collaboration partners to support them in sustained efforts of making their lives more sustainable. According to students, investing in building long-term customer relationships would increase customer loyalty and trust and improve the opportunities to create shared value. Students expected retailing companies to stop greenwashing, and to achieve this, they proposed that companies invest in professional and fair marketing and communication practices and expertise. These selected quotations illustrate students' ideas about how to increase consumer engagement and improve its impact:

> Educating consumers on how to avoid food waste. This could be done through information materials or campaigns that emphasize the importance of avoiding waste, encourage customers to make conscious shopping decisions, and guide them to store and handle their food properly. (Company 10, R27)

> They should promote a waste-free lifestyle and planetary diet. In this way, they can inspire consumers to participate in the company's goals towards a waste-free society. One way of engaging with the consumers is to start a dialogue on social media. (Company 15, R13)

> To guide and inform consumers, our proposal is the implementation of ecopoints in stores. Ecopoints are a fantastic way of recycling but sometimes it's difficult to find them in some areas. (Company 11, R24)

> The company should put an end to greenwashing and try to differentiate itself in the market through honesty. We suggest: 1) setting a realistic date for the end of production from plas-

tic materials; 2) termination of greenwashing and adopting 100% transparency regarding the materials used; 3) focusing sales campaigns on promoting products made from natural materials that are not harmful to the planet and decompose rapidly; 4) running educational campaigns of explaining the problems of polyester clothing production and promoting awareness of the problem. (Company 16, R28)

Our recommendations from the consumer point of view are: to be more honest and transparent; to address and take responsibility for the issues that the company has; to share a detailed series of action or objectives that they are planning to achieve; to listen to and communicate with their customers and stakeholders; and to focus on improving the channels through which they are keeping the consumers informed about sustainable implementations. (Company 1, R23)

Second only to the environment, students' expectations and development ideas were heavily focused on **community involvement and development (CID).** As retailing companies' business activities are integrated with people's everyday lives and they often occupy visible and prominent roles in communities, students also expected them to be actively involved in developing the communities around them. Students expected companies to invest in innovations and initiatives that promote sustainability and social inclusion and to form partnerships that leverage their impact on making the surrounding communities and ecosystems more sustainable and better equipped to serve the needs of all stakeholders. Students also expected the retailers to use their brand, size, and power to promote sustainable practices and social initiatives not only among citizens but also among business sector players. As one concrete form of community involvement, students expect retailing companies to be active in partnering with NGOs, nonprofits, and charities to arrange volunteering opportunities and donations to serve the less advantaged groups of the population and to contribute to environmental protection.

Students also felt that retailing companies should widen their community impact by sharing some responsibilities usually attributed to the public sector or municipal organisations, including responsibilities and partnerships contributing to the promotion of public health, education and skills development, and technology development and access. For example, students suggested that retailers invest in the promotion of healthy living by providing health information, sponsoring local fitness events, partnering with local farmers to offer natural and biological products, and supporting farmers' markets to sell healthy and sustainable products through joint campaigns, apps, promotions, and discounts. Further, students suggested that retailers could partner with local educators to improve the employability and skills development of young people. In terms of technology investments, students suggested that retailers continue to invest in electronic car charging stations and reverse vending machines to promote the green transition and circular economy.

It would be good of the company to get in touch with local universities and conduct a workshop on 'Healthy Eating on a Low Budget', bringing someone who works at the company and an influential person with expertise in dietetics as extra inspiration and motivation for students to participate in the session. For young adults, it is also important to provide educational campaigns on healthier and safer drinking culture. (Company 8, R33)

> To promote healthier lifestyles to their clients, the company could search for local farmers near each store and negotiate with them to offer natural and biological products at lower prices than regular ones. (Company 2, R22)

> In our opinion the company should continue with their project that involves the placing of electric car charging stations in car parks (Company 5, R6)

> The profit made on one specific day each month could be donated to a local social institution. This would strengthen the company's commitment to SDG1 No Poverty, SDG2 Zero Hunger, SDG10 Reduced Inequalities, and SDG16 Peace, Justice and Strong Institutions. (Company 8, R31)

> The company has an obligation to assist and interact with the communities in which it conducts business as a significant corporation. This can entail establishing fresh alliances with neighborhood nonprofits and charities, advocating for inclusion and diversity in the workplace, and funding local economic development projects. It can also help the community by building positive work conditions, improving work-life balance, enforcing ethical code of conduct, and making healthy food more accessible to everyone (e.g. for families on a budget or in financial distress). (Company 10, R26)

> The company could contribute to the establishment and operations of a job training and skills development center focused on the inclusion of vulnerable populations such as people with disabilities, immigrants, and low-income families facing barriers to employment and economic mobility. This could be done in collaboration with diverse vocational service providers. (Company 10, R27)

To sum up, the data analysis above was carried out by conducting two rounds of content and thematic analysis, the first with the help of TBL categorisations and the second by applying the organisational governance approach and six core subjects of ISO 26000. Through the analysis, we sought to illustrate the diversity and focus of our international students' stakeholder expectations concerning retailing companies' responsible business practices and impacts. Even though the ISO 26000 categorisations overlap with the more general TBL categorisations, their application helped further demonstrate and classify the different types of students' stakeholder concerns. The categorisations, thematic subcategories, and direct quotations from student reports illustrate the richness and abundance of students' ideas and the holistic nature of their thinking and perceptions. Even if not all development ideas proposed were immediately realisable, they are a powerful testimony to the growing stakeholder interest in and concern about the impact of companies' CSR and sustainability practices.

8 Discussion and Conclusion

We analysed 166 university students' CSR and sustainability expectations towards retailers in the light of TBL and ISO 26000. There was some thematic overlap in the data between the content analysis categories applied, as students formulated complex and integrative expectations and intertwined development ideas. Nevertheless, our content and thematic analysis allowed us to qualitatively explore and demonstrate the diversity, specific areas of interest, and holistic focus of students' expectations and ideas. Our quantification of prevalent themes in the data can be used to outline the most popular content categories, but overall, the data points towards students' diverse, balanced, and holistic take on the subject. In addition, diverse high-quality development ideas were also proposed outside of the most prevalent categories.

Of all the CSR and sustainability dimensions, our students expressed the greatest variety of development ideas related to the environment, community involvement and development, and labour practices. This seems to indicate that their stakeholder expectations are rather strongly directed towards broader societal and planetary concerns. All in all, the majority of their expectations and development ideas related to the TBL and ISO 26000 categories were tilted towards issues larger than themselves, with emphasis on (1) sustainable resource use and waste management, (2) engaging in community development and charity work, (3) promoting responsibility in the value chain, (4) transparent, inclusive, and fair policies and working conditions, and (5) guiding and educating consumers for sustainable lifestyles. In addition, concrete and high-quality development ideas were proposed, for example, concerning stakeholder collaboration, social inclusion, and the ethical consideration of the needs of diverse vulnerable and disadvantaged stakeholder groups.

Furthermore, the necessity of forming partnerships to broaden the impact of retailers' CSR and sustainability activities was reinforced throughout the data. Students saw that it was imperative for retailing companies to form collaboration partnerships with a wide range of private and public players, including environmental organisations, certification bodies, governmental and civil organisations, NGOs, charities, educational institutions, worker unions, technology providers, health experts, and so on. Students also expect companies to set ambitious targets and milestones for CSR performance, regularly monitoring progress and benchmarking impacts against international industry standards and best practices.

Fatima and Elbanna (2023) argue that "as CSR continues to establish a stronger foothold in organizational strategies, understanding its implementation is needed for both academia and industry". They move on to advocate an integrative birds-eye view on the many facets and contexts of CSR implementation. Along these same lines, our sample of university students expected that companies adopt a holistic approach to CSR because in so doing, they are better equipped to drive stakeholder engagement, prioritise their expectations, and address their concerns. Students saw that by increasing the transparency of their CSR operations and communication both internally and externally, companies can build trust and accountability. Through sharing information on CSR progress and impacts, applying feedback mechanisms, and involving stakeholders in decision-making, companies can signify that stakeholder voices are heard and their needs are valued.

The pathways of organisational learning are intricate and iterative, often marked by progress followed by setbacks when faced with new demands. According to Zadek (2004), companies usually go through five stages as they navigate the learning curve needed for them to cope in the face of complex changes and stakeholders' volatile expectations about companies' ability to address societal problems. During the defensive stage, companies encounter unexpected criticism, often from civil activists and the media but occasionally also from direct stakeholders such as customers, employees, and investors. In this defensive stage, their response to growing stakeholder pressure and criticism is denial and refusal to assume responsibility. Transitioning to the compliance stage involves establishing and adhering to corporate policies, typically in ways that can be visibly demonstrated to critics ("We ensure compliance with agreed-upon standards"). Advancing to the managerial stage, companies recognise the need to address long-term issues in ways that go beyond mere compliance or public relations tactics. To reach the strategic stage, companies need to understand how aligning their strategies with responsible business practices can generate competitive advantage and contribute to long-term success.

In the final stage of their learning curve, the civil stage, companies advocate for collective action to address societal concerns, sometimes integrating broader participatory measures into their strategic planning. The Civil-Learning Tool discussed by Zadek can assist companies in assessing their position, as well as that of their competitors, regarding demanding societal issues. The tool can aid organisations in devising strategies aligned with societal expectations, which fosters acceptance among stakeholders. Some organisations extend their focus even further, planning their future operations on the level of "metastrategy", anticipating the evolving role of businesses within societal structures and the need to contribute towards greater "stability and openness" in society and the global community (Zadek, 2004).

The university students' ideas brought up in this study demonstrate the kind of stakeholder expectations that require from retailing companies capacity to develop higher-level organisational learning, strategic agility, and willingness to engage in collective action. To reach the civil stage in their learning curve, companies may have to take short-term risks, be prepared to commit to long-term partnerships, and aim at achieving wider economic benefits through, for example, broad industry participation or collaboration initiatives with educational and civil institutions.

Universities can help companies to progress in their learning curve in several ways. First of all, pedagogical approaches can be developed in ways that actively engage students in addressing societal challenges and equip them with skills, competences, attitudes, and values that prepare and empower them for change and collaboration. Such "key sustainability competences" (e.g. Wiek et al., 2011; Brundiers et al., 2021) include, among others, transversal skills related to systemic and anticipatory thinking, values thinking, strategic thinking, creative implementation, critical analysis and reflection, self-awareness, and intercultural and interpersonal communication and collaboration. Overall, moral and ethical "reflexivity" (Cunliffe, 2016) is needed to reassess and redesign our ways of living on this planet in harmony with the environment and each other.

Secondly, universities can join forces with companies to establish strategic partnerships that contribute to effective knowledge sharing, co-creation, co-innovation, and co-research aiming at facilitating collaborative learning. Dialogical interaction between professionals in enterprises and university students, teachers, and researchers can generate innovative and impactful business, management, and stakeholder collaboration processes and solutions. Despite the many practical challenges of upholding long-term and strategic university–industry collaboration (Rybnicek & Königsgruber, 2019), it is imperative that an increasing number of university teachers reach out across disciplinary borders and build partnerships with a broad range of industry partners (Stewart et al., 2022) to create purpose-driven efforts to improve the reach and impact of CSR and sustainability initiatives.

Finally, university staff can take an even bolder step out of their traditional professional roles as teachers and researchers and extend both their own and their students' expertise, creativity, continuous learning capacity, and time resources more actively for the disposal of different social innovation networks and initiatives (McDonnell-Naughton & Păunescu, 2022). In the European context, social innovation refers to joint initiatives between public authorities, civil society, academia, and enterprises that aim to find solutions to societal problems and seek to "advance European life through improving working conditions, education, community development or health, or through tackling critical problems such as poverty or discrimination" (European Commission, n.d.). Judging from our data, university students strongly expect retailing companies to embark on a path of social innovation. Students are also eager and ready to engage in such efforts of social innovation aiming to contribute to changing the world we live in (Graham & Moir, 2022) and to our planetary survival (Sterling, 2021).

This study contributes to research into university students' sustainability conceptions. According to Zeegers and Francis Clark (2014), university students tend to have "an environmentally focused perspective of sustainability" at the expense of social sustainability. Kagawa (2007) identified the same focus and argued that there was a knowledge gap regarding students' conception of the social and economic aspects of sustainable development. The present study, in contrast, demonstrates that the students of our sample have a holistic and balanced view of sustainability, addressing both environmental and social dimensions and with due attention to both the economic potential and the investment requirements engrained in effective and impactful CSR and sustainability activities. Our students embraced a long-term perspective on sustainability, recognising the interconnectedness of environmental, social, and economic factors and striving for holistic solutions that create value to all stakeholders. They wanted to foster a culture of continuous improvement and innovation across all aspects of CSR, underlining the need to exchange best practices, share knowledge, and drive collective action to attain shared sustainability goals.

Due to our convenience sampling technique and qualitative approach, the generalisability of our results is limited. Our multicultural and multigender project groups did not allow us to systematically compare the informants' background variables and their sustainability expectations. To develop the research design, a mixed-method approach could be applied to investigate the potential differences in students' sustainability expectations according to, for example, their nationality, academic discipline, age, gender, or socioeconomic status. A quantitative survey could be used to measure students' expectations before and after the collaborative course project and the creation of qualitative reflections. In this study, we found no significant differences in students' expectations vis-à-vis different retailer types. Further research into the most prevalent categories of responses, differences among informants based on their background variables, and variations in informants' expectations across different types of retailers would provide more comprehensive insights and deeper understanding.

References

Andrew, M. B., Dobbins, K., Pollard, E., Mueller, B., & Middleton, R. (2023). The role of compassion in higher education practices. *Journal of University Teaching & Learning Practice, 20*(3).

Bansal, P., & DesJardine, M. R. (2014). Business sustainability: It is about time. *Strategic Organization, 12*(1), 70–78.

Berelson, B. (1952). *Content analysis in communication research.* The Free Press.

Braun, V., & Clarke, V. (2006). Using thematic analysis in psychology. *Qualitative Research in Psychology, 3*(2), 77–101. https://doi.org/10.1191/1478088706qp063oa

Brundiers, K., Barth, M., Cebrián, G., Cohen, M., Diaz, L., DoucetteRemington, S., Dripps, W., Habron, G., Harré, N., Jarchow, M., Losch, K., Michel, J., Mochizuki, Y., Rieckmann, M., Parnell, R., Walker, P., & Zint, M. (2021). Key competences in sustainability in higher education—Toward an agreed-upon reference framework. *Sustainability Science, 16*, 13–29. https://doi.org/10.1007/s11625-020-00838-2

Burritt, R. L., & Schaltegger, S. (2010). Sustainability accounting and reporting: Fad or trend? *Accounting, Auditing & Accountability Journal, 23*(7), 829–846. https://doi.org/10.1108/09513571011080144

Carroll, A. B. (1979). A three-dimensional conceptual model of corporate social performance. *Academy of Management Review, 4*, 497–505.

Carroll, A. B. (1991). The pyramid of corporate social responsibility: Toward the moral management of organizational stakeholders. *Business Horizons, 34*(4), 39–48.

Chadwick, B. A., Bahar, H. M., & Albrecht, S. L. (1984). Content analysis. In B. A. Chadwick et al. (Eds.), *Social science research methods*. Prentice-Hall.

Chandler, D. (2020). *Strategic corporate social responsibility* (5th ed.). Sage.

Clarkson, M. B. (1995). A stakeholder framework for analyzing and evaluating corporate social performance. *Academy of Management Review, 20*(1), 92–117.

Cording, M., Harrison, J. S., Hoskisson, R. E., & Jonsen, K. (2014). Walking the talk: A multistakeholder exploration of organizational authenticity, employee productivity, and post-merger performance. *Academy of Management Perspectives, 28*(1), 38–56. https://doi.org/10.5465/amp.2013.0002

Cunliffe, A. L. (2016). Republication of "On becoming a critically reflexive practitioner". *Journal of Management Education, 40*(6), 747–768. https://doi.org/10.1177/1052562916674465

Donaldson, T., & Preston, L. E. (1995). The stakeholder theory of the corporation: Concepts, evidence, and implications. *Academy of Management Review, 20*(1), 65–91.

Du, S., Bhattacharya, C. B., & Sen, S. (2010). Maximizing business returns to corporate social responsibility (CSR): The role of CSR communication. *International Journal of Management Reviews, 12*(1), 8–19. https://doi.org/10.1111/j.1468-2370.2009.00276.x

Elkington, J. (1997). *Cannibals with forks: The triple bottom line of 21st-century business.* Capstone.

Elo, S., & Kyngäs, H. (2008). The qualitative content analysis process. *Journal of Advanced Nursing, 62*(1), 107–115. https://doi.org/10.1111/j.1365-2648.2007.04569.x

European Commission. (2024). *Internal market, industry, entrepreneurship and SMEs/retail.* Retrieved from https://single-market-economy.ec.europa.eu/single-market/services/retail_en

European Commission. (n.d.). *European social fund plus/social innovation and transnational cooperation.* Retrieved from https://european-social-fund-plus.ec.europa.eu/en/social-innovation-and-transnational-cooperation
Falck, O., & Heblich, S. (2007). Corporate social responsibility: Doing well by doing good. *Business Horizons, 50*(3), 247–254. https://doi.org/10.1016/j.bushor.2006.12.002
Fatima, T., & Elbanna, S. (2023). Corporate social responsibility (CSR) implementation: A review and a research agenda towards an integrative framework. *Journal of Business Ethics, 183*(1), 105–121.
Freeman, R. E. (1984). *Strategic management: A stakeholder approach.* Pitman Publishing.
Freeman, R. E., Wicks, A. C., & Parmar, B. (2004). Stakeholder theory and "the corporate objective revisited". *Organization Science, 15*(3), 364–369. https://doi.org/10.1287/orsc.1040.0066
Freeman, R. E., Harrison, J. S., & Wicks, A. C. (2007). *Managing for stakeholders: Survival, reputation, and success.* Yale University Press.
Graham, C. W., & Moir, Z. (2022). Belonging to the university or being in the world: From belonging to relational being. *Journal of University Teaching & Learning Practice, 19*(4).
Gray, R. (2006). Social, environmental and sustainability reporting and organizational value creation? Whose value? Whose creation? *Accounting, Auditing & Accountability Journal, 19*(6), 793–819.
Gupta, S., & Pirsch, J. (2008). The influence of a retailer's corporate social responsibility program on re-conceptualizing store image. *Journal of Retailing Consumer Services, 15*, 516–526.
Hart, S. L., & Milstein, M. B. (2003). Creating sustainable value. *Academy of Management Executive, 17*(2), 56–67.
Haski-Leventhal, D. (2020). *The purpose-driven university: Transforming lives and creating impact through higher education.* Emerald.
Haski-Leventhal, D. (2022). *Strategic corporate social responsibility: A holistic approach to responsible & sustainable business.* Sage.
Henriques, A., & Richardson, J. (2004). *The triple bottom line: Does it all add up? Assessing the sustainability of business and CSR.* Routledge.
Holsti, O. R. (1968). Content analysis. In G. Lindzey & E. Aronson (Eds.), *The handbook of social psychology* (Vol. II, 2nd ed., pp. 596–692). Amerind Publishing.
Hou, M., Liu, H., Fan, P., & Wei, Z. (2016). Does CSR practice pay off in east Asian firms? A meta-analytic investigation. *Asia Pacific Journal of Management, 33*(1), 195–228.
ISO 26000. (2018). *ISO 26000 and SDGs* (2nd ed.). International Organization for Standardization (ISO).
Kagawa, F. (2007). Dissonance in students' perceptions of sustainable development and sustainability: Implications for curriculum change. *International Journal of Sustainability in Higher Education, 8*(3), 317–338. https://doi.org/10.1108/14676370710817174
Krippendorff, K. (1980). *Content analysis: An introduction to its methodology.* Sage.
López-López, C., Martínez-Rodríguez, F. M., & Fernández-Herrería, A. (2021). The university at the crossroads of eco-social challenges: Pedagogy of care and the community of life for a transformative learning. *Frontiers in Sustainability, 14.*
Lozano, R. (2015). A holistic perspective on corporate sustainability drivers. *Corporate Social Responsibility and Environmental Management, 22*(1), 32–44.
Lozano, R., & Barreiro-Gen, M. (Eds.). (2021). *Developing sustainability competences through pedagogical approaches: Experiences from international case studies.* Springer. https://doi.org/10.1007/978-3-030-64965-4
Margolis, J. D., Elfenbein, H. A., & Walsh, J. P. (2007). Does it pay to be good? A meta-analysis and redirection of research on the relationship between corporate social and financial performance. *Journal of Management, 33*(6), 958–984.
Marrewijk, M. (2003). Concepts and definitions of CSR and corporate sustainability: Between agency and communion. *Journal of Business Ethics, 44*(2–3), 95–105.

Matten, D., & Moon, J. (2008). "Implicit" and "explicit" CSR: A conceptual framework for a comparative understanding of corporate social responsibility. *Academy of Management Review, 33*(2), 404–424.
McDonnell-Naughton, M., & Păunescu, C. (2022). Facets of social innovation in higher education. In C. Păunescu, K. L. Lepik, & N. Spencer (Eds.), *Social innovation in higher education: Innovation, technology, and knowledge management*. Springer. https://doi.org/10.1007/978-3-030-84044-0_2
Mitchell, R. K., Agle, B. R., & Wood, D. J. (1997). Toward a theory of stakeholder identification and salience: Defining the principle of who and what really counts. *Academy of Management Review, 22*(4), 853–886.
Murray, A., Haynes, K., & Hudson, L. J. (2010). Collaborating to achieve corporate social responsibility and sustainability? Possibilities and problems. *Sustainability Accounting, Management and Policy Journal, 1*(2), 161–177. https://doi.org/10.1108/20408021011089220
Orlitzky, M., Schmidt, F. L., & Rynes, S. L. (2003). Corporate social and financial performance: A meta-analysis. *Organization Studies, 24*(3), 403–441.
Parmar, B. L., Freeman, R. E., Harrison, J. S., Wicks, A. C., Purnell, L., & de Colle, S. (2010). Stakeholder theory: The state of the art. *Academy of Management Annals, 4*(1), 403–445. https://doi.org/10.5465/19416520.2010.495581
Păunescu, C., Lepik, K.-L., & Spencer, N. (Eds.). (2022). *Social innovation in higher education: Landscape, practices, and opportunities*. Springer. https://doi.org/10.1007/978-3-030-84044-0
Pereira, T. H. M., & Martins, H. C. (2021). People, planet, and profit: A bibliometric analysis of triple bottom line theory. *Journal of Management and Sustainability, 11*(1), 64. https://doi.org/10.5539/jms.v11n1p64
Peterson, R. A., & Merunka, D. R. (2014). Convenience samples of college students and research reproducibility. *Journal of Business Research, 67*(5), 1035–1041.
Porter, M. E., & Kramer, M. R. (2006). Strategy and society: The link between competitive advantage and corporate social responsibility. *Harvard Business Review, 84*(12), 78–92.
Porter, M. E., & Kramer, M. R. (2011). Creating shared value. *Harvard Business Review, 89*(1/2), 62–77.
Rahdari, A., Sheehy, B., Khan, H. Z., Braendle, U., Rexhepi, G., & Sepasi, S. (2020). Exploring global retailers' corporate social responsibility performance. *Heliyon, 6*, e04644.
Ramasamy, B., & Yeung, M. (2009). Chinese consumers' perception of corporate social responsibility (CSR). *Journal of Business Ethics, 88*, 119–132.
Rybnicek, R., & Königsgruber, R. (2019). What makes industry–university collaboration succeed? A systematic review of the literature. *Journal of Business Economics, 89*, 221–250. https://doi.org/10.1007/s11573-018-0916-6
Schaltegger, S., Burritt, R., & Petersen, H. (2017). *An introduction to corporate environmental management: Striving for sustainability*. Routledge. https://doi.org/10.4324/9781351281447
Stempel, G. H. (1989). Content analysis. In G. H. Stempel & B. H. Westley (Eds.), *Research methods in mass communications*. Prentice-Hall.
Sterling, S. (2021). Concern, conception, and consequence: Re-thinking the paradigm of higher education in dangerous times. *Frontiers in Sustainability, 2*, 743806. https://doi.org/10.3389/frsus.2021.743806
Stewart, I. S., Hurth, V., & Sterling, S. (2022). Re-purposing universities for sustainable human progress. *Frontiers in Sustainability, 3*, 859393.
Svensson, G., & Wagner, B. (2015). Implementing and managing economic, social and environmental efforts of business sustainability. *Management of Environmental Quality: An International Journal, 26*, 195–200. https://doi.org/10.1108/MEQ-09-2013-0099
Wiek, A., Withycombe, L., & Redman, C. L. (2011). Key competencies in sustainability: A reference framework for academic program development. *Sustainability Science, 6*, 203–218.
Zadek, S. (2004). The path to corporate responsibility. *Harvard Business Review, 82*(12), 125–132.
Zeegers, Y., & Francis Clark, I. (2014). Students' perceptions of education for sustainable development. *International Journal of Sustainability in Higher Education, 15*, 242–253.

Dr Tanja Vesala-Varttala works as Principal Lecturer in Marketing and Communication at Haaga-Helia University of Applied Sciences, Helsinki, Finland. She teaches sustainability marketing and acts as a researcher and project manager in European-funded RDI projects at the Research Unit of Entrepreneurship and Business Development. She has been developing marketing and communication studies in industry-university projects for more than 15 years. She is actively involved in Erasmus+ project and research networks, recent projects including LEARN&CHANGE—Collaborative Digital Storytelling for Sustainable Change (coordinator 2021—2023) and CORALL—Coaching-oriented Online Resources for Autonomous Learning of LSP (partner 2019–2022). She has published in the fields of higher education teaching and learning, sustainability competence development, sustainability marketing, multicultural corporate communication, collaborative autonomous learning, digital team and community building, digital storytelling, and the ethics of narrative.

Dr Carmina Nunes is Invited Assistant Professor and teaches courses in corporate social responsibility (CSR), ethics, and sustainability at the University of Coimbra, Portugal. She is the author of scientific articles published in specialized journals in the same fields, and she is integrated with the research center of GOVCOPP (Competitiveness, Innovation and Sustainability Research Group). She was involved as an associated partner in the Erasmus+ RDI project LEARN&CHANGE, coordinated by Haaga-Helia University of Applied Sciences, Helsinki, Finland, where students of Coimbra University developed their stakeholder collaboration skills and worked as change agents co-innovating CSR solutions that inspire and drive sustainable change in the international retail sector. Presently, she is involved in the project Interreg-Europe CHEERS4EU—Circular Hubs, promoting to accelerate the transition to a circular economy in Europe by focusing on Circular Hubs.

Dr Susana Garrido is Associate Professor at the University of Coimbra, Portugal. Her research interests lie in the circular economy, sustainability, supply chain, lean, green, and logistics management. She has published over 300 scientific books, chapters, articles, and conference proceedings. She is a member of the Scientific Evaluation Board of the ERA-MIN European Funding Program on “Raw Materials for Sustainable Development and the Circular Economy” and the Portuguese Agency for Assessment and Accreditation of Higher Education. She has participated in several projects in the area of sustainability and circular economy, such as CHEERS4EU—Circular Hubs serving as an Engine for European Regions to foster a Strong Green Economy 4 Europe (Interreg-Europe); BioAgroFloRes—Sustainable Supply Chain Management Model for Residual Agro-forestry Biomass supported in a Web Platform (FCT, PCIF/GVB/0083/2019, 2019–2023); Lean, Agile, Resilient and Green Supply Chain Management LARGeSCM (MIT-Pt/EDAM-IASC/0033/2008, Universidade Nova de Lisboa Unidade de Investigação e Desenvolvimento em Engenharia Mecânica e Industrial, Portugal); and Optimization of Organic Residues Fermentation Processes (CAPES-PRINT-UNESP, Brazil. 2021-2023).

The Importance of the Family Businesses for Fundamental Sustainable Improvements: Cases from Bulgaria, Romania, and Uzbekistan

Daniel Pavlov, Silvia Puiu, and Deniza Alieva

1 Introduction

Many national governments play an important role in sustainable development in their national plans. But the balance between the three basic elements (nature, society, and business) is not easy. The core of the marketing economy is based on financial flows and the money is recognized as an instrument to facilitate the trade between the marketing participants. As bigger the trade, as bigger the profit. Trade is based on production with a focus on different production factors; some of them are labor, technologies, and resources.

In a period of crisis like the COVID-19 pandemic, these three factors get into risky situations, due to the related sources of risks—demographic conditions, technological development, access to natural resources, political environment, level of competition, etc. From 2020 through 2022, all governments recommended that people "Stay home," which created problems for employees working onsite (Coombs, 2020; Dawson et al., 2021; Ng & Stanton, 2023; Willcocks, 2020; McGaughey, 2022). As a result of these restrictions, most of the firms reduced their personnel, searching for production lines and services which can operate with much less human resources. This process of dehumanization of the companies is not new. In the second half of the twentieth century, many agricultural producers changed the technology of grain harvesting—from manual to combine harvesters. A huge

D. Pavlov (✉)
Entrepreneurship Center of the University of Ruse "Angel Kanchev", Ruse, Bulgaria
e-mail: dpavlov@uni-ruse.bg

S. Puiu
Faculty of Economics and Business Administration, University of Craiova, Craiova, Romania

D. Alieva
Management Development Institute of Singapore in Tashkent, Tashkent, Uzbekistan

S. O. Idowu, S. Vertigans (eds.), *Sustainability in Global Companies*, CSR, Sustainability, Ethics & Governance, https://doi.org/10.1007/978-3-031-77971-8_2

number of reapers lost their jobs. Thanks to the increased demand of the labor force in the factories, many of these jobless people got employed in the industrial plants.

The technological development reached further improvements and the automation and robotization of production processes deepened towards the end of the twentieth century, leading to a second wave of workforce replacement, but this time in production processes with automated production lines. In the twentieth century, production technologies made the workforce easily replaceable, both in agriculture and in industrial enterprises. But, with the creation of the computer, humanity entered a new phase of its development—the information society. The rapid growth of digital businesses has allowed many new firms to hire staff, thus reducing the social strain on the unemployed, also emphasizing the need for new skills and knowledge (Kehinde & Olatunde, 2022; Simić, 2019; Barna & Epure, 2020; Guitert et al., 2020; Pirzada & Khan, 2013; Stofkova et al., 2021). At the beginning of the twenty-first century, various digital companies pushed technological development to even higher levels by creating artificial intelligence to replace their staff. Again, the workforce can be replaced (Nguyen & Vo, 2022; İŞCAN, 2021; Bordot, 2022; Mutascu & Hegerty, 2023; Bruun & Duka, 2018).

Through all these transformations, one participant remains constant—the owner of the firm or the related shareholders. If in the twentieth century, the owner of the company had to hire 100 people to perform the relevant activity manually, then in the twenty-first century the same amount of work can be done with 2–3 people and the necessary technologies. These few people may be his family members.

In the period of COVID-19 restrictions, the companies have reduced their staff (especially those who are not their relatives) creating different economic and psychological problems for the people, especially the single ones. In contrast, the family of each owner was in a safer position, because they shared their responsibilities for family stability and managed the different risks (Rivo-López et al., 2022; Amore et al., 2022).

Nowadays the green transition creates new restrictions for companies (Licastro & Sergi, 2021; Rizos et al., 2016; Rahman et al., 2020) and thanks to the digital transformation the managers are able to reduce part of their staff and optimize the way they organize their businesses. Other firms relocate their production facilities to other countries with fewer green restrictions and better access to natural resources. According to Bakracheva et al. (2020) family businesses help people to go through economic crises on the base of mutual support of relatives. Therefore, the purpose of this chapter is to reveal some benefits of family businesses in terms of sustainability.

1.1 Contribution of the Intergenerational Family Businesses to the Improvement of Sustainable Regional Development

The new theoretical fundamentals for the intergenerational family business under INTERGEN were designed in 2017 (Pavlov et al., 2017). This theory states that when younger generations start their new businesses, they integrate the businesses of the previous generations. The classic example is:

- The first generation (grandparents) produces grapes.
- The second generation (parents) starts to produce wine, including the grapes of the grandparents.
- The third generation creates an online shop and sells the grapes and wine of their relatives.

Thus, all three generations could support each other and have sustainable development in their families.

The following example of classic INTERGEN businesses follows the classification of Toffler (1980) for agricultural, industrial, and information society:

- The first generation is representative of the agricultural society. Their wine production activity has been based on a good balance between nature and renewable natural resources (biomass), which has been the main factor for the sustainable development of mankind for millennia.
- The second generation is representative of the industrial society, and mankind has two main choices—to use renewable sources, like biomass in the given classical example, or to organize the production activity from exhaustible resources such as oil, coal, ores, and others. Therefore, we can recognize the green transition concept as an effort to stop using scarce resources, because sooner or later they will not be available anymore. Therefore, the green transition will lead to a reduction in production in some sectors and the closure of a variety of jobs. International Labour Organization (2018) emphasizes the creation of jobs in the green economy, also stating that "The job-creating potential of environmental sustainability is not a given: the right policies are needed." From a management point of view, people are considered just "human resources," which makes them replaceable when they are no more useful.
- The third generation is representative of the information society. Many young people have lost their contact with the land and they do not have the skills to deal with agricultural production, which has been the typical human activity for millennia. Living in cities makes these people highly dependent on the store. Many of them do not want to work in factories, which creates problems for the industry to find the required workforce. In many countries, a solution to this problem is sought by importing from other countries the necessary employees. A new moment for global business is that digital businesses could be done from home.

Situation in Bulgaria

Due to COVID-19 restrictions, many Bulgarian families have changed their living places—from the city to a village. The good transport infrastructure facilitates their travel to nearby bigger cities and they have access to different services. Another important factor for their decision to stay in the rural areas after the start of the COVID-19 pandemic is the high speed of the Internet in Bulgaria. Thus, digital entrepreneurs manage their activities from a home office. Family members have more time together and they could manage a variety of risks and take advantage of the new opportunities.

But are the family members capable of doing business together? In industrial societies, individuals were stimulated to leave their families and spend their fruitful time as employees, which affected the future of the family as an institution (Oppenheimer, 1994). The promotion of career development was among the main goals of the education process; the schools were proud of their graduates who worked at a high career level. As Gibb (1987) states, education taught in universities and entrepreneurship are sometimes different things, being at opposite poles. The educational system is still very much focused on creating employees for the companies (Harry, 2007), while the idea of family entrepreneurship is not so well integrated into the curriculums.

The information society creates new opportunities for the family members to be together and some of the professors should be able to explain to the students a simple business model of how different generations could deal together. This is one of the main goals of the professors from the INTERGEN international academic network—to encourage their students through specific course assignments to develop family business ideas, too.

The University of Ruse "Angel Kanchev" has different courses where Bulgarian students are encouraged to deepen their business relations with some family members:

- *Risk management* course—Students have the freedom to choose to develop some risk analyses for existing firms or their ideas to start some new family business.
- *Planning and forecasting* course—Students have as a course assignment a business plan for a new small company, using the TEHNOSTART template of the Bulgarian Ministry of Economics. Also, the students are stimulated to export their production, and in accordance with a recent study by Gueorguiev and Kostadinova (2021), they have to be well informed about the different quality management ISO standards, too.
- *Small business management* course—Students develop a course assignment on a business plan to start a family business. They are encouraged to integrate their relatives from the previous generations. The experience has shown that over 80% of the course assignments are about family businesses in production using renewable sources (biomass), which also proves the outcomes of a Bulgarian study that more entrepreneurs are interested in the food processing industry (Todorova et al., 2018).

Situation in Romania

The **University of Craiova** and especially the *Faculty of Economics and Business Administration* has specific courses to encourage Romanian students to go for family business and a more sustainable form of entrepreneurship. Some of them are:

- *Management* course in which professors can direct some topics into a discussion about managing your own business that can start as a small family unit.
- *Ethics in Business* is another course that shows students how to create sustainable, responsible, and ethical businesses.
- *Marketing* course is useful for teaching students how to promote their businesses which is especially important for small family businesses.
- *Entrepreneurship* course is a good opportunity for professors to give students assignments regarding the development of a business plan for a family business which is usually small at the beginning and as a project, it can be more easily understood by students.
- *B to B Marketing* and *Consumer Behavior* are marketing-related courses which can be of tremendous help for students starting their own family businesses.
- *Sales Management, Human Resources Management,* and *Strategic Management* are courses taught to students in Management and can be easily applied to small businesses.
- *Entrepreneurship and Business Administration* is an optional course that students wanting to become entrepreneurs might choose which proves to be useful for family businesses.
- *Business Risk Management* is a course taught to graduate students in Business Management. This helps them to manage risk when occurs and implement adequate strategies for mitigation.
- *Managerial Simulations* course helps students better understand a specific business context, and family businesses are a good example for these simulations considering their small dimensions, at least at the beginning.
- *Social Responsibility of Organizations* is a course taught to graduate students in Human Resources Management. The course is useful for understanding the role played by many organizations, family businesses included, in the community and the impact they have or can have.
- Other courses taught to graduate students in Marketing and Business Communication can be useful for family businesses in the way they manage their marketing activities to gain more visibility on the market. Such courses are *Client Relationships Management; E-marketing; Strategic Marketing; Brand Management;* and *Creative Writing in Marketing.*

These courses are not focused only on family businesses but the professors can and usually give task assignments that are more easily understood by students in the context of a small family business. The fact that many students have such businesses is also helpful for them to understand some theoretical aspects. Some students have their parents or grandparents in rural areas and they already started small businesses. The young generation can help with the knowledge they gain in faculty about marketing and management to help their family to be more sustainable and also visible on the market.

The Faculty of Economics and Business Administration has been the organizer of a yearly competition for students and pupils called *The Best Business Idea* since 2014. This is a great opportunity for students to think of innovative businesses and some of them can be successful family businesses with a great impact on their communities.

Situation in Uzbekistan

The **Management Development Institute of Singapore in Tashkent** offers a comprehensive range of courses specifically created to inspire and equip Uzbek students with the necessary skills and knowledge to succeed in family business ventures and to instill a deep appreciation for nature preservation. The courses included in programs of partner universities from the UK (Bangor University, Teesside University, and University of Sunderland) foster a conducive environment where students can develop a profound understanding of the intricacies involved in managing family enterprises while simultaneously cultivating a strong sense of environmental stewardship. The blend of theoretical insights with practical applications empowers students to become adept leaders and entrepreneurs in their respective family businesses. At the same time, there is a nurturing of a sense of responsibility towards sustainable practices and the conservation of natural resources.

The following courses are provided by the University:

- *Principles of Business Management* explores foundational concepts and theories essential for understanding the fundamental principles and practices of managing a business effectively.
- *Entrepreneurship, Capital, and Firm* delves into the acquisition and management of capital within firms, fostering an understanding of the entrepreneurial process from inception to growth.
- *Leadership and HRM* equips students with the skills needed to lead and manage diverse teams within organizational settings.
- *Business Management Project* guides students through the process of conceiving, planning, and executing a business project, emphasizing project management principles and techniques.
- *International Business Competency* develops students' understanding of the complexities of conducting business in the global marketplace, covering topics such as cross-cultural management, international trade, and global strategy.
- *Corporate Governance and Ethics* addresses issues related to accountability, transparency, and ethical behavior within organizations.
- *Business Organizations and Society* examines the impact of businesses on society and vice versa.

- *Understanding Organizations* provides insights into the structures, processes, and dynamics of organizations, enabling students to comprehend how organizations function and evolve over time.
- *Projects and Organizations* explores how projects are initiated, executed, and integrated within organizational frameworks.
- *Global Business Dynamics* examines economic, political, technological, and cultural factors influencing international business operations.
- *Sustainability, Strategy, and Society* explores how businesses can create value while addressing environmental and social challenges.
- *Current Issues in Business Ethics and CSR* examines emerging ethical dilemmas and strategies for promoting responsible business practices.

The Relevance of the Courses Taught

In examining the courses offered across educational institutions in Bulgaria, Romania, and Uzbekistan, it becomes evident that there exists a profound interconnection between the curricula and the overarching goal of promoting sustainable development through family businesses. Beyond the mere dissemination of theoretical knowledge, these courses serve as dynamic platforms for fostering interdisciplinary understanding, cultivating entrepreneurial spirit, and instilling a deep-seated commitment to ethical entrepreneurship and environmental stewardship.

At the heart of this interconnected educational landscape lies a shared recognition of the evolving role of family businesses in the contemporary socio-economic milieu. In Bulgaria, where the COVID-19 pandemic has prompted a reevaluation of lifestyle choices, the courses at the University of Ruse "Angel Kanchev" are tailored to equip students with the skills necessary to harness the potential of family enterprises in rural settings. By integrating risk management, strategic planning, and sustainability principles, these courses empower students to leverage familial bonds and local resources to drive sustainable economic growth and community resilience.

Similarly, in Romania, the Faculty of Economics and Business Administration at the University of Craiova places a strong emphasis on ethical entrepreneurship and social responsibility within the context of family businesses. Through courses that explore topics such as business ethics, marketing, and strategic management, students are encouraged to envision family enterprises not merely as vehicles for profit but as agents of positive change within their communities. By fostering a culture of innovation and responsible business practices, these courses pave the way for the emergence of family businesses that are not only economically viable but also socially and environmentally conscious.

In Uzbekistan, where the Management Development Institute of Singapore in Tashkent collaborates with international partners to offer a comprehensive curriculum, there is a deliberate focus on integrating business management principles with environmental sustainability. Through courses that delve into topics such as corporate governance, sustainability strategy, and global business dynamics, students are equipped with the knowledge and skills necessary to navigate the complexities of family enterprises in an increasingly interconnected world. By fostering a deep appreciation for the interplay between business operations and environmental preservation, these courses inspire students to become stewards of both economic prosperity and ecological integrity.

Beyond the geographical boundaries that separate these educational institutions, there exists a common thread that binds them—a shared commitment to preparing future generations of entrepreneurs who are not only adept at navigating the intricacies of family businesses but also cognizant of their role in driving sustainable development. By fostering interdisciplinary collaboration, promoting ethical entrepreneurship, and instilling a sense of environmental stewardship, these courses lay the foundation for a new breed of business leaders who are poised to address the complex challenges of the twenty-first century with creativity, compassion, and resilience. In essence, the connection between these courses extends far beyond the confines of the classroom, serving as a catalyst for transformative change within families, communities, and societies at large.

1.2 Examples from Bulgaria, Romania, and Uzbekistan of Family Firms with Foreign Clients and an Approach to Sustainability

Examples from Bulgaria

There are a variety of family firms in the region of Ruse, Bulgaria, which could be recognized as using the INTERGEN approach with linkages between at least two generations. Here are some examples from the city of Ruse. The information was collected by direct meeting with the owners of these firms in 2022 and 2023.

The first example is that of a **furniture** company with production facilities situated in Ruse. The owners are the parents while their children have a furniture shop in the center of the city where they sell the furniture of their family and also of other business partners. The children do not deal with the physical production of the furniture, but they prefer to trade; they have received financial support from their parents to develop the shop. As better the shop operates, more family furniture will be sold. Thus, the parents have a financial interest in the shop being well-established in the regional market. Unfortunately, not all owners of furniture factories in Ruse have children who want to develop similar business linkages with their parents and those firms were closed or sold after the parents got too old to keep the business.

Another example is that of a mother who is well-known as a good **doctor** in Ruse city. Her medical skills are well-known among many patients and they bring more ill people to be treated by her. At the very beginning of her medical career, she got the financial support of her husband and later she became the main source of income for the family. Their daughter has seen how many times her mother was called by her patients at inappropriate times, strongly disturbing the family's pleasure. Therefore, the daughter does not want to become a medical doctor to avoid the negatives of having a private medical cabinet. After some discussions, the daughter has decided to study to become a dentist. She has financial support from both of her parents and the plans are for some of the mother's patients to use the dental services of her daughter. The daughter likes the idea of being a dentist and takes advantage of the good reputation of her mother to become well-known among potential patients. The daughter studies medicine very hard to gain the proper skills and keep her patients for a long time.

Stiliyana Stoyanova and Nikola Kynchev have graduated from the Faculty of Business and Management at the University of Ruse "Angel Kanchev." They got married and have become happy parents of two small children. At the same time, they have established a business, which is typical for the INTERGEN approach:

- The first generation (grandparents) have a cow farm in a village near the city of Ruse. They have used this type of work and both grandparents have been happy with what they have done for over 25 years.
- The second generation (the father) decided to create a dairy company in the yard of the cow farm. The milk produced by the first generation is primarily used in it.
- The third generation (Stiliyana and her husband) set up a grocery store where they sell the milk of the first generation and the dairy products (cheese, yellow cheese) of the second generation. They also created an online opportunity for the clients to contact them via the Facebook profile of their grocery store—https://www.facebook.com/magazinotfermata. Stiliyana and Nikola decided to allow some of their strategic partners to use this grocery store to sell other types of food, too. Nikola is very critical regarding quality and he allows only those food products, which meet his high criteria for quality. As he says, "In time we had to interrupt our relations with some of the business partners because they didn't meet our criteria, but we also had the opportunity to make stronger and sustainable contacts with other business partners." The grocery store has managed to keep enough sustainable clients and be competitive with the big stores, which are a few meters away. Stiliyana and Nikola have invested their time in creating a reliable reputation for their grocery store and are therefore welcomed by many business partners and clients.

Also, both Stiliyana and Nikola go to the cow farm to help the grandparents with the cow farming. In 2022, Stiliyana was invited to share her experience with students and pupils. She was captivating and all attendees listened to her with interest. At the end of the lesson, she said "Look at me—I have good clothes, a nice face and perfumes, but after 30 minutes I will be on the cow farm to help my grandparents. I want my family to be free and, Yes, I am ready to pay the price—to work in the grocery store and at least 2–3 days a week to go and help my grandparents. Freedom is all we have as entrepreneurs, and I am ready to keep it, despite the nice dresses and perfumes I have. Is there anyone among you, who would like to come with me and work in the cow farm?". Then, 80% of the attendees raised their hands to join her in cow farming. For them, Stiliyana was a model of a free person with dreams, a parent with children and a spouse, and a businesswoman with well-developed intergenerational family linkages. She had previously worked in the administration of some companies and she did not want to be employed again, but to be a free person—forever independent and she paid that price every day in her life. The capacity of their cow farm is small and it targets only the local market, but the family has the potential to go for some limited export, too, reaching clients abroad. This business is based on a renewable source (biomass to produce milk), which makes it sustainable and a positive influence on regional development, too. Each succeeding generation has received financial and other forms of assistance from the previous generations.

Examples from Romania

There are many family businesses of different sizes and in different domains in Romania, some of them engaged in trying to be more sustainable, while also expanding their activity and addressing international markets. In the following paragraphs, I will present a few family businesses I had the opportunity to know more about through the interviews I had with their founders in March 2023. The interviews were online considering that they are in several parts of the country. The interviews were structured and sent via a Google Forms to 10 family businesses, but only two accepted to be included in this research.

Alia's Unicorn (www.aliasunicorn.studio) is a small but ambitious family business in the domain of photography founded in 2022 after the COVID-19 pandemic when the number of events increased and thus the potential for this type of business. This business identified as an opportunity the fact that there was a need for photography as an art on the market. The founders are a married couple, and each partner brought their unique strengths and qualities into the business. Of course, as for any business at its beginning, there are some barriers too: the difficulty in finding qualified employees for marketing and photography. Still, they overcame these limits by training them at work, costs covered by the family business. Investments in marketing are important and needed for gaining visibility in the market, especially in a domain where people do not always understand the difference between an average good image taken with your phone and one taken by a professional photographer in a studio. Alia's Unicorn tries to change these ideas and raise awareness of the value of photography as a form of art.

Still, the owners are not discouraged and see these limits as opportunities to overcome themselves, and try new techniques and new approaches in terms of photography, but also in terms of marketing to reach a higher number of customers that understand the value of a high-quality photo. The competition for this type of business might seem high, but if things are approached at the level of the quality provided by Alia's Unicorn, the competition is rather low. The challenge is to meet the demand at this level. Unfortunately, the prices dropped a lot and many small businesses and professional photographers struggle. One of the main strengths of the business is the passion they put into this project, a passion that transpires in their posts on social media where they try to also educate people on what quality in this domain means and show them the difference between a low-quality image and a high one provided by a professional in a photo studio with the right equipment.

Still, at their beginning, the business does not have collaborations with partners in other countries yet, but they mention in the interview they consider this opportunity. Another important aspect revealed during the interview is the sustainable approach of this small family business. They use green forms of energy for heating, they try to reduce their energy consumption as much as they can (which helps in cutting costs but also protects the environment) and they also recycle which shows the social responsibility of this business in their community. As Iulia, one of the founders, mentions: "It is just a matter of time, effort and being consistent" to reach the top. My opinion after the interview is that passion is a good driving force to keep the family business on the market and create new opportunities for development and expansion.

Another example of a family business is ***RawLala Raw Vegan*** (https://www.rawlala.ro/), a cake laboratory and shop in Craiova, Romania, which focuses on creating raw vegan desserts with natural ingredients without adding sugar. It was founded in 2023 by two brothers. Even if financial difficulties are something experienced by many small firms in their beginnings, the founders consider that this is a domain in which they can improve themselves constantly while being creative. Asked about competition, the founders said: "More important than competition is to outrun yourself daily, trying to do things better and tastier." This shows us their desire to satisfy the needs of their customers by offering them healthy and tasty products that also look amazing. Their business targets a niche of customers but this segment is rising, with more and more people choosing healthier desserts. The customers are not only vegans or vegetarians but also people with some allergies and food sensitivities people during religious fasting (mainly before Christmas and Easter) or simply people who love to try new, healthy, and delicious products. The passion put in the laboratory for creating high-quality sweets with wonderful decorations shows that this is also an art. As in the previous example, the energy of the founders is seen not only in their products but also in the way they post on social media.

Asked about their motivation to start a family business, the founders say that this is a good opportunity to make a profit while also enjoying what they do. Practically, this is passion and creativity turned into profit and customers are enjoying their products so there are benefits for everyone.

They also had foreign customers but no collaboration with partners in other countries at the moment. Still, the founders mention their openness toward this approach and opportunity in the future. Their experience so far is that things might be more complicated at the beginning, but they become easier with the experience they gain on the market. In terms of sustainability, they focus only on desserts produced with natural ingredients without adding sugar and thus products that help people be healthier. They also attended a local event to have more opportunities to meet new partners and customers. The founders said in the interview that people who tried their products at the event were surprised pleasantly after trying their raw vegan cakes.

These are examples of small but ambitious family businesses, but there are also family businesses that reached important dimensions, expanding their activities in the entire country and with great success. But this happened after many years. A few examples are ***Dedeman*** in construction retail which has 60 stores in Romania and a turnover of more than 11 billion lei in 2022 founded by two brothers in 1992; ***FAN Courier***, another business started by brothers together with a friend in 1998 that now became a leader in the domain of courier services; ***Autonom***—A car rental business started by two brothers in 2006 that now has branches not only in Romania but also abroad (Ghitulescu, 2024).

Family businesses in Romania are still in their first generation, most of them started after the fall of the communist system in the 90s. Only 9% of family businesses reached a second generation, a level considered low in comparison with other European countries (Club Antreprenor, 2024).

Examples from Uzbekistan

Family businesses play a significant role in the economic landscape of Uzbekistan, contributing to employment, innovation, and overall economic growth (Tolipov, 2024). With a rich cultural heritage and a long tradition of entrepreneurship, Uzbekistan's family businesses span various industries, ranging from agriculture to manufacturing, retail, services, and beyond (Mardievna & Oblokulovich, 2021). One of the defining features of family businesses in Uzbekistan is their resilience and adaptability (Mardievna & Oblokulovich, 2021; Karajanova & Rakhimboev, 2023), often passing down entrepreneurial skills and knowledge from one generation to the next. These businesses are deeply ingrained in the fabric of society, serving as pillars of stability and sources of employment within their communities.

In Uzbekistan, family businesses are typically structured around kinship ties, with ownership and management closely held within the family unit (Karajanova & Rakhimboev, 2023). This familial bond often fosters a strong sense of loyalty, trust, and commitment among family members (Welter et al., 2017), which can be advantageous in navigating the complexities of running a business in a developing economy.

Agriculture remains a cornerstone of Uzbekistan's economy, and many family businesses are engaged in farming activities, cultivating crops such as cotton, wheat, fruits, and vegetables. These agricultural enterprises often operate on a small to medium scale, with family members working together to manage the land, harvest crops, and market their produce. One of the examples of this type of enterprises might be a family business in the Khorezm region of Uzbekistan, Khorazm Anor, renowned for its cultivation and export of pomegranates. Benefiting from the region's rich soil and favorable climate, the enterprise has flourished as a multi-generational endeavor, deeply rooted in familial tradition and agricultural expertise.

Founded decades ago, by Abdulla, the business has been passed down through three generations, with ownership and management closely held within the family unit. Today, the business is overseen by Abdulla's sons and grandsons. Each family member brings their unique talents and skills to the table, whether it is agricultural expertise, orchard management, marketing savvy, or logistical know-how, ensuring the smooth functioning of the enterprise from orchard to export.

The **Khorazm Anor** activities extend beyond national borders. Leveraging Uzbekistan's strategic location and growing reputation as a source of high-quality agricultural products, the enterprise actively engages in exporting its pomegranates to international markets. Through strategic partnerships, export agreements, and participation in trade fairs, Khorezm Pomegranate Gardens has successfully penetrated markets in Central Asia, Russia, and the Caucasus. This expansion not only brings economic prosperity to the family and the region but also elevates Uzbekistan's reputation as a leading exporter of premium agricultural products.

In addition to agriculture, family businesses in Uzbekistan are active in manufacturing, producing a diverse range of goods including textiles, garments, furniture, and processed foods. These manufacturing enterprises may start as small workshops or cottage industries, gradually expanding over time as they reinvest profits and attract new customers. **Rakhim-ota furniture workshop**, located in the heart of Tashkent, the capital of Uzbekistan, was founded over five decades ago. It traces its roots back to the artisanal traditions of Uzbekistan, where woodworking skills have been passed down through generations. Hasan, a master carpenter with a passion for preserving Uzbek cultural heritage, established the workshop with a vision to create timeless pieces of furniture that blend traditional craftsmanship with contemporary design. Over the years Rakhim-ota furniture workshop has faced many challenges, including competition with mass-produced furniture, economic crises, and environmental concerns. However, it continues operating in the market, ensuring a sustainable approach to work. Through meticulous planning and efficient production processes, the workshop minimizes waste generation, repurposing scraps and off-cuts for smaller projects or donating them to local artisans for use in their crafts. Investment in energy-efficient equipment and technologies, such as solar panels and LED lighting, helps reduce energy consumption and reliance on non-renewable resources. Finally, engaging with the local community through educational workshops, seminars, and exhibitions raises awareness about sustainability issues and fosters a sense of collective responsibility for environmental stewardship.

Retail is another thriving sector for family businesses in Uzbekistan, with many families operating shops, markets, and trading businesses selling a variety of goods to local consumers. From neighborhood grocery stores to specialty boutiques, these retail establishments play a vital role in meeting the everyday needs of Uzbekistan's population (Isokova, 2023). The services sector also sees a significant presence of family businesses in Uzbekistan. These may include family-owned restaurants, cafes, hotels, transportation services, construction firms, and professional services such as legal, accounting, and consulting firms (Elo, 2016). In many cases, these businesses are passed down through generations, with each successive cohort building upon the legacy of those who came before them.

Governmental support for family businesses in Uzbekistan plays a crucial role in fostering entrepreneurship, economic growth, and job creation (Rutkauskas & Ergashev, 2012; Tadjiev et al., 2023). Recognizing the significance of family businesses in driving the economy, the Uzbek government has implemented various policies and initiatives to support their development and sustainability. One key area of governmental support is in the realm of financial assistance (Welter et al., 2017). The government provides access to subsidized loans, grants, and financial incentives tailored specifically for family businesses. These financial resources help entrepreneurs overcome initial capital constraints, invest in expansion opportunities, and navigate economic uncertainties. Additionally, the government often collaborates with financial institutions to offer favorable lending terms and credit facilities to family-owned enterprises.

In addition to financial support, the Uzbek government offers regulatory incentives and streamlined procedures to facilitate the establishment and operation of family businesses (Olimjanova, 2024). Simplified registration processes, reduced bureaucratic red tape, and tax incentives for small and medium-sized enterprises (SMEs) make it easier for family businesses to formalize their operations and comply with legal requirements. Moreover, the government frequently reviews and updates business regulations to create a more favorable environment for entrepreneurship and innovation (Welter et al., 2017).

Another crucial aspect of governmental support is the provision of educational and training programs tailored to the needs of family businesses. The government partners with academic institutions, industry associations, and vocational training centers to offer workshops, seminars, and skill development courses covering various aspects of business management, leadership, marketing, and technology adoption (Rutkauskas & Ergashev, 2012). These programs equip family business owners and managers with the knowledge, tools, and resources needed to enhance their competitiveness, adapt to market changes, and sustain long-term growth.

In recent years, the Uzbek government has intensified its efforts to support family businesses as part of broader economic reforms aimed at stimulating private sector development and diversifying the economy. By fostering an enabling environment conducive to entrepreneurship and innovation, the government seeks to unleash the full potential of family businesses as engines of economic growth, job creation, and social development.

However, still there are challenges facing family businesses in Uzbekistan, which include access to financing, bureaucratic hurdles, limited access to modern technology and managerial expertise, and succession planning (Ostonov, 2023). Many family businesses struggle with the transition from one generation to the next, particularly in terms of leadership succession and strategic planning.

Despite these challenges, family businesses in Uzbekistan continue to thrive and adapt to changing market conditions. With the right support and enabling environment, these businesses have the potential to continue serving as engines of growth and sources of prosperity for years to come.

In examining the landscape of family entrepreneurship in Bulgaria, Romania, and Uzbekistan, several interconnected themes emerge, illuminating the dynamic relationship between familial enterprises and the aspirations of the younger generation.

Firstly, it is essential to recognize the enduring significance of family businesses within these nations' economies. Bulgaria, Romania, and Uzbekistan, like many other countries, have long relied on family-owned enterprises as cornerstones of economic activity. These businesses often embody the values of tradition, continuity, and resilience, passed down through generations. The cultural and historical contexts of these countries have shaped the prominence of family entrepreneurship, fostering a deeply ingrained entrepreneurial spirit that transcends societal and economic shifts.

Against this backdrop, the study on student intentions sheds light on the evolving attitudes and aspirations of the youth toward family businesses. By exploring the perceptions, motivations, and career preferences of students in Bulgaria, Romania, and Uzbekistan, the study offers valuable insights into the future of family enterprises within these contexts. It reveals not only the extent to which family businesses remain influential but also how they adapt to attract and retain the next generation of leaders.

One key connection lies in the intersection of tradition and innovation. Family businesses in Bulgaria, Romania, and Uzbekistan often grapple with the tension between preserving heritage and embracing modernization. Similarly, students contemplating careers in family enterprises navigate a similar dichotomy, weighing the allure of tradition against the desire for innovation and growth. The study elucidates how these tensions manifest in students' career preferences, reflecting a nuanced understanding of the evolving dynamics within family businesses.

In conclusion, the study on student intentions regarding family businesses in Bulgaria, Romania, and Uzbekistan serves as a compelling lens through which to explore the intricate interplay between tradition, innovation, succession, and socioeconomic trends within the realm of family entrepreneurship. By contextualizing the findings within the broader landscape of familial enterprises in these countries, a nuanced narrative emerges, elucidating the complex dynamics shaping the future of family businesses and the aspirations of the youth. Moving forward, fostering dialogue and collaboration between academia, industry, and government will be crucial in harnessing the potential of family entrepreneurship as a driving force for economic growth and societal development in Bulgaria, Romania, Uzbekistan, and beyond.

1.3 Student Intentions About Family Businesses in Some Universities in Bulgaria, Romania, and Uzbekistan in 2021 and 2023

Methodology Design

In the previous parts of this chapter, we have pointed out the importance of entrepreneurs to developing business family culture and we may propose the next thesis: *A family business could be sustainable by many factors and one of them is the good business connections of the entrepreneurs with some of their relatives and friends.* From this thesis, we may derive two hypotheses:

H1: The respondents are ready to turn to their relatives and friends for advice when they make decisions.
H2: The respondents rely on their friends and relatives in difficult times.

We checked the two hypotheses by using the database of the INTERGEN international academic network. The international survey was conducted using a questionnaire, which was originally designed in English and then translated into the official languages of each involved country. The main purpose of the questionnaire is to study intergenerational family businesses as a stress management instrument for entrepreneurs. Most answers are based on the Likert scale: 1-No; 2-Rather No; 3-N/A; 4-Rather Yes; 5-Yes.

In 2021, the INTERGEN academic network had collected the answers of 4001 respondents from 20 universities, in 8 countries (Table 1): Albania, Bulgaria, Iran, Poland, Romania, Russia, Serbia, and Uzbekistan. This diversity has empowered the INTERGEN academic network to prepare different comparisons.

This chapter is focused on the answers from three universities: Bulgaria, the University of Ruse "Angel Kanchev"; Romania, the University of Craiova; and

Table 1 Respondents from the second survey (2021–2022) under the INTERGEN project by countries and gender

University	Females	Males	Total
Albania—University of Tirana	137	16	153
Bulgaria—University of Ruse "Angel Kanchev"	220	69	289
Bulgaria—"Konstantin Preslavski" University of Shumen	93	54	147
Bulgaria—Svishtov Academy of Economics "D.A. Tsenov"	106	132	238
Bulgaria—"St. Cyril and St Methodius" University of Veliko Tarnovo	84	16	100
Bulgaria—Technical University of Gabrovo	44	98	142
Bulgaria—University of Economics—Varna	124	75	199
Iran—Allameh Tabatabaei University—Tehran	79	97	176
Poland—"Jan Kochanowski" University in Kielce	148	86	234
Romania—Timisoara Politehnica University	167	148	315
Romania—UBB University Center of Resita	153	49	202
Romania—University of Craiova	154	46	200
Romania—West University of Timisoara	160	53	213
Russia—Chelyabinsk State University	195	84	279
Russia—Lomonosov Moscow State University	81	62	143
Russia—Orel State University	81	126	207
Russia—University of Tyumen	198	46	244
Serbia—University of Belgrade, Technical Faculty in Bor	74	30	104
Uzbekistan—Bukhara Engineering-Technological Institute	64	136	200
Uzbekistan—Management Development Institute of Singapore in Tashkent	65	151	216
Total	**2427**	**1574**	**4001**

Uzbekistan, Management Development Institute of Singapore in Tashkent. These three universities have contributed to the INTERGEN database in 2021 and 2023 by developing Google Forms to obtain answers from their students and alumni. The respondents received a link to the online INTERGEN questionnaire by e-mail or Messenger.

We checked the two hypotheses based on the following INTERGEN questions:

For H1, it will be Question 6 *When I make decisions, I turn to my relatives and friends for their advice.*
For H2, it will be Question 7 *When I have a problem, I share it with my relatives and friends.*

Findings

Tables 2 and 3 present the findings, ordered by countries and the year of responses. The related two figures present the data in percentages.

Figure 1 gives a visualization of the responses to Question 6 in percentages. In 2021, most of the Bulgarian respondents (69.6%) answered Yes and Rather Yes (25.3% and 44.3%, respectively), which shows they are quite positive about turning to their relatives and friends for advice when making decisions. After the COVID-19 period, this share has been reduced to 58.5% (which is 35.1% for Yes and, respectively, 23.4% for Rather Yes). In 2021, most of the Romanian respondents (66.5%, which is 37.5% for Yes and, respectively, 29% for Rather Yes) turn to their relatives and friends for advice when making decisions, while in 2023, this positive attitude goes down to 61.2% (which is 33.8% for Yes and, respectively, 27.4% for Rather Yes). In Uzbekistan, the respondents show opposite results—51.4% (which is 24.1%—Yes and 27.3%—Rather Yes) in 2021 and 58.8% (which is 25.1%—Yes and 33.7%—Rather Yes) in 2023.

Figure 2 gives a visualization of the responses to Question 7 in percentages. In 2021, most of the Bulgarian respondents (77.9%) answered Yes and Rather Yes (32.5% and, respectively, 45.3%), which is an indicator of the high level of trust in their relatives and friends during a period with many challenges. After the COVID-19

Table 2 Responses in 2021 and 2023 to Question 6 "When I make decisions, I turn to my relatives and friends for their advice"

Responds to Q6 "When I make decisions, I turn to my relatives and friends for their advice."	BG 2021	BG 2023	RO 2021	RO 2023	UZ 2021	UZ 2023
No	14	44	17	30	47	45
Rather No	57	67	39	32	36	43
N/A	17	131	11	16	22	22
Rather Yes	128	136	58	55	59	90
Yes	73	204	75	68	52	67
Total responses	**289**	**582**	**200**	**201**	**216**	**267**

Table 3 Responses in 2021 and 2023 to Question 7 "When I have a problem, I share it with my relatives and friends"

Responses to Q7 "When I have a problem, I share it with my relatives and friends."	BG 2021	BG 2023	RO 2021	RO 2023	UZ 2021	UZ 2023
No	9	36	21	38	53	63
Rather No	37	65	49	44	39	52
N/A	18	117	13	22	25	31
Rather Yes	131	129	52	37	67	72
Yes	94	235	65	60	32	49
Total responses	**289**	**582**	**234**	**201**	**216**	**267**

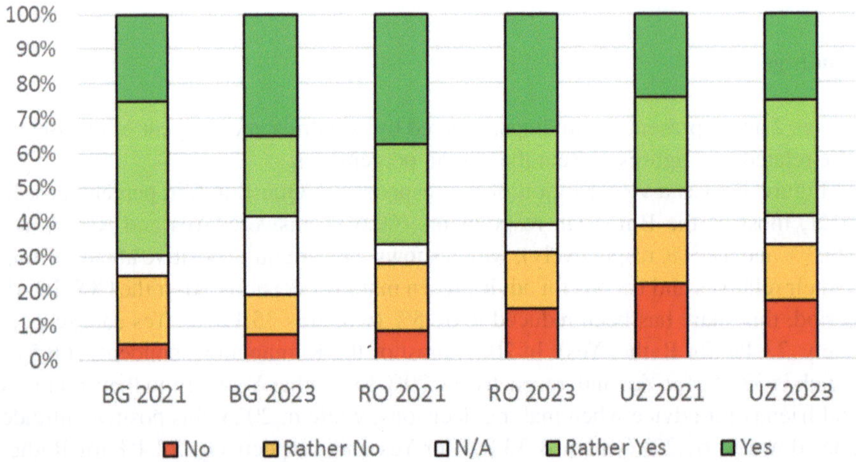

Fig. 1 Responses in 2021 and 2023 to Question 6 "When I make decisions, I turn to my relatives and friends for their advice"

period, this share has been reduced to 62.6% (which is 40.4% and, respectively, 22.2%). In 2021, most of the Romanian respondents—58.5% (which is 32.5% and, respectively, 26%) turn to their relatives and friends to solve problems, while in 2023 this positive attitude goes down to 48.3% (which is 29.9% and, respectively, 18.4%). In Uzbekistan, the respondents show similar results—45.8% (which is 14.8% and, respectively, 31%) in 2021 and 45.3% (which is 18.4% and, respectively, 27%) in 2023.

Discussion

The majority of the respondents turn to their relatives and friends for advice when they make some decisions (Fig. 1). It is a good indicator for keeping some proper long-term relations with them. Both Bulgarian and Romanian students state that in 2021 (the COVID-19 period) they searched for their relatives and friends, while in

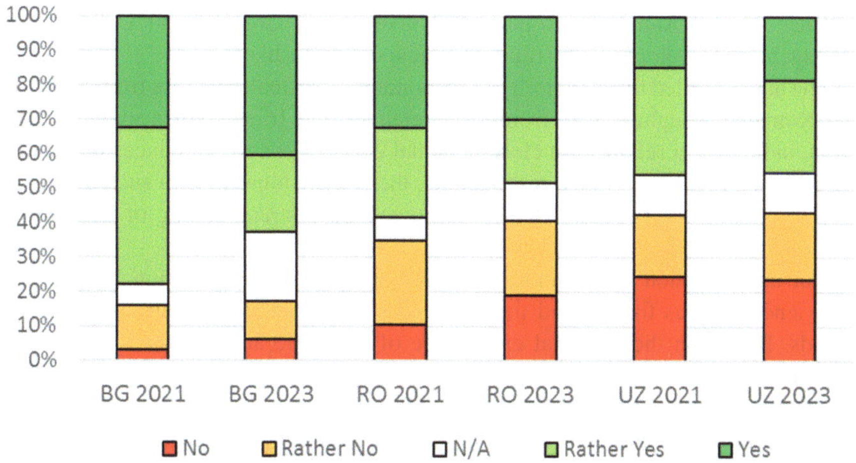

Fig. 2 Responses in 2021 and 2023 to Question 7 "When I have a problem, I share it with my relatives and friends"

2023 this attitude is reduced, but still high. In the context of sustainability, it is not good that some representatives of these two groups of respondents reduce their meetings to discuss important issues. Another crisis could come.

The number of Uzbek respondents who contact their relatives and friends for advice has increased after the COVID-19 period (Fig. 1). It is a good indicator for the development of sustainable relations between them because entrepreneurs do need partners when keeping and developing their businesses. The responses to Question 6 confirm Hypothesis 1—The respondents are ready to turn to their relatives and friends for advice when they make decisions.

Some of the answers to Question 7 show that some Bulgarian and Romanian respondents search for their relatives and friends mainly in times of trouble (Fig. 2). The Uzbek responses are quite similar for the two periods, which means that they have already made clear with whom to stay in contact in good and bad times. The responses to Question 7 confirm Hypothesis 2—The respondents rely on their friends and relatives in times of problems.

2 Conclusion

In conclusion, the findings presented in this research paper underscore the significance of entrepreneurs fostering a business family culture for the sustainability of family businesses. Through an analysis of the role of good business connections with relatives and friends, two hypotheses were formulated and subsequently confirmed by responses from students across three universities: Bulgaria's University of Ruse "Angel Kanchev," Romania's University of Craiova, and Uzbekistan's Management Development Institute of Singapore in Tashkent.

The data revealed that a majority of respondents, particularly those from Bulgaria and Romania, sought advice from their relatives and friends when making decisions, indicating a reliance on close personal connections for guidance. While this trend persisted among Uzbek respondents, there was an increase in such consultations following the COVID-19 period, suggesting a strengthening of ties during times of uncertainty.

Our interpretation is that under conditions of green transformation, people will experience various threats and they will seek assistance from their relatives and friends. Based on the personal experience of the three authors, there is still not enough attention in these three universities paid to training dedicated to strengthening family relationships.

In addition to theoretical contributions, the research provides some ideas for practical application. The insights from this study can help entrepreneurs in family businesses, especially those in the first generation to better understand the younger generations and their opinions and attitudes related to family businesses, including their desire to involve parents, siblings, or grandparents in a business. Building and nurturing these relationships can serve as a valuable asset in decision-making processes and provide support during challenging times. Policymakers can use the findings of this research to inform the development of policies and initiatives aimed at supporting family businesses. By recognizing the role of interpersonal relationships in business sustainability, policymakers can implement measures to facilitate networking opportunities, access to resources, and support mechanisms for entrepreneurs. Business associations and chambers of commerce can organize networking events and platforms specifically tailored to facilitate connections among family business owners, relatives, and friends. These initiatives can provide opportunities for knowledge sharing, collaboration, and mutual support, ultimately enhancing the resilience and sustainability of family enterprises. Finally, family business owners can integrate the insights from this research into their long-term strategic planning efforts. By prioritizing the cultivation of strong relationships with relatives and friends, entrepreneurs can create a supportive network that contributes to the longevity and success of their businesses across generations.

In essence, by recognizing the significance of interpersonal relationships in business decision-making and resilience during challenging times, entrepreneurs can leverage their familial and social networks to enhance the sustainability and longevity of their enterprises. This research provides valuable insights for practitioners and policymakers alike, emphasizing the integral role of personal connections in the pursuit of sustainable business practices within the context of family-owned enterprises.

References

Amore, M. D., Pelucco, V., & Quarato, F. (2022). Family ownership during the Covid-19 pandemic. *Journal of Banking & Finance, 135*, 106385. https://doi.org/10.1016/j.jbankfin.2021.106385

Bakracheva, M., Pavlov, D., Gudkov, A., Diaconescu, A., Kostov, A., Deneva, A., Kume, A., Wójcik-Karpacz, A., Zagorcheva, D., Zhelezova-Mindizova, D., Dedkova, E., Haska, E., Stanimirov, E., Strauti, G., Taucean, I., Jovanović, I., Karpacz, J., Ciurea, J., Rudawska, J., Ivascu, L., Milos, L., Manciu, V., Sheresheva, M., Tamasila, M., Veličković, M., Damyanova, S., Demyen, S., Kume, V., & Blazheva, V. (2020). *The intergenerational family businesses as a stress management instrument for entrepreneurs* (Vol. 1). Academic Publisher University of Ruse "Angel Kanchev".

Barna, C., & Epure, M. (2020). Analyzing youth unemployment and digital literacy skills in Romania in the context of the current digital transformation. *Review of Applied Socio-Economic Research, 20*(2), 17–25. Accessed from https://www.reaser.eu/ojs/ojs-3.1.2-1/index.php/REASER/article/view/61

Bordot, F. (2022). Artificial intelligence, robots and unemployment: Evidence from OECD countries. *Journal of Innovation Economics & Management, 37*, 117–138. https://doi.org/10.3917/jie.037.0117

Bruun, E., & Duka, A. (2018). Artificial intelligence, jobs and the future of work: Racing with the machines. *Basic Income Studies, 13*(2), 20180018. https://doi.org/10.1515/bis-2018-0018

Club Antreprenor. (2024). *Viitorul afacerilor de familie în România – Ediția a II-a, 30 ianuarie 2024*. Accessed March 23, 2024, from https://www.clubantreprenor.ro/2024/01/29/viitorul-afacerilor-de-familie-in-romania-editia-a-ii-a-30-ianuarie-2024/

Coombs, C. (2020). Will COVID-19 be the tipping point for the intelligent automation of work? A review of the debate and implications for research. *International Journal of Information Management, 55*, 102182. https://doi.org/10.1016/j.ijinfomgt.2020.102182

Dawson, N., Williams, M. A., & Rizoiu, M. A. (2021). Skill-driven recommendations for job transition pathways. *PLoS One, 16*(8), e0254722.

Elo, M. (2016). Typology of diaspora entrepreneurship: Case studies in Uzbekistan. *Journal of International Entrepreneurship, 14*, 121–155. https://doi.org/10.1007/s10843-016-0177-9

Ghitulescu, R. (2024). *Afaceri de familie în tandem: Cine sunt frații care au avut succes în România*. Accessed March 23, 2024, from https://www.capital.ro/afaceri-de-familie-in-tandem-cine-sunt-fratii-care-au-avut-succes-in-romania.html

Gibb, A. A. (1987). Enterprise culture—Its meaning and implications for education and training. *Journal of European Industrial Training, 11*(2), 2–38.

Gueorguiev, T., & Kostadinova, I. (2021). ISO standards do good: A new perspective on sustainable development goals. In *Proceedings of the 13th international joint conference on KDKEKM-2021*, Vol. 3, pp. 133–137.

Guitert, M., Romeu, T., & Colas, J. F. (2020). Basic digital competences for unemployed citizens: Conceptual framework and training model. *Cogent Education, 7*(1), 1748469. https://doi.org/10.1080/2331186X.2020.1748469

Harry, W. (2007). Employment creation and localization: The crucial human resource issues for the GCC. *The International Journal of Human Resource Management, 18*(1), 132–146. https://doi.org/10.1080/09585190601068508

International Labour Organization. (2018). *World employment and social outlook 2018: Greening with jobs*. ILO.

Işcan, E. (2021). An old problem in the new era: Effects of artificial intelligence to unemployment on the way to industry 5.0. *Yaşar Üniversitesi E-Dergisi, 16*(61), 77–94. https://doi.org/10.19168/jyasar.781167

Isokova, M. (2023). Doing small business and private entrepreneurship of each family in Uzbekistan. In *Ethiopian international multidisciplinary research conferences*, pp. 39–42.

Karajanova, R., & Rakhimboev, M. (2023). Family business development in the Republic of Uzbekistan: Features and challenges. *Innovative Development in Educational Activities, 2*(12), 16–26.

Kehinde, E. E., & Olatunde, A. M. (2022). Digital skills needed by business education graduates for unemployment reduction in the 21st century. *Nigerian Journal of Business Education (NIGJBED), 9*(1), 134–148.

Licastro, A., & Sergi, B. S. (2021). Drivers and barriers to a green economy. A review of selected Balkan countries. Cleaner. *Engineering and Technology, 4*, 100228. https://doi.org/10.1016/j.clet.2021.100228

Mardievna, S. G., & Oblokulovich, K. S. (2021). Methodology for determining the role of family business in the economy. *European Business and Management, 7*(6), 199. https://doi.org/10.11648/j.ebm.20210706.16

McGaughey, E. (2022). Will robots automate your job away? Full employment, basic income and economic democracy. *Industrial Law Journal, 51*(3), 511–559. https://doi.org/10.1093/indlaw/dwab010

Mutascu, M., & Hegerty, S. W. (2023). Predicting the contribution of artificial intelligence to unemployment rates: An artificial neural network approach. *Journal of Economics and Finance, 47*(2), 400–416. https://doi.org/10.1007/s12197-023-09616-z

Ng, E., & Stanton, P. (2023). Editorial: The great resignation: Managing people in a post COVID-19 pandemic world. *Personnel Review, 52*(2), 401–407. https://doi.org/10.1108/PR-03-2023-914

Nguyen, Q. P., & Vo, D. H. (2022). Artificial intelligence and unemployment: An international evidence. *Structural Change and Economic Dynamics, 63*, 40–55. https://doi.org/10.1016/j.strueco.2022.09.003

Olimjanova, S. O. (2024). The features of the organization of family entrepreneurship in the agricultural sector of Uzbekistan. *Journal of Modern Educational Achievements, 3*(1), 26–32.

Oppenheimer, V. K. (1994). Women's rising employment and the future of the family in industrial societies. *Population and Development Review, 20*, 293–342. https://doi.org/10.2307/2137521

Ostonov, O. (2023). The situation of handicrafts and family business in Uzbekistan (on the example of the pandemic period). *Society and Innovations, 4*(3/S), 369–379.

Pavlov, D., Sheresheva, M., & Perello, M. (2017). The intergenerational small family enterprises as strategic entities for the future of the European civilization - A point of view. *Journal of Entrepreneurship & Innovation, Issue 9*, 121–133. Accessed from https://jei.uni-ruse.bg/Issue-2017/10.%20Pavlov_Sheresheva_Perello.pdf

Pirzada, K., & Khan, F. (2013). Measuring relationship between digital skills and employability. *European Journal of Business and Management, 5*(24).

Rahman, T., Ali, S. M., Moktadir, M. A., & Kusi-Sarpong, S. (2020). Evaluating barriers to implementing green supply chain management: An example from an emerging economy. *Production Planning & Control, 31*(8), 673–698. https://doi.org/10.1080/09537287.2019.1674939

Rivo-López, E., Villanueva-Villar, M., Vaquero-García, A., & Lago-Peñas, S. (2022). Do family firms contribute to job stability? Evidence from the great recession. *Journal of Family Business Management, 12*(1), 152–169. https://doi.org/10.1108/JFBM-06-2020-0055

Rizos, V., Behrens, A., Hofman, E., Ioannou, A., Kafyeke, T., Flamos, A., Rinaldi, R., Papadelis, S., & Topi, C. (2016). Implementation of circular economy business models by small and medium-sized enterprises (SMEs): Barriers and enablers. *Sustainability, 8*(11), 1212. https://doi.org/10.3390/su8111212

Rutkauskas, A. V., & Ergashev, A. (2012). Small business in Uzbekistan: Situation, problems and modernization possibilities. In *7th international scientific conference on business and management*.

Simić, I. (2019). Digital competencies in the function of reducing unemployment. *Tourism International Scientific Conference Vrnjačka Banja - TISC, 4*(1), 288–306. Accessed from https://www.tisc.rs/proceedings/index.php/hitmc/article/view/257

Stofkova, J., Poliakova, A., Stofkova, K. R., Malega, P., Krejnus, M., Binasova, V., & Daneshjo, N. (2021). Digital skills as a significant factor of human resources development. *Sustainability, 14*(20), 13117. https://doi.org/10.3390/su142013117

Tadjiev, B. U., Ataev, J. E., Akhmetshin, E. M., Vasilev, V. L., & Kukhar, V. S. (2023). Assessment of the effectiveness of the reforms to support entrepreneurship in Uzbekistan. In *E3S web of conferences* (Vol. 449, p. 03002). EDP Sciences.

Todorova, M., Ruskova, S., & Kunev, S. (2018). Research of Bulgarian consumers' reactions to organic foods as a new product. In *The 6th International Conference Innovation Management, Entrepreneurship and Sustainability (IMES 2018)*. https://doi.org/10.18267/pr.2018.dvo.2274.0

Toffler, A. (1980). *The third wave*. William Morrow Publisher.

Tolipov, F. (2024). Family business in Uzbekistan and its role in public life of people (historical and social analysis). *Oriental Journal of Social Sciences, 4*(01), 10–37547. https://doi.org/10.37547/supsci-ojss-04-01-03

Welter, F., Smallbone, D., & Mirzakhalikova, D. (2017). Women entrepreneurs between tradition and modernity – The case of Uzbekistan. In *Enterprising women in transition economies* (pp. 59–80). Routledge.

Willcocks, L. (2020). Robo-apocalypse? Response and outlook on the post-COVID-19 future of work. *Journal of Information Technology, 36*(2), 188–194. https://doi.org/10.1177/0268396220978660

Websites

https://www.rawlala.ro/
www.aliasunicorn.studio

Daniel Pavlov, PhD, is Associate Professor at the University of Ruse "Angel Kanchev," Bulgaria. His PhD thesis is in the field of the management of economic complexes, and his monograph is related to the business models under Osterwalder's Canvas. In 2004, he had a 4-month PhD specialization under Project 44745/A001—Special Educational Partnership Program in Regional Development Management with Bulgaria ECA/A/S/U-03-22—Cornell University, Ithaca, New York, USA. In 2009, he had a 6-day specialization in European entrepreneurship colloquium (Netherlands) with professors from Harvard Business School, Stanford University, the European Forum for Entrepreneurship Research, Katholike Universiteit, and RWTH Aachen University. For the period 2005–2019, he has been ERASMUS Lecturer in Risk Management at Karel-de-Grote Hogeschool, Antwerp, Belgium, with students from all EU countries; Lecturer of the courses: Small Business Management, Business Models for Entrepreneurs, Risk Management, and Social Entrepreneurship; Head of the Entrepreneurship Centre at the University of Ruse, 2008 to present; Coordinator of the Master Program in Entrepreneurship and Innovation at the University of Ruse, 2009 to present; Member of the RESITA net (Academic Network in Southeast Europe in Entrepreneurship and Innovation), 2008–2018; and Co-founder of INTERGEN academic network "The intergenerational family businesses as a stress management instrument," 2018 to present (http://intergen-theory.eu/). Since 2013, he has been an expert at the European Commission on the HEInnovate topic. This online self-assessment tool helps the academic community identify academic entrepreneurship opportunities. He was invited to the HEInnovate events as a speaker or an expert in Brussels in 2014, 2015, 2016, 2017, 2018, 2019, 2021 (online), 2022, and 2023. He has experience in the following international academic projects: (1) (2023–2026): 101087248-GET-AHED-ERASMUS-EDU-2022-PI-FORWARD. "Green Education and Transition - A Higher Education Digital Buddy" (GET-AHED). Position: Technical coordinator of the University of Ruse. (2) (2019–2022): 612887-EPP-1-2019-1-AT-EPPKA3-PI-FORWARD. "Developing the

Organisational Capacity of Higher Education Institutions using the HEInnovate platform to facilitate peer learning and a pan-European community of practice" (BeyondScale). Position: Technical Coordinator of the University of Ruse. (3) (2013–2017): 544573-TEMPUS-1-2013-1-BG-TEMPUS-JPHES (2013-4571/001-001) MATcHES "Towards the Modernization of Higher Education Institutions in Uzbekistan." Position: Coordinator of the entire consortium. (4) DNTS 02-26/01.10.2010 "Comparative study regarding the training needs for development of entrepreneurial competencies in the context of E.U. Post-integration," funded by Bulgarian Ministry of Education and Romanian Ministry of Education, with EU funds. Position: Coordinator of the Bulgarian team.

Silvia Puiu is Associate Professor PhD Habil. at the Department of Management, Marketing and Business Administration within the Faculty of Economics and Business Administration, University of Craiova, Romania. She has a PhD in Management and teaches Management, Marketing, Ethics Management in Business, and Creative Writing in Marketing. In 2014, she conducted research on *Ethics Management in Higher Education System of Romania*. In 2015, Silvia Puiu completed her postdoctoral studies on *Ethics Management in the Public Sector of Romania*. During the last few years, Silvia Puiu has published more than 60 articles in national and international journals or in the proceedings of international conferences. Her research covers topics from corporate social responsibility to ethics management, sustainability, marketing, and management. She is also the founder of an NGO in Romania—Building Hopes—which organizes workshops for non-formal education of youngsters.

Deniza Alieva, is Docent of School of Business and Management of the Management Development Institute of Singapore in Tashkent, Uzbekistan. She holds a PhD degree in Human Resources Psychology and a DSc degree in Economics. Her recent years of research focused on a variety of topics such as sustainability, community development, tourism and governance, education, and HRM. Deniza's practical and academic expertise permits her to take part in various projects, such as ADJACENT (Ministry of Science and Innovations of Spain), UNDP's Climate Promise: Policy Action for climate security in Central Asia (UNDP), Feasibility Study on Scaling Innovation (KIX EAP), and Development of Tourism Diversification Strategy for the Khorezm Region of Uzbekistan (UNWTO, EBRD, and Ministry of Tourism and Cultural Heritage of Uzbekistan). Currently, Deniza acts as an editorial member, member of the advisory board, and reviewer in such journals as *REDES: Revista hispana para el análisis de redes sociales*, the *Journal of Community Psychology*, and *SN Social Sciences*.

Sustainability in Aviation: CSR and Air Transportation

Panagiotopoulos Φ. Ioannis

1 Introduction to Sustainability and CSR

Sustainability is an international trend that is suddenly overused in an increased number of fields of economic activities. Its concept as per a popular and initial description given by United Nations (UN) consists of this kind of development that covers today's needs without compromising future needs of next generations (Brundtland, 1987). Sustainability focuses on an equilibrium between human activities and economic development on the first side with environmental preservation and social welfare on the other side. Several side effects of intense human economic activity (that mainly targets to profit) like climate change, pollution of oceans, soil, and atmosphere, melting of ice in poles, and in total the phenomenon of environmental degradation which is developed with continuously increased pace has signaled civil society, politicians, scientists, and, finally, business people and companies to consider corporations as corporate citizens or social enterprises. The concept of corporate citizenship is a theory that allocates responsibilities to companies either for environmental or social issues considering them as entities with moral responsibilities to their internal and external environment (Carroll, 1998; Crowther & Seifi, 2021; Dahl, 1972; Whitehouse, 2003).

Corporate citizenship is a part of Corporate Social Responsibility (CSR) ideological complex. CSR is described as an idea whereby companies incorporate social and environmental concerns in their decision-making processes and everyday business operations, considering their interaction with their stakeholders on a voluntary basis (European Commission, 2001). CSR reflects the formation of a vehicle for the

P. Φ. Ioannis (✉)
Department of Business Administration, Mediterranean College - Derby University, Athens, Greece

S. O. Idowu, S. Vertigans (eds.), *Sustainability in Global Companies*, CSR, Sustainability, Ethics & Governance, https://doi.org/10.1007/978-3-031-77971-8_3

implementation of sustainable principles in business field by corporations through the design and implementation of CSR strategy.

The CSR policy should recognize the critical stakeholders of a corporation and provide a constructive equilibrium among their interests and company's everyday operations. The term stakeholders represent every party inside or outside a corporation that could potentially be affected by the everyday activity of it like its employees and their families, its suppliers, the local community, etc. These stakeholders keep an interest in corporation's decisions and in many cases have a level of influence over the shaping of those decisions, too (Crowther & Seifi, 2021: 19–20).

2 Introduction in Aviation

Aviation is one of the most globalized sectors of international economy. It is responsible for the transportation of 65.6 million cargo tones annually that correspond to 35% of global trade by value or to 1% of global trade by volume. This quantity equals with $8.5 trillion worth of goods annually, or $23.2 billion worth of goods every day (IATA, 2024b). Regarding the transportation of passengers, a hundred million passengers were using aircrafts in 1960, while this number became 4.5 billion in 2019 (EESI, 2022).

The main factors in aviation are the air carriers, the airports, the air navigation service providers (ANSPs), and the control centers (ACCs) along with the International Air Transportation Association (IATA) and the International Civil Aviation Organization (ICAO). Aviation industry consists of several airline business models, e.g. business aviation, traditional, low cost, regional, holiday, and cargo airlines (Efthymiou & Papatheodorou, 2018b). These abovementioned factors are also considered stakeholders of aviation along with the passengers, the employees like pilots, cabin crew, ground support personnel, engineers and technicians, administrative people at the offices, the employees' families, the citizens who live in immediate vicinity of airports, etc.

The most important effects of air transportation to environment are (Panagiotopoulos, 2020):

- Noise pollution
- Atmospheric pollution through emissions of air pollutants and micro-particles
- Exhaust of non-renewable resources (oil)

ICAO (2024) has agreed with its member states to direct its efforts to three main fields where it is critical for air carriers to take action:

- Climate change and aviation emissions
- Aircraft noise
- Local air quality, in areas in close proximity to the airport

This is the reason that ICAO (2008) has created analogous environmental indices for the measurement of:

- Number of people who are exposed to high level of noise caused by aircrafts based on a 3 years' variable average
- Production of air emissions like CO_2, NOx, H_2O, and micro-particles
- Efficiency of the use of air fuel proportionally with the earnings of the aircraft and the performed distance based on a 3 years' variable average

The burn of air fuel in the turbines of the aircrafts causes the emission of air pollutants which are by 70% CO_2, by 29% vapors, and the rest is a mix of several gases:

- NO_x, nitrous oxides in mixtures with (NO_2) nitrogen dioxide and monoxide (NO)
- N_2O, dinitrogen monoxide
- CH_4, methane
- VOC, hydrocarbons without methane (NMHC)
- CO, carbon monoxide
- PM, particulate matter, the most dangerous are those with average aerodynamic diameter less than 10 μm (PM10) and 2.5 μm (PM2.5)
- SOx, sulphur oxides

According to the International Council on Clean Transportation (ICCT), global CO_2 from commercial aviation was 707 million tons in 2013. In 2019 that value reached 920 million tons, having increased approximately 30 percent in 6 years. Although the collective usage of private cars, the power generation, the industrial and rural sectors contribute more to climate change phenomenon than air transportation, passenger air travel was resulting in the highest and fastest growth of individual emissions in the period prior of coronavirus, even though there is a significant improvement in efficiency of aircraft and flight operations over the last 6 decades (Graver et al., 2020).

Many of the abovementioned gases are considered Green House Gases (GHGs). CO_2 is categorized as the main GHG. Jet fuel consumption produces CO_2 at a defined ratio (3.16 kilograms of CO_2 per 1 kilogram of fuel consumed), regardless of the phase of flight. However, there are more gases that contribute to this phenomenon whose impact is under investigation. The effect of these gases in the Green House phenomenon varies. For example, CO_2 remains for a long time in the atmosphere (30 percent of a given quantity of the gas is removed from the atmosphere naturally over 30 years, an additional 50 percent disappears within a few hundred years, and the remaining 20 percent stays in the atmosphere for thousands of years) while vapors for a limited period. Aircrafts emit nitrous oxides in very small quantities and only during the phase of non-fuel-efficient operation of their turbines during flight (EESI, 2022; Muhammad-Azfar et al., 2015).

Water vapor has been recently recognized as a very important gas that counts for almost 30 percent of the exhaust. It has a short lifespan in the atmosphere since it is part of the water cycle, and minimal direct warming impact. However, it is responsible for the phenomenon of contrails of aircrafts. The condensation of the water vapor in the exhaust when the ambient temperature is cold enough, as particulates in the exhaust form the nucleus of ice crystals, creates the contrails in the tail of aircrafts. When the ambient atmosphere is sufficiently humid and cold, these small ice crystals increase their size as they attract water vapor from the atmosphere and are sustained as contrails that can spread horizontally and vertically to form contrail-induced cirrus clouds. These microcrystals cover 0.1% of the earth surface contributing in the clouding of the planet. However, their allocation is not characterized by uniformity since it depends on the local meteorological parameters and the regional air-traffic (Williams et al., 2002). These lingering contrails and contrail-induced cirrus clouds trap infrared rays, producing a warming effect up to three times the impact of CO_2. Even though these cirrus clouds have a relatively short life span, usually a few hours, their collective influence, produced by thousands of flights, has a serious warming effect. The effect is intense and as per the latest researches it is considered as having more contribution in the total warming than all the CO_2 emitted by aircrafts (EESI, 2022).

It is important to be highlighted that the emissions of these gases happen in high altitudes and this characteristic changes the impact of them on the atmosphere comparing with the same emissions on the ground. This is why scientists have changed their opinion regarding their negative effect to the environment and global warming since more studies gradually have contributed to a wider examination of the issue.

The noise caused by aircrafts is another negative impact of air transportation. This impact is intense when airplanes perform a flight in low altitude, thus, during the take-off and the landing phases of the flight when they are close to airports and their surrounding areas. The aircraft noise could be of high volume and duration proportionally with the size of the airport and its traffic. The factors that could enhance this problem are (ICAO, n.d.):

- The types of aircrafts that are in use in the airport
- The total annual number of take-offs and landings, and the maximum number of them in specific periods of high traffic
- The busiest period of the day when the majority of the flights are performed
- The most frequent air routes (flight paths) on the sky that are expected to be in use for the reduction of noise and whether they are followed
- The runways used in the airport
- The average weather conditions in the area of the airport
- The topography in the area where the airport is located
- The position and the size of the urban areas that surround the airport
- The processes for the everyday operation of the airport
- The general conditions that affect the everyday operations in the airport

The development of technology of jet turbines and in aerodynamics has contributed significantly in the reduction of aircraft noise pollution. For example, a Boeing 747-8 has by 30% less noise footprint than the previous model while its CO_2 emissions are reduced by 15% (https://www.boeing.com/commercial/747-8). Modern aircrafts are in average 20dbs less noisy than previous generation of aircrafts and this difference corresponds to 75% reduction in their noise footprint. For the further control of aircraft noise pollution, aircrafts follow special instructions, processes, and limitations during their landing and take-off modes. As an overview, modern aircrafts are 50% less noisy than those which were in the forefront of technology ten year earlier and it is projected that an extra 50% reduction will be achieved during the third decade of the twenty-first century (Dickson, 2014).

3 Sustainability and CSR in Aviation Sector

In global level, the concerns about aircraft noise pollution and pollution of planet's atmosphere, and their impacts on the health and quality of life have increased the environmental regulations and restriction in aviation. These measures have increased the operational and fixed costs of air carriers and airports. Furthermore, the abovementioned situation put at risk their ability to manage the continuous development of the aviation industry (Neufville & Odoni, 2003).

The majority of actions and policies that have been applied at the end of the twentieth century and at the beginning of the twenty-first century in favor of environment had had a "prohibition character" like new stringent standards for aircraft engines and limitations for the flights. In the last years, the interest has been shifted to initiatives that propose more incentives than restrictions and fines. Under this consideration could be categorized policies like taxes on fuels and on air-pollutants, and a trading system for these emissions. Sweden levies a prototype taxation on the number of landings of aircrafts. Initially, the criterion for this regulation was related with the noise pollution in the areas around the airports and gradually its concept was developed more including the emission of specific air pollutants like hydrocarbons (HC) and nitrous oxides (NO_x) (Carlsson & Hammar, 2002).

In 2007, the stakeholders of aviation agreed mutually in a strategy with four main pilons that was attested by ICAO. These four pilons concern fields where the stakeholders should direct their efforts: (a) improved technology, (b) efficient processes, (c) positive financial measures, and (d) effective infrastructure.

Technology is the field among the abovementioned four axes which keeps the best prospects to be faster developed within next years, especially through the design and manufacturing of new aircrafts with modern engines that emit less air pollutants than their predecessors. New lighter materials, improvements in engines, new sustainable aviation fuel (SAF), etc., provide ground for optimism that aviation will reduce impressively its environmental footprint within next decades. Furthermore, the design and adoption of new operational procedures during flight could result in less air pollutant emissions. Such processes could comprise more efficient processes for the implementation of flight itself, measures for the decrease of the aircraft weight, optimization of the net of itineraries, and holistic flight management that minimizes empty seats and adjusts proportionally the frequency of flights (Neufville & Odoni, 2003).

Improved infrastructure with more efficient premises and modern systems that control energy consumption, adjust air traffic, and regulate the operation of airports could contribute to the battle against pollution. However, among these four axes, only the positive financial measures have the power to imply results in short term until the rest efforts show their positive impact in the minimization of aviation environmental footprint. Thus, financial movements like taxes, fines, incentives for fleet renewal, a system for trading transactions, etc., could tackle the negative impact of aviation on society and environment in short to mid-term.

4 Initiatives by European Union-European Trading System (ETS)

European Union (EU) is at the forefront of tackling the negative impact of aviation in the environment in global level. The two major initiatives by EU are: a) the European Union Emissions Trading System (EU ETS) and b) the unification of the European air space (Single European Sky—SES).

The Kyoto Protocol in 1997 was the first prestigious initiative for air pollutants in aviation. It established binding emission reduction targets for thirty-seven (37) industrialized countries and economies in transition and the EU. One of the first goals was the decrease by 5% in emissions compared with 1990 levels the period 2008–2012. Specifically for aviation, that goal embraces only the internal flights of the states. The international flights that were not covered by the Kyoto protocol were included in the EU ETS in 2008. ETS is a system for trading rights for emissions that is not exclusively designed for aviation. It embraces several sectors of economy, mainly industrial sectors, power generation, aviation, etc. It is focused on the trading of rights for CO_2 emissions without considering other critical air pollutants like nitrous (NOx) or sulphur oxides (SOx). The way of its function is described by the Moto "cap & trade". This means that each corporate entity in the selected sectors that is a producer of CO_2 due to its everyday core activities has obtained an up limit of emissions of CO_2. On the same time, it keeps the right to trade rights for emissions either by selling its rights to other companies whether it considers that could achieve reduction in its annual emissions or by buying rights from other companies whether it considers that it is going to overpass its allowed threshold. Furthermore, there is provision for fines when the regulation is violated. ETS does not include special categories of flights like military flights, special purpose flights, etc.

Several big air carriers and countries outside EU like China, Russia, and the USA expressed their objection to this legislation and asked to be exempted by these rules because they are not state-members of EU. Even though they did not manage to be exempted as per the decision made by the European Court, EU has decided to exempt from ETS those flights with taking-off or landing in airports outside EU or outside from the European Economic Area (EEA) in 2013. However, there are air carriers which although had the chance to escape from the compliance with ETS, they preferred to follow its application like Korean Air, Fed Ex, and Nippon Air because they recognized the environmental advantages of its application (Efthymiou & Papatheodorou, 2019).

The second very important initiative that came up from EU is SES. European air space is extremely congested since the daily average of flights is more than thirty-three thousand. In addition, it is also a complicated air space considering the management of flights since it consists of several smaller national air spaces. The complexity is leveling due to the existence of thirty-seven (37) Air Navigation Service Providers (ANSPs) and sixty-three Area Control Centers (ACCs). This segmentation of European air space has some negative results in the flights since aircrafts have to perform as average forty-two (42) kilometers more per flight in comparison with the existence of a single European air space. This means increased fuel consumption, increased pollutants, delays, and increased pricing for the passengers. Comparing the European example with the situation in the USA, where there is an air space of approximately equal size with the European air space, there are double flights per day that are managed by a single ANSP with reduced operational cost.

The realization of these inherent difficulties that could be transformed into inherent disadvantages in the international competition created the background for the creation of SES in 1999 which has as target the continuous improvement of European air traffic management (Efthymiou & Papatheodorou, 2018a). European Commission is responsible for SES and has divided it into nine (9) parts aiming to its efficient operation. The expected advantages are: a) the increase of safety by ten (10) times, tripling of available air space, the reduction of cost or air traffic control by fifty percent (50%), and the reduction in air pollutants by ten percent (10%).

SES application is gradual because there are many stakeholders that should be cooperated. In general, these two initiatives prove that EU wishes to be a leader in the efforts against climate change and for the further development of aviation in compliance with the utmost environmental standards. EU has been in negotiations with ICAO for the establishment of a global trading system for the emissions by all air carriers. However, EU had announced that the failure in the design and application of a global trading system by ICAO that could satisfy EU's targets could result in the application of ETS in its full range on 1/1/2024. However, the current status in 2024 is that EU is going to continue to apply the EU ETS solely to intra-EU flights, but a review will extend it to all flights into and out of the EU after 2026, unless the Carbon Offsetting and Reduction Scheme for International Aviation (CORSIA) is "positively evaluated". Thus, the previous date of 2024 has been postponed for 2027 unless ICAO manages to fulfill the European requirements for such a trading system. In the meantime, EU continues negotiations with other states, ICAO and IATA for the proper scheme of a trading system or an equivalent system that will operate as a vehicle toward the major target of net zero emission by 2050. If the outcome of these discussions negative, EU's intention is to apply ETS in its extended full scope after 2026 (Schvartzman, 2023).

5 Initiatives by ICAO

ICAO has taken very important initiatives in its effort to alert air carriers for the negative impacts of air transportation to environment. ICAO implements studies, researches and issues reports that are announced in its members providing analysis of the best practices in the field of aviation. Main target for ICAO is the production of air pollutants in 2050 to be the half of that of 2005 and the stabilization of improvement pace in fuel efficiency at the level of 1.5–2% up to 2050 (IATA, 2013).

ICAO has oriented its interest to new lighter materials for the manufacturing of aircrafts, new technologies for engines, more efficient processes, and the commercialization of alternative fuels for aircrafts. A major issue for ICAO is the establishment of a measure that could control the production of CO_2 while it is a market-based measure (MBM). Such a measure should include a common method for the estimation of air pollutants of aircrafts internationally, processes for monitoring, reporting and verification of emissions, and a mechanism for the determination of the offsetting programs quality.

ICAO has created a series of tools for the environmental protection like the ICAO Carbon Emission Calculator which measures the CO_2 emissions during a flight for the design of corresponding offsetting programs. There is also the ICAO Green Meetings Calculator which estimates the potential production of CO_2 in the decision-making process for the selection of the venue of a conference. ICAO offers the Fuel Savings Estimation Tool that supports small air carriers and developing countries to design energy management programs and justifies the expected advantages (ICAO, 2024).

The most promising program is CORSIA (Carbon Offsetting and Reduction Scheme for International Aviation) that is designed to control CO_2 in all international flights. CORSIA consists of an international initiative for controlling air emissions from flights that goes beyond regional or national interests and points of view. It is considered as an international market-based measure, brand-new, and prototype for any industrial or business field. It takes care of the smooth operation of air transportation market avoiding the deterioration of competition principles and respects the particularities of ICAO Member States. CORSIA tackles the CO_2 emissions that could not be eliminated with measures and initiatives like SAF, technological and operational improvements, by offsetting them with emissions units from the carbon market (ICAO, 2023a, 2024). A first main target is air carriers to manage to offset the extra quantity of CO_2 that may generate in comparison with the levels of 2020 through environmental initiatives and offsetting actions. It has been estimated that the first years of application, the target of zero net emissions was achieved partially. The participation in the program is initially voluntary. It is worth to be mentioned that all member states of EU participate in this program (European Commission, 2024).

ICAO has designed relevant initiatives to tackle aircraft noise pollution based on decisions made by the organization like ICAO[a]-Doc 9829—Guidance on the balanced approach to aircraft noise management, Appendix C of Assembly Resolution

A35-5, Appendix C of Assembly Resolution A36-22, etc., taking into consideration proper scientific research (Environmental Assessment Standards Manual (IESM), 2015). As per ICAO, the four cornerstones for facing this negative impact of aviation are:

- Reduction of the volume of noise in the source that produces it. This principle has led ICAO to establish specific standards for the noise concerning the construction of new types of aircrafts.
- Proper design and management of operations on the ground. This means that the area around the airports should not have communities with residents or other activities that are considered of high noise pollution that could act cumulatively.
- Design and application of operational functions to limit noise. This policy includes instructions and technics for the selection of runways, air routes, and trajectories of the aircrafts. The particularities of each airport, the geographical allocation of the populated areas in proximity with airports, and the stringent application of all safety regulations should be considered too.
- Establishment of restrictions for the use and operation of aircrafts. There are some countries which have banned the operation of specific aircraft types in specific airports which are already congested with passengers and aircrafts facing high levels of noise pollution. This measure is particularly tough for air carriers either of small size or from developing countries which cannot easily renew their fleet with brand-new aircrafts. This is why ICAO suggests this measure to be examined after having applied all other possible measures which do not burden disproportionally such vulnerable air carriers threatening the fair play status in the industry (IATA, 2015)

Since 1981, ICAO has suggested charges either to airports with high aircraft noise pollution or to air carriers which occupy aircraft types that produce significant noise in comparison with the relevant average. ICAO has encouraged states to consider even the application of both measures. However, these charges should not be high enough to make unprofitable the use of specific aircraft types because this equals with intervention in the competition of the air transportation and could be of favor of bigger air carriers of developed and wealthier countries (ICAO, 2008).

6 Initiatives by IATA

IATA has also adopted as main part of its mission to tackle the phenomena of climate change and global warming. A program for offsetting CO_2 is in use having as target to calculate the total quantity of CO_2 that is produced during a flight and support proportional offsetting policies by the air carriers in order the balance to become ideally zero (IATA, Aviation Carbon Offset Programmes, IATA guidelines and toolkit).

Another program of IATA is the IATA Environmental Assessment for air carriers that initiated in 2015 and the participation in that is voluntarily. It includes the assessment of air carriers by specialized external independent organizations regarding their compliance with modern environmental standards and the adoption of the best environmental practices of the industry (IATA, 2015). This program is in alliance with Eco-Management and Audit Scheme (EMAS) and ISO 14001 that are both popular environmental tools. It is divided into two stages. At the first stage, there is an assessment process for the several phases and parts of a flight like the start-up of engines, the take-off and landing, the potable water usage, etc. The second stage focuses on corporate level and examines activities like heating, ventilation, and air conditioning (HVAC) in corporate offices, logistics and supply chain procedures, etc., regarding their environmental footprint (Environmental Assessment Standards Manual (IESM), 2015).

IATA has also developed an electronic platform called "FRED+" taking into consideration the introduction of CORSIA by ICAO. This initiative has as target to facilitate air carriers to implement reporting based on the standards of CORSIA. Thus, the users of this program are the air carriers themselves either they are members of IATA or not. Through this way, it is facilitated the compliance of both air carriers and ANSPs with CORSIA by assuring the direct exchange of data and information without any cost about fuel consumption, air pollutants, etc. (IATA, 2024a).

7 Initiatives by United Nations

The Global Compact consists of a cornerstone in the history of CSR. It is a global effort by the United Nations (UN) to join the efforts by corporations, labor unions, organizations made by simple citizens, etc., in the sectors of human rights protection, working rights, elimination of corruption, and environmental protection. Global Compact is the main initiative of UN for the spread of CSR throughout business world and industry and consists of a call for every modern company to adopt its ten (10) main sustainable principles. The initiative was announced in Davos by the General Secretary Kofi Annan in 1999. In 2015, there were four thousand five hundred (4500) national or international business and non-business participants while nowadays this number has been six (6) times bigger with twenty-four thousand six hundred twenty-five (24,625) participants from one hundred sixty-seven (167) countries. The Global Compact has founded national networks in each country where the local enterprises can join and through them, they become automatically participants in the global effort. There are currently sixty-one (61) local networks. Many air carriers have been participants in this initiative like Lufthansa, KLM Royal Dutch, Air France, Emirates, etc. (United Nations, 2024).

The restructured concept behind the UN Global Compact is to increase business awareness and action in favor of fulfilling the Sustainable Development Goals by 2030. United Nations (UN) established the 2030 Agenda for Sustainable Development in 2015 as a frame of targets that could direct humanity toward sustainability. The 2030 for Sustainable Development operates as a compass for governments, international organizations, non-governmental organizations (NGOs), and corporations having seventeen (17) main sustainable goals called Sustainable Development Goals (SDGs) These goals comprise one hundred sixty-nine (169) more analytical targets and two hundred thirty-two (232) indicators. Aviation is an industry that is directly related with Goal 7 about energy and Goal 13 about Climate action based on the environmental axis of CSR. However, aviation could be connected with more SDGs like those related with gender equality and women's empowerment, economic growth, infrastructure, industrialization, etc. In general, CSR in aviation could potentially be linked with all these SDGs because aviation could possibly either be directly or indirectly related with them and air carriers have the economic power to interfere positively through specialized CSR programs in the relevant fields.

8 Case Study: Presentation of CSR Policy Designed by Emirates

Emirates Airlines is at the forefront in the battle for the mitigation of negative impacts of aviation on environment and quality of life, with a really extensively developed CSR strategy. These initiatives are characterized of high financial and technological demands and embrace investments in Research and Technology (R & D), specialized software for energy management during flight, new equipment, new aircrafts, new processes and regulations. Emirates participates in the Global Compact since 2023, too.

The air carrier has invested in the re-design of flights and the optimization of their trajectory. A program called Flextracks includes the evolution of technology that could provide freedom in the selection of proper trajectory for a specific flight instead of the existence of pre-selected flight trajectories. New routes are designed each time by using a specialized software that needs as input meteorological data and suggests the best route having as priority to keep the flight safe from bad weather conditions, to achieve time savings and optimization of fuel usage. Through this orientation, the control and decrease of air emissions is achieved, too. The initial application of this program for one year in the itinerary Sidney-Melbourne-Dubai with a total of five hundred ninety-two (592) flights was the cause for savings of six hundred twenty-eight (628) tons of fuel and fifty-seven (57) hours of flight. Every minute of flight corresponds to sixty-two (62) liters of air fuel and one hundred sixty (160) kilograms of CO_2. The savings per flight as average was estimated as six (6) minutes of flight, approximately half tone of fuel and approximately a tone of CO_2 (Emirates, 2019).

Some states have allowed the company to use a special software that improves the trajectory of the aircraft amid the flight (Re-routing en route). This is effective especially in the intercontinental itineraries where an airplane flies in high altitudes for many hours and the meteorological data are updated periodically every few hours. This innovative software brings improvements in the air route after take-off. The final outcome includes reduced flight time, less fuel consumption, and reduced air pollutant emissions.

Emirates Airlines has also invested in a new technology called "iFlex" which has similar targets and calculates alternatives air routes for the flights. This program has been applied initially in the itinerary Dubai-Sao Paolo and it is calculated that 18 minutes of flights and seven thousand seven hundred (7700) kilograms of CO_2 are saved per flight.

A third program that is developed (Tailored Arrivals) focused especially on the arrivals of aircrafts in the hub airport of Dubai where the base of the company is located. This initiative offers a high level of coordination for the arrivals of the aircrafts avoiding traffic congestions and delays. The continuous update of information in the control center of the airport for the aircrafts that are approaching it is an expected and ordinary process. This initiative goes one step beyond by sending the same data to the approaching aircrafts which have the ability to adjust their flight plan to arrive to their base. This means that the aircraft can change the parameters of its flight like its speed and arrive later or faster in its destination without causing any further congestion to the airport and avoiding doing circles while waiting its turn for landing. This is a program that also achieves optimization of fuel consumption, extensive delays, and less emissions by the aircrafts (Emirates, 2019).

Furthermore, the company has active involvement in the design of its aircrafts by requesting the aircraft manufacturers to be oriented to new models of aircrafts that are lighter than previous models, demanding less fuel for the same distance and emit less air pollutants. For example, Emirates Airlines cooperated with Airbus, the biggest European aircraft manufacturer and the output was the further reduction of the weight of the model A380 by five (5) tons. A-380 is the most popular aircraft in the fleet of Emirates Airlines and it is considered the model with the lowest noise footprint globally (Emirates, 2019).

The company has not stopped to find out new ways to improve its CSR policy toward environment:

- It has adopted the electronic ticket avoiding printing papers and this equals with important savings in paper consumption annually.
- It has established strict rules for the allowed weight limit in passengers' baggage based on the fact that every reduction in the aircraft weight by one-kilogram entails in saving of thirty-four thousand (34,000) liters of fuel annually.
- The company is keen to renew often its fleet with new aircrafts and a representative example of this policy is that it was one of the first air carriers to obtain the Boeing 777-300ER which was a model with very advanced jet engines. The average age of its fleet is nine (9) years considerably lower than the industry average. Emirates Airlines had ordered one hundred fifty (150) aircrafts Boeing 777X which is an improved version of 777 with reduced noise and fuel footprint amid the Dubai Air Show of 2013 achieving a great advertisement for its CSR policy.

- During landing, pilots use only one engine saving fuel.
- It uses aircrafts that are taxing with one engine when conditions allow that. It is estimated that one more minute of taxing with one engine could save four hundred thirty thousand liters of fuel annually.
- The aircrafts are connected with the ground grid for their electricity needs like air condition and ventilation when they are parked on the airport. The usual practice so far by the majority of air carriers was covering their needs all the time through the auxiliary power unit (APU) which works with the jet fuel. By doing so, the consumption of jet fuel from the aircraft tanks when airplane is still is reduced by 85%.
- The cabin crew has as task to separate the trash that could be recycled like those that are made by aluminum and paper. The company recycles more than one hundred tons of trash every month. The amount of recycled paper equals with the protection of thirty-five thousand (35,000) tons of trees annually.
- Emirates Airlines applies a program of scheduled tactic preventive maintenance for its fleet. This is based on the principle that a well maintained and cleaned engine is more efficient and emits less air pollutants than a dirty and not well-maintained engine. Especially the process of cleaning pays particular attention to the cleaning of compressors engines from dirt deposits to optimize performance. The average efficiency of the fleet regarding fuel consumption is above the international average of IATA as per 2013 measurements.
- Emirates has created the Green Standard Operating Procedures (Green SOPs) which is implemented on a daily basis by the pilots of the company. This program contains results from statistical processing of thousands of flights that provide useful directions about important issues of a flight like: minimization of engine taxi, idle reverse, prudent judgement on extra fuel, optimized flap landing, inflight speed management to optimize fuel consumption, and use of direct routing opportunities (Emirates, 2024).

The period 2022–2023 Green SOPs along with all the rest CSR programs resulted in fuel savings of fifty thousand (50,000) tones and reduction in carbon emissions of more than one hundred sixty (160,000) tones. Green SOPs include the training of pilots to use the abovementioned useful tools for the re-routing of aircraft amid flight, the proper way of performing take-off and landing, etc. The most important is that the air carrier has understood that the pilots consist of a key factor for the successful implementation of a variety of CSR programs, thus, it is valuable to achieve their awareness that their flying techniques influence not only the safety of the flight but additionally the fuel consumption, the noise pollution, and the air pollutants of flight.

9 Critical Analysis of CSR Policies in Aviation

Modern air carriers have realized that are checked by their customers for their environmental footprint, their social profile, and their economic practices in parallel with their ticket pricing and the quality of the offered services. Thus, the competition among air carriers has extended into the field of CSR and sustainability. This reality explains why many air-carriers have created specialized departments for CSR policy, they spend significant parts of their budget for green initiatives, and they circulate annually CSR reports.

Although the abovementioned described activities by corporations in the fields of CSR are absolutely positive, a critical analysis is required for gaining spherical review and for providing feedback for potential improvements. This could be considered imperative since it pertains corporate activities that could impact on corporate brand (Duchin et al., 2010; Panagiotopoulos, 2020) and budget.

Many times, there is a variation between the annual CSR policy announced by an air carrier and the real CSR activities that will have been implemented by it after the completion of a year. This deviation could come up due to several reasons like a combination of really ambitious targets and poor financing of CSR initiatives, incomplete design of activities, lack of experienced personnel, etc. It seems that there are companies which invest in their CSR policy in order to improve their green marketing and enhance their branding without really caring for their CSR performance. A strong advertising policy without a proportionally strong CSR program could disorient customers, convince and attract those who are environmentally sensitive.

The annual CSR reporting usually includes announced policies and initiatives about CSR without a critical review of the application of them and of course avoiding any comparison between the established aims and the fulfilled achievements. The above reality entails the conclusion that the expected economic advantage of a CSR strategy is much more desired than the expected positive impact into the environment and the society.

There is also a contradiction between the announced interest of air carriers for the society and the environment on the first side and their actions on the other side. It is widely known that the decrease of aircraft weight equals with reduced fuel consumption, thus with reduced emissions of gases (Panagiotopoulos, 2020). Air carriers have reduced the allowed baggage weight limit and charge for any extra kilogram as a measure to discourage passengers to carry on more weight in favor of less total gross weight for the aircraft. However, the same companies continue to burden their aircrafts with extra unnecessary weight like products for duty free sales on flights and luxurious facilities for A class passengers. On the same time, the average comfort for the economic class is continuously reduced with smaller, cheaper, and lighter seats.

It is important to be mentioned that a complete CSR strategy should include beyond design, funding and implementation, the steps of critical review, and extraction of conclusions periodically. These processes demand the existence of specialized tools that could assist an air carrier to assess its Corporate Social Performance (CSP). However, there is a scarce in such type of specialized CSR assessment tools that are tailor made for the specific industry of aviation. The ETS cannot consider a CSP assessment too. The same applies for ISO 26000 for CSR because it is more a guide for the design of CSR policy and not an assessment tool that certifies a good performance. This is why CORSIA by ICAO is of significant importance since it is one of the first which was introduced in a single specific industry and could be used as a tool for CSP assessment (Panagiotopoulos, 2020).

10 Conclusions: Flight Toward Sustainability

If the target of public policy is to eliminate emissions while avoiding unfair socio-economic results and aggressive taxation, aviation would be a particularly appropriate field for interventions. There are currently some recently introduced carbon pricing systems for aviation like the EU Emissions Trading System and ICAO's CORSIA. Both are based on total fuel consumption independently of the number of passengers. It is on the discretion of the air carriers to allocate the extra cost to their passengers. An idea is the increase of tickets with focus on premium and frequent travelers and no horizontal increases to all passengers. This method could bring more money to air carriers for environmental initiatives (Graver et al., 2020).

At the 41st Assembly of ICAO, states agreed to support the target of aviation to succeed net-zero carbon emissions by 2050. ICAO has brought back the baseline for the CORSIA to the level of 85% of 2019 emissions along with the statement that this is going to be the only financial measurement for the control of emissions in air transportations. ICAO expects governments to design strategies facilitating sustainability in energy management and securing CORSIA future. However, the majority of the states so far have not managed to establish successfully incentives for initiatives like the introduction of SAF and this increases the risk states to select to create more environmental taxes for the sake of the environment undermining CORSIA and the financial sustainability of the sector (ICAO, 2023b). This is the main reason why EU is thinking to apply the inclusion of all flights into and out of the EU after 2026 in the EU ETS. However, this intention is perceived as a lack of support to CORSIA initiative from EU's side by ICAO and IATA.

It is very important to be highlighted the importance of the high level of involvement of the personnel of air carriers into these CSR policies. The training of pilots and the increase of their awareness and sensitivity in environmental issues beyond safety issues which undoubtful is very important, should be always an internal target for air carriers. This applies to the rest specializations too. Cabin crew and technicians are very important in the abovementioned CSR initiatives and good practices, and the quality of their work could determine the percentage of success of these efforts.

References

Brundtland, G. H. (1987). *Report of the World Commission on environment and development: "Our common future"*. United Nations.

Carlsson, F., & Hammar, H. (2002). Incentive-based regulation of CO2 emissions from international aviation. *Journal of Air Transport Management, 8*, 365–372.

Carroll, A. B. (1998). The four faces of corporate citizenship. *Business and Society Review, 100–101*(1), 1–7. https://doi.org/10.1111/0045-3609.00008

Crowther, D., & Seifi, S. (Eds.). (2021). *The Palgrave handbook of corporate social responsibility*. Palgrave Macmillan. https://doi.org/10.1007/978-3-030-42465-7

Dahl, R. (1972). A prelude to corporate reform. *Business and Society Review, 1*, 17–23.

Dickson, N. (2014). *Balanced approach to aircraft noise management*. ICAO.

Duchin, R., Ozbas, O., & Sensoy, B. A. (2010). Costly external finance, corporate investment, and the subprime mortgage credit crisis. *Journal of Financial Economics, 97*, 418–435.

EESI. (2022). *The growth in greenhouse gas emissions from commercial aviation*. Environmental and Energy Study Institute.

Efthymiou, M., & Papatheodorou, A. (2018a). Environmental considerations in the single European sky: A Delphi approach. *Transportation Research Part A: Policy and Practice, 118*, 556–566. https://doi.org/10.1016/j.tra.2018.09.024

Efthymiou, M., & Papatheodorou, A. (2018b). *The Routledge companion to air transport management: Evolving business models*. Routledge.

Efthymiou, M., & Papatheodorou, A. (2019). EU emissions trading scheme in aviation: Policy analysis and suggestions. *Journal of Cleaner Production, 237*(117734). https://doi.org/10.1016/j.jclepro.2019.117734

Emirates. (2019). *The Emirates Group environmental performance report 2017-18* (p. 191). Retrieved from https://cdn.ek.aero/downloads/ek/pdfs/report/annual_report_2018.pdf

Emirates. (2024). *Our planet - Reducing emissions [official website]*. Emirates. Retrieved from https://www.emirates.com/ae/english/about-us/our-planet/reducing-emissions/

European Commission. (2001). *Green paper: Promoting a European framework for corporate social responsibility*. Commission of the European Communities. Retrieved from https://ec.europa.eu/commission/presscorner/detail/en/DOC_01_9

European Commission. (2024). *EU emissions trading system (EU ETS)*. Climate Action. Retrieved from https://climate.ec.europa.eu/eu-action/eu-emissions-trading-system-eu-ets_en

Graver, B., Rutherford, D., & Zheng, S. (2020). *CO2 emissions from commercial aviation 2013, 2018 and 2019*. The International Council of Clean Transportation.

IATA. (2013). *Reducing emissions from aviation through carbon-neutral growth from 2020*. ICAO.

Environmental Assessment Standards Manual (IESM), 2nd edn. (2015). IATA.

IATA. (2015). *Environmental assessment program manual (IEPM)*.

IATA. (2023). *Our commitment to fly net zero by 2050*. Retrieved from https://www.iata.org/en/programs/environment/flynetzero/

IATA. (2024a). *IATA - Sustainability [Official IATA website]*. Retrieved from https://www.iata.org/en/programs/environment/

IATA. (2024b). *The value of Air Cargo—Air Cargo makes it happen [IATA]*. Retrieved from www.iata.org/cargo

ICAO. (2008). *Manual on global performance of the air navigation system. (DOC 9883)*.

ICAO. (2023a). *ICAO environment: Climate change*. Retrieved from www.icao.int/environmental-protection/pages/climate-change.aspx

ICAO. (2023b). *International Air Transportation Association: Annual review 2023*.

ICAO. (2024). *ICAO - Environment [official website]*. Retrieved from https://www.icao.int/environmental-protection/Pages/default.aspx

ICAO. (n.d.). *Aircraft noise*. ICAO. Retrieved April 29, 2024, from https://www.icao.int/environmental-protection/Pages/noise.aspx

Muhammad-Azfar, A., Boon-Cheong, C., & Syaiful-Rizal, H. (2015). Benchmarking key success factors for the future green airline industry. In *6th international research symposium in service management*. UiTM Sarawak.
Neufville, R., & Odoni, A. (2003). *Airport systems: Planning, design and management.*
Panagiotopoulos, F. I. (2020). *Energy management and corporate social responsibility: The case of air-transportation.* University of the Aegean.
Schvartzman, R. (2023, April 18). *EU ETS reform destabilizes international consensus for aviation carbon reductions.* IATA. Retrieved from https://www.iata.org/en/about/worldwide/europe/blog/eu-ets-reform-destabilizes-international-consensus-for-aviation-carbon-reductions/
United Nations. (2024). *UN global compact* [United Nations Official]. Retrieved from https://unglobalcompact.org/
United Nations. (n.d.). *Sustainable development goals - The sustainable development agenda.* UN Official Website. Retrieved March 27, 2023, from https://www.un.org/sustainabledevelopment/development-agenda-retired/#:~:text=On%201%20January%202016%2C%20the,Summit%20%E2%80%94%20officially%20came%20into%20force
Whitehouse, L. (2003). Corporate social responsibility, corporate citizenship and the global compact: A new approach to regulating corporate social power? *Global Social Policy, 3*(3), 299–318. https://doi.org/10.1177/1468018103003002

Dr. Panagiotopoulos Φ. Ioannis is a Greek Electrical & Electronic Engineer who holds a Bch in Electrical Engineering mainly in Power Generation, a Bch in Electrical & Electronic Engineering mainly in Telecommunications, a MBA with specialization in Innovation & Technology Management, a Master of Arts in Lighting Design & Multimedia, and a PhD in Business Administration. His PhD thesis was focused on Energy Management and CSR in Aviation. He is an independent researcher and writer with more than 45 papers in scientific conferences and journals with double blind peer review in the fields of sustainability, CSR, ESG, energy management, sustainable lighting design, etc. He has introduced the concept of Operational Ethical Dilemma for Corporations (F. I. Panagiotopoulos, 2020; I. Φ. Panagiotopoulos, 2022) and a prototype Assessment Tool for the CSR programs of Energy Management in aviation industry (F. I. Panagiotopoulos, 2020). He has also introduced the term of Sustainable Lighting Design (F. I. Panagiotopoulos, 2018, 2019) and a specialized tool for the assessment of CSR programs focused on Energy Management of Buildings. He is an academic lecturer and a Sustainable Supply Chain/Procurement Manager in Construction/Oil and Gas industry. He is married with one son and lives in the UAE.

Frontier Markets Compliance with Global ESG Practice: Sample of Baltic Countries

Ilze Zumente, Jūlija Bistrova, Nataļja Lāce, and Ilja Arefjevs

1 Introduction

The increasing emphasis on ESG practices across the global corporate landscape has underscored the imperative for transparent and comprehensive disclosure of sustainability efforts. This chapter delves into the emerging domain of ESG implementation within the corporations of Central and Eastern Europe (CEE), with a specific focus on the Baltic states—Estonia, Latvia, and Lithuania.

The chapter starts off with an examination of sustainability trends within the broader CEE context, analyzing the evolution of corporate mission statements in light of shifting sustainability paradigms. This segment assesses the degree to which companies have reoriented their mission statements toward sustainability, stakeholder engagement, and long-term value creation, reflecting broader trends in corporate governance and strategic focus.

Next, acknowledging the challenges posed by the limited availability of ESG scores across CEE corporations (Dumrose et al., 2022; Zumente & Lāce, 2021), this study employs alternative methodologies to gauge the current ESG landscape and its integration within corporate practices. Through a thorough examination of stock-listed corporations in the Baltic region, this research presents an in-depth analysis of ESG disclosure levels, leveraging qualitative content analysis of sustainability reports and their equivalents from the year 2020 and extending the investigation to the developments observed in 2022. Furthermore, the study casts light on the

I. Zumente · J. Bistrova (✉) · N. Lāce
Faculty of Engineering Economics and Management, Riga Technical University, Riga, Latvia
e-mail: julija.bistrova@rtu.lv; natalja.lace@rtu.lv

I. Arefjevs
BA School of Business and Finance, Riga, Latvia
e-mail: ilja.arefjevs@ba.lv

S. O. Idowu, S. Vertigans (eds.), *Sustainability in Global Companies*, CSR, Sustainability, Ethics & Governance, https://doi.org/10.1007/978-3-031-77971-8_4

sectoral and industry-specific patterns of ESG disclosure, offering insights into the variances and progressions in ESG transparency.

In addition to assessing disclosure and strategic orientations, this chapter also examines the external pressure for wider ESG implementation extending to the role of financial investors and asset managers. A detailed survey of financial market participants in the Baltic region is offered comparing the stance at two time points—2019 and 2024 offering unique insights into the trend evolvement over time. This section explains the perceptions, challenges, and drivers of ESG integration within the investment evaluation process, highlighting the critical role of capital owners and regulatory frameworks in fostering sustainable investment practices.

Finally, the chapter devotes attention to the readiness of introducing ESG principles in investing process among the pension fund managers, clients' readiness to demand ESG layer, and the readiness of the regulators to supervise the compliance with this regard. The conclusions are based on the analysis of the expert interview representing three types of interested parties coming from Latvia, Lithuania, and Estonia.

Overall, this chapter aims to contribute to the growing field of ESG research within the CEE and the specific Baltic region context, offering a comprehensive examination of current practices, challenges, and prospects for enhancing sustainability disclosure and practices among corporations. Through this multi-layered analysis, the study aims to provide valuable insights for policymakers, investors, and corporate leaders alike, in navigating the complexities of ESG integration and promoting a more sustainable and transparent corporate ecosystem in the Baltic region and beyond.

1.1 Sustainability Trend in the CEE: Mission Statement Analysis

The concept of shareholder value has long been a cornerstone of corporate strategy (Friedman, 1970), guiding the actions and policies of companies worldwide. This principle suggests that companies should primarily focus on increasing the value for their shareholders, which is often reflected in the company's mission statement and operational strategies. However, as global priorities shift toward sustainability and ethical governance (Jensen, 2002; Moir et al., 2007), there has been a significant interest in understanding how these evolving values are integrated into corporate practices, especially in regions undergoing rapid economic development (UN Global Compact, 2020). CEE with its unique economic landscape presents an interesting case study in this context. In 2012, Bistrova and Lace validated a model of shareholder value by analyzing stock market data from the CEE countries. Furthermore, they investigated whether companies in the region emphasized long-term value creation for their shareholders in their mission statements. After reviewing the mission statements of 85 publicly listed companies from the CEE region, they found that only approximately 30% mentioned a commitment to their shareholders. Instead, themes like customer focus, quality, leadership, and market position were more frequently highlighted (Bistrova & Lace, 2012).

As the world trends toward more sustainability, one might anticipate that this will show up in the revisions of the mission statements in the recent decade, which offers a special chance for a trend analysis based on the original data set. Mission statements play an important role in setting the company's course toward a comprehensive sustainability strategy (Analoui & Karami, 2002). The importance of a strategically aligned mission statement can be crucial for successful sustainability performance not only for large global corporations but also for small and medium size entities (Duygulu et al., 2016). Moreover, referrals to the fundamental business drivers of the company in their mission statements are proven to influence their financial performance positively (Barth et al., 2001), thus underlying the importance that the mission statement can have on the overall business performance (Bartkus et al., 2006).

A text search software was used to perform a frequency analysis of 20 concepts to examine if these factors were also reflected by the companies in the CEE region. The analysis mainly focused on the concepts related to the stakeholders themselves and their interests, to assess if the corporate mission statements were giving less priority to the shareholders' interests and more to the interests of other stakeholders. The results (see Fig. 1) indicate that the highest focus is put on sustainability-related metrics, including references to responsibility, innovations, environment, long-term orientation, and community. More than 90% of the companies had at least one reference to these topics in their mission statements. The second highest priority was consumers—more than half of the companies referred to their customers in their mission statements. References to the stakeholders were found more frequently than shareholders. Shareholder commitment and financial performance were mentioned comparatively rarely. While the categories in the analysis have been added mostly for illustrative purposes, it can be argued that all three first categories (sustainability, consumer, and stakeholders) generally relate to a wider non-financial dimension of companies on their path toward sustainable development and are constituents of the general dimensions usually comprised by the ESG factors.

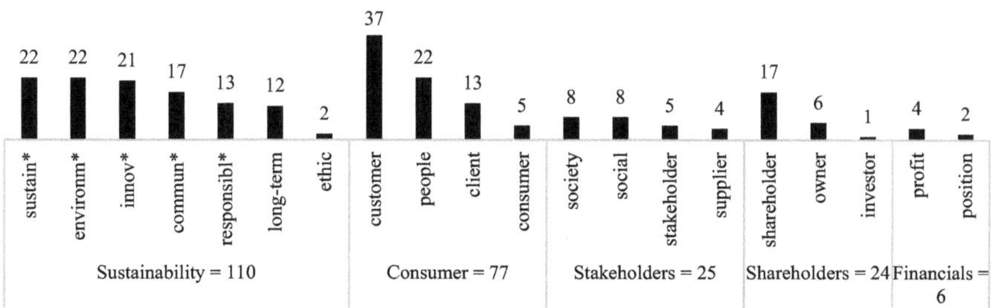

Fig. 1 Results of the mission statement analysis of CEE companies. Created by the authors

To analyze the mission statement changes over the last decade, the sample data of Bistrova and Lace (2012) used for the mission statement analysis was retrieved, and the updates to the mission statements of the companies used in 2012 were added either from the web pages or annual statements of the companies as of February 2021. By doing so, a database of the mission statements that the sample of companies had in 2012 and 2021 was created allowing the author to explore how significantly the companies have altered their purpose and reason for existence in the indicated time frame. In order to allow for direct comparison, the same companies were chosen for this analysis ("like-for-like" sample). From the previously analyzed sample of 122 companies and 85 available mission statements, 70 updates as of 2021 were available due to some companies undergoing restructuring or liquidation. In the like-for-like comparison of the same companies, the following topics (as summarized in Fig. 2) underwent the most significant changes.

While in 2012, the content analysis of the offered mission statements showed that a third of the companies mentioned their commitment to the shareholders, this number over the decade has decreased to only 17%. Surprisingly, the reference to shareholders in the mission statements has experienced the most dramatic decrease, followed by similar terms describing the financial orientation as the position (e.g., market position) and profit. On the other hand, the concentration has increased toward contributions to society (characterized by terms such as "people," "society," "community") and sustainable operations ("sustainability," "responsibility," "long-term," "environment," and "innovation"). Bistrova and Lace's analysis of 2012 data showed that companies highlighting shareholder value also emphasized profitability and financial performance. However, over the decade, there has been a shift; companies now prioritize societal, environmental contributions, and sustainable performance for long-term value creation in their mission statements. The noticeable decline in mentions of profitability and market focus aligns with a broader trend: achieving long-term shareholder value through high ESG scores and improved financial performance (Aguinis & Glavas, 2012; Duygulu et al., 2016), rather than a narrow focus on short-term financial outcomes. This pivot toward a more inclusive and sustainable focus in mission statements sets the stage for investigating the actual implementation of ESG practices within corporations.

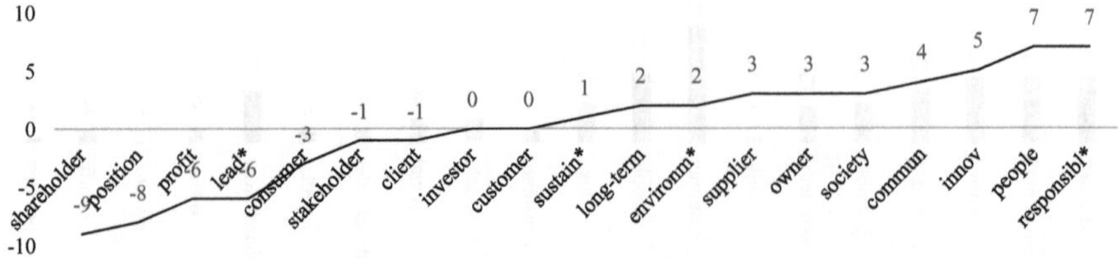

Fig. 2 Results of the like-for-like mission statement analysis summarizing the changes in topic occurrence in the company mission statements over the period from 2012 to 2021 (count). Created by the authors

1.2 ESG Disclosure Level in the Baltics

The observed shift in corporate mission statements toward sustainability and social responsibility naturally leads to questions about how these commitments are reflected in tangible ESG practices. The lack of a global ESG reporting standard leads to varied disclosure practices among companies, even within the same industry or country. This variation, influenced by specific regulations and norms, hinders the comparability of ESG reports across organizations. Consequently, assessing a company's ESG performance becomes challenging, creating uncertainty for investors and stakeholders (Ioannou & Serafeim, 2012). However, as the levels of ESG-compliant assets rise, there is a noticeable positive trend in overall ESG disclosure level (Chen & Xie, 2022).

Due to the scarcity of third-party ESG scores for corporations in CEE, alternative approaches are necessary to assess the actual level of ESG practices. This analysis focuses on a select group of CEE corporations, specifically those listed on the Baltic stock exchange.

A detailed qualitative content analysis was conducted on the sustainability information initially found in their reports (or equivalent documents) for 2020. Recognizing the dynamic nature of corporate sustainability and evolving regulatory frameworks, this study also aims to capture changes over time by including data up to 2022. The research encompasses a sample of 38 companies from the NASDAQ Baltic stock exchange as of August 2022. This includes 32 companies on the prime list and 6 on the secondary or alternative markets, all of which have published dedicated ESG or Corporate Social Responsibility (CSR) reports within their annual disclosures. From the original 2020 cohort, 3 companies were delisted and 5 new entities were listed, leading to slight changes in the sample composition. Excluding three prime-listed companies for not publishing their 2021 annual reports, the updated sample maintains an 85% similarity to the 2020 group, comprising 16 companies from Lithuania, 17 from Estonia, and 5 from Latvia. Due to the previously noted limited availability of corporate sustainability evaluations for companies in the Baltic region, we adopted the ESG disclosure score computation method initially developed by Roca and Searcy (2012). This methodology has also been applied in studies by Bakar et al. (2019) and aligns closely with Bloomberg's disclosure score calculation. The approach utilizes content analysis to review disclosures, verifying the inclusion of specific measures and factors by the companies in question. In the 2020 study, the evaluation checklist comprised 106 factors, drawing from the NASDAQ ESG Reporting guidelines, GRI Reports, the NASDAQ Corporate Governance Code, and the UN Sustainable Development Goals (SDGs). By 2022, the maximum possible score increased to 119, reflecting additional indicators and measures companies were expected to report, such as EU taxonomy related information and whistleblowing policies.

Qualitative content analysis was performed on the non-financial reports of the companies for the year 2020, as well as on reports and disclosures in the regulatory filings available on the stock exchange's website. Information and sustainability disclosures located elsewhere were not considered in this analysis to ensure consistency in comparison. The forms of reports analyzed included Sustainability Reports, ESG Reports, Social Responsibility and Governance Reports, Non-financial Reports, and specific sections within the Management Reports of annual disclosures, accommodating the variances in reporting formats.

The split between the E, S, and G metrics was found to be approximately similar—35 indicators corresponding to the environmental factors, 45 indicators revealing information on the social facet, and 38 indicators reporting on the corporate governance practices. One point was added to the checklist for each case the company had reported on the specific ESG indicators.

Correspondingly, the ESG disclosure score was calculated by dividing the sum of individual disclosure items by 119 (the max score according to the checklist) in 2022 sample and by 106 in 2020 sample. Given that the computation method does not provide information on the quality or performance of the specific disclosures, the result must be interpreted as the relative degree of ESG transparency rather than an overall level of ESG performance or corporate sustainability achievement. The result is expressed in percentage terms to allow for easier comparability.

By evaluating the total reported information volume against the maximum attainable transparency level, the percentage of the disclosed ESG information was obtained. As visible in Fig. 3, the average ESG disclosure score had improved by 7 p.p. from 40% in the 2020 sample to 47% in the 2022 sample. While there are still companies that disclose insufficient non-financial information (12% minimal disclosure level), there is also a significant improvement in the best performers—one corporation achieving even 93% transparency level. The results of the study reveal that in line with expectations, Baltic stock-listed companies have improved their overall ESG affinity and are indeed more transparent in their corporate sustainability achievements.

When split between the industries, as depicted in Fig. 4, the highest ESG disclosure level is achieved by companies in the utility sector, identically as noted in 2020. The average level of disclosure in this sector has increased from 61% in 2020 to 63% in 2022. The lowest disclosure scores demonstrate companies in the real estate segment—presenting only half of the transparency level achieved by their peers in the utilities segment.

The results across the pillars as explained in Fig. 5 have remained consistent with the patterns observed in 2020—the highest transparency level is achieved across the

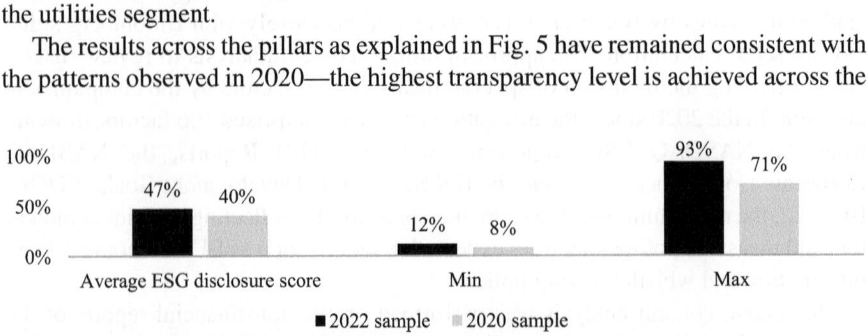

Fig. 3 ESG disclosure score in the Baltic public listed companies—comparison between 2020 and 2022 data. Created by the authors

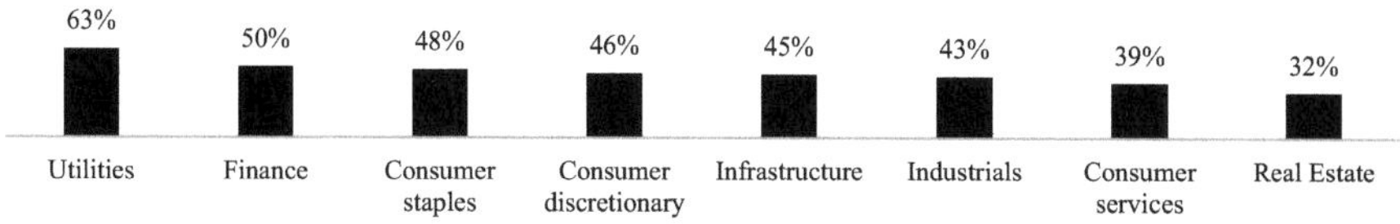

Fig. 4 Average ESG disclosure level in the Baltic stock-listed companies by industries (from max 100%). Created by the authors

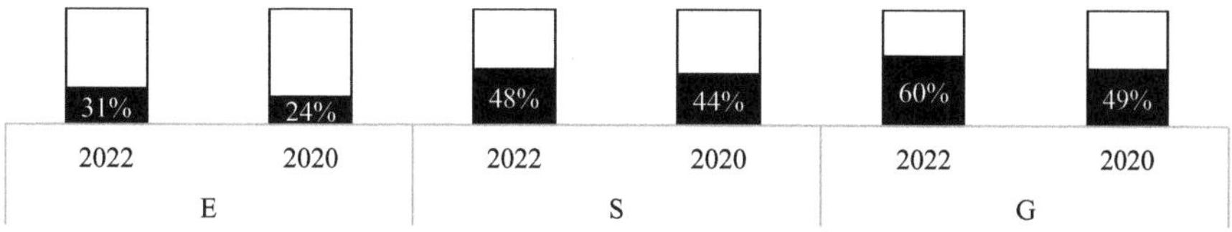

Fig. 5 Disclosed ESG information level by factors in the Baltic stock-listed companies (% of 100%). Created by the authors

governance pillar (60%), followed by social disclosures at 48% level and environmental pillar of 31%. The average disclosure level has increased since 2020 across all the ESG factors.

While the disclosures in the corporate governance pillar are strongly driven by the requirement for the stock-listed companies to publish a dedicated Corporate Governance Report, the environmental pillar is still relatively underrepresented driven by more complex data measurement. Nevertheless, with the overall trend of higher ESG transparency in the Baltic region, also improvements in environmental data disclosure have been documented. The specific improvement areas relate, for example, to emission level measurement (57% of the sample companies report on their emission levels at least in Scope 1 and 2), as well as more quantified information on water consumption, waste generation levels, etc. All in all, however, particularly the Environmental data availability remains the largest pitfall for the Baltic corporations confirmed by the vast number of companies disclosing insufficient environmental data (Fig. 6).

Overall, the results show a moderate level of ESG disclosure across the stock-listed companies of the Baltic countries once again signaling the need for additional focus on this topic for the examined companies. Stock exchanges are generally in a unique position to contribute to a wider implementation of ESG practices in company reporting standards and therefore higher overall transparency of the capital markets (Bizoumi et al., 2019). Sectoral specificity and emphasis on the material disclosures in the stock exchange-issued guidelines shall help to promote the focus on the right sustainability drivers and, by doing so, to increase the ESG disclosure value for the investors.

Simultaneously, financial investors, with their considerable influence on ESG adoption and the regulatory requirements they impose, serve as pivotal drivers for inserting ESG considerations within the investment landscape. This synergy between stock exchanges and financial investors highlights a multi-layered approach

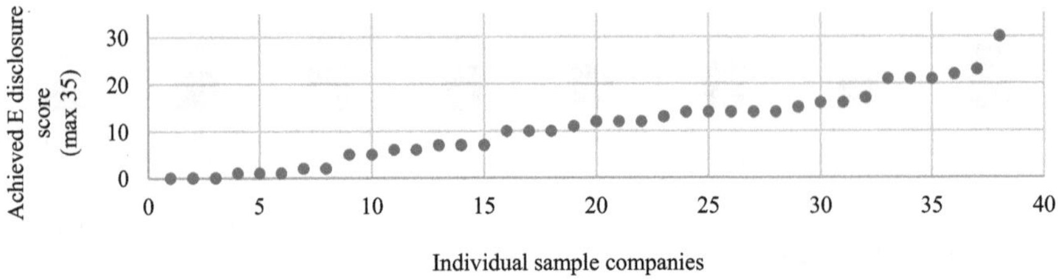

Fig. 6 Environmental disclosure level across the Baltic listed companies (from max thirty-five points). Created by the authors

to advancing ESG practices, further explored through a detailed examination of the role of financiers, asset managers, and pension funds in the Baltic region's financial market.

1.3 Financial Investors and Their Impact on ESG Adoption

Financial investors including asset managers yield considerable influence over ESG adoption, acting as catalysts by extending regulatory compliance requirements to their investment entities, thereby amplifying ESG integration across markets. It is found that that sustainable investing does not compromise financial returns and can positively affect investment outcomes, reinforcing the role of investors as pivotal to the widespread adoption of ESG practices (Amel-Zadeh & Serafeim, 2017; Friede et al., 2015; Schramade, 2016).

2 Current Adoption Status of ESG Investing

ESG is a major trend observed in the asset management industry now exerting pressure on each asset management industry player to offer products with ESG filter. As the data show, sustainably invested assets rapidly grow and more companies involve themselves in the topic. According to Bloomberg Intelligence's latest ESG 2021 Midyear Outlook report, by 2025 ESG assets under management to reach USD 50 trillion, which is going to be a third of total assets under management globally. To compare, in 2020 ESG assets totaled USD 35 trillion and in 2016—USD 23 trillion (Bloomberg, 2021). ESG adoptions across institutional asset managers rise globally: in 2021 more than 70% offered ESG strategies as compared to more than 60% in 2019. Given such a major trend and client demand, asset managers and particularly pension fund managers obviously have societal obligations to reach ESG goals in terms of climate change goal achievements and improvement in corporate culture. Index Industry Association conducted an international survey of 300 asset managers and found out that the main factor driving adoption of ESG products and services is client demand, by far outpacing desire for increased return, investment policy, or ESG factor concern (Index Industry Association, 2022). Interestingly, regulatory and reputation risks were mentioned as least important. The challenges faced by the asset managers primarily mentioned were the lack of data standardization and its availability as well as lack of regulatory certainty.

3 ESG Investing Adoption Barriers

Global study by Capital Group of 1000 investors in 16 countries (Capital Group, 2023) indicated that the most significant barrier in ESG adoption is lack of trustworthy ESG data. ESG score provided by rating agencies differ quite substantially and the access to ESG data often is limited. The problem of ESG rating inconsistency is pointed out in many research papers: (Berg et al., 2019; Chatterji et al., 2016; Dumrose et al., 2022; PwC, 2021). In addition, asset managers cite inconsistencies in corporate reporting (Capital Group, 2023). The major barrier for the clients to demand sustainable investment products is consistent lack of knowledge. Additionally, the clients and financial advisors mention a mess in terminology and the quality of ESG data, which is not standardized yet. Lack of clarity in regulatory and reporting standards is found to be the most significant barrier to adopt ESG (Doyle, 2018). Another obstacle, which does not stimulate asset managers to adopt ESG investing standards is lack of robust evidence of ESG value adding to the investment and financial performance as addressed in several research papers (Bruno, 2018; Dorfleitner et al., 2015). In addition, in string theory based research suggests that managers' conformance to ESG criteria will not lead to more sustainable investing behavior (Sakuma Keck & Hensman, 2013).

4 ESG Investing Adoption Drivers

ESG topic is being widely discussed and adopted at all levels by the financial market participants: financial regulators, fund managers and their customers. Regulation is one of the major forces driving investments in the sustainable direction. It stimulates not only the fund management industry but also the investment universe, which is the main variable in the whole ESG value chain. As admitted by Ioannou and Serafeim (2012), political, legal, and labor market institutions are significant factors affecting variation in the corporate social performance. A study of over 1963 large-cap companies headquartered in 49 countries, found that firm characteristics explain most of the variation in firm ESG disclosure, while differences in country factors such as corruption and political rights explain less (Yu & Luu, 2021). Regulatory institutions have an ultra-important role in adopting ESG principles in investment decision-making by asset managers as admitted by 91% of respondents in Barnett Waddingham survey (2022) (Barnett Waddingham, 2022).

Legislation factor substantially dominates over other factors considered in the research such as risk management, client demand, and peer influence (Karadima, 2022). Both asset managers and asset owners confirm that in future ESG regulation will continue to dominate influencing power, and it will be substantially expanded further, therefore driving also the costs of the market participants. Regulations are so heavily imposed that European Union regulatory bodies have deserved overregulation critics. Its US peer, Securities and Exchange Commission, allows certain freedom to financial market participants in interpretation of ESG principles for fund managers to decide on adequate underlying criteria and approaches and whether the investment universe discloses these sufficiently.

Despite the critics, it seems that regulation is efficient enough to advance ESG principles and to follow up with the arising problems, particularly in Europe (OECD, 2020). Lately, European regulator started to actively address greenwashing of financial products by introducing Sustainable Finance Disclosure Regulation, while the focus of US regulator SEC is toward improvement of the climate disclosures by the US companies (Morrison & Foerster LLP, 2020). ESG factors' importance to the performance improvement is named as the first-choice motivation by the mainstream investment organizations (Amel-Zadeh & Serafeim, 2017). Relevance to the performance is followed by client demand, product strategy, and ethical considerations.

Speaking about the valuation being influenced of ESG factor, Schramade (2016) in his research believes that only 5% should be assigned to ESG factors, and states that anyway ESG leads to better analysis results due to additional more in-depth look (Schramade, 2016). One should admit that it is not only the regulation stimulating evolvement of responsible investment from niche to mainstream investing, but also the client demand per se. The survey conducted by Institutional Shareholder Services admits that the financial investors offer respective ESG products as stimulated both by their clients and regulators (Sauders, 2020). Invesco, in its survey of 161 financial advisors and 201 investors, brings to the attention the fact that virtually every client is considered to be an ESG client and for 79% of clients sustainability is an important aspect of investment. In addition, the study results confirm stronger interest in responsible investments exhibited by younger generations (Invesco, 2022). The primary motivation of introducing ESG investing is brand reputation—59% of respondents admitted, while external stakeholder requirement—46%, improving returns—45%, and decreased investment risk—39%. External stakeholders', namely clients' and regulators', requirement is mentioned as the secondary motivation for 329 institutional investors from 19 countries in 2021 as indicated in the BNP Survey (BNP Paribas, 2021). Interestingly, over the last 2 years, the motivation coming from external stakeholders substantially increased, while motivation created by the necessity to generate higher returns substantially decreased. According to van Duuren et al. (2016), conventional asset managers actively integrate ESG investing also for risk management and for red flag detection. ESG matters should be particularly important for the pension funds given their large share in total managed assets and, therefore, high bargaining power, as well as long-term horizon and societal concerns of various ESG issues (van Duuren et al., 2016). Also, the clients of pension funds indicate that it is important for them to have their funds invested in sustainable companies. So, the pension funds have to comply with the societal mandate assigned by the society. According to SEB pension survey, it is imperative for 75% future Baltic pensioners that their pension capital is invested sustainably (SEB, 2021). What is more, 15% of the respondents said that they would agree to choose financial products with sustainability being a sole aim. Obviously, these results show a very high consciousness about ESG matters among the customers.

5 Baltic Financiers Survey

Moving from a global outlook to a closer look at the Baltic region, we shift focus from the general motivations behind ESG investing to how these factors play out locally. After examining why institutions worldwide are turning toward ESG for reasons like brand reputation and risk management, we now explore the specific attitudes and practices toward ESG within the Baltic financial sector. This closer examination aims to uncover how local financial players prioritize, implement, and perceive the challenges of ESG, providing a deeper understanding of ESG's role in the Baltic investment landscape.

To gather insights on the adoption of ESG factors in the Baltic financial sector, a survey targeting 50 entities, including financial investors, banks, asset managers, venture capital, and private equity funds as well as mezzanine lenders was distributed digitally. The anonymous survey aimed at investment managers and decision-makers was conducted from January 4 to February 24, 2024, and sought their perspectives on ESG's importance, implementation practices, and challenges, alongside the ESG considerations in portfolio management and the SFDR impact. Achieving responses from 25 institutions, the survey consisted of 15 open and closed questions focusing on the Baltic investor's opinion on (1) ESG factor importance in their investment evaluation process, (2) the methods and practices applied in the evaluation process, as well as (3) current obstacles in ESG implementation. To offer insights and findings with a temporal dimension, the results were compared with a similar study conducted in 2021 (Zumente & Bistrova, 2021). The sample split according to the operation types is presented in Table 1.

The various operation types were underlined by the differences also in the average investment tickets—the size comprised 3.52 million EUR, ranging from 0.1 million EUR up to even 10 million EUR. Also, the count of the assets in the portfolio varied based on the respondent type—9 respondents reported 6 to 15 companies, while 7 investors reported over 40 business cases.

The initial part of the survey was designed to gauge investor sentiment regarding the incorporation of sustainability into financial decision-making. When participants were prompted to share their views on which financial market players ought to integrate ESG data into their investment evaluation processes, two clusters emerged.

Table 1 Sample split of the Baltic financiers. Created by the authors

Operation type	Quantity	Sample proportion
Venture capital fund	7	28%
Asset management company	6	24%
Private equity fund	5	20%
Bank	4	16%
Alternative investment fund	1	4%
Early stage / accelerator	1	4%
VC fund / crowdfunding	1	4%

Fig. 7 Share of respondents who believe that ESG factors should be considered by the mentioned type of investors. Created by the author based on survey results

As depicted in Fig. 7, the vast majority named banks (96%), pension funds (92%), PE funds (88%), and asset managers (79%). The share of proponents was relatively smaller for the second cluster—venture capital funds (50%), alternative lenders (42%), and early-stage funds (38%). Interestingly, when comparing these results to the 2021 data, the share of the proponents has increased for the first cluster, while slightly decreased for the second, suggesting that the focus on the ESG factors is believed to be most important for the most institutional and financially impactful financial market players.

When asked about their own experience in ESG due diligence, 88% of the respondents answered positively (up from 81% in 2021), whereof 52% perform ESG evaluation for all their investments and 36% do that in limited scope or for companies representing specific industries. Only 12% do not perform any ESG evaluation prior to investment. This progression highlights a practical application of ESG principles, underlining a broader acceptance and implementation in the financial decision-making landscape.

The large share of ESG-integrating financiers and generally the positive sentiment toward the ESG inclusion goes in line with the previous conclusion that capital owners can be one of the primary drivers ensuring that certain level of ESG compliance is achieved by the investment portfolio (Eurosif, 2016). The specific long-term and active relationship between the financial investors and the companies ensures that the private equity and venture capital companies are particularly well suited to integrate and improve the ESG standards in their portfolio companies (Invest Europe, 2021). As highlighted by a recent study about investment funds in Latvia, even after more than a decade after the first risk capital funds were launched in

Latvia, the funding is still largely dependent on local or international public resources (government, EU funds, EBRD, etc.). As found out at the time of the study, there were no VC funds in Latvia without public capital (Matisone & Lāce, 2017). This finding partly explains the results—as a significant share of the sample companies manage capital, which is based on public resources, they have an implied requirement of at least a high-level sustainability risk evaluation in their investment process. In addition, the results of the survey imply that also the private capital managers are similarly minded.

When asked about the primary drivers directly, most of the respondents (52%) cited regulation as the main reason to perform ESG evaluation whereby 72% of the responses indicated that the recent SFDR has further enhanced the ESG integration. Global tendencies (24%), risk avoidance (16%), and client's requirements (8%) were also among the most frequently selected answers. Also, the macroeconomic changes and market turbulences have further positively impacted the ESG adoption whereby 40% of the financiers admitted extended focus on some of the ESG dimensions.

In line with the previous answers, 64% of the respondents believed that ESG performance can be a value driver for the investment. Interestingly, this belief has decreased over the last three-year gap from 72% in 2021 to 64% in 2024. This implies that there are investors, who have lost the belief in the ESG as a value driver. This result comparatively is close to the one reported by Deloitte for Central and Eastern Europe investment funds, where 62% of the respondents considered ESG factors as a value driver. The finding generally also goes in line with the global evidence as summarized by, e.g., Eccles et al. (2019).

An important section of the study concentrated on the obstacle determination allowing to potentially explore the ways on how to solve them in a meaningful manner by the policy makers. With respect to the current ESG challenges, only 8% of the respondents believed that there are no current obstacles in ESG data application in the investment process. The majority (76%) cited data quality issues, lack of ESG awareness (36%), and insufficient knowledge (12%) as well as lack of materiality focus (20%). 4% mentioned greenwashing risk as a potential hurdle.

Also globally, data availability, which goes in line with the general ESG application by the invested companies, is one of the most commonly cited obstacles in ESG application. With respect to the Baltic companies, the situation is even dimmer given the fact that only a handful of companies have an external ranking ESG score available, most of which are rather large, publicly listed, or state-owned companies, which mostly are outside of the investment scope for the local financial investors (except certain largest banks and a slight share of the asset managers). The lack-of-data barrier is supported by the survey results, which suggests that around 86% of the respondents are not satisfied with the volume and quality of non-financial data that the companies can offer. The largest gap seems to occur specifically in environmental data. The lack of proper benchmark data has been found as another meaningful obstacle by the literature—as the privately held, mostly SMEs can hardly be comparable to the global listed peers, the financial investors are frequently struggling to understand the reasonable level of the metrics measured (Kotsantonis &

Serafeim, 2019). Also, as noted by several respondents—due to the different reporting approaches, industries, and materiality, the ESG data among the portfolio companies are rarely comparable, leading to an overall benchmarking problem in the market. Arguably, with the new regulations coming into force, this problem shall be at least partially lifted.

Correspondingly to the poor level of general data availability, when asked about the ESG data sources used, majority 72% of the respondents admitted using in-house research data has significantly increased the third-party data usage (from 22% of the respondents in 2021 to 40% in 2024, especially among the pension funds and asset managers who majorly invest outside of the Baltic region). Also, the importance of external consultants has increased (Fig. 8).

To assess the impact of ESG considerations on investment strategies, the study asked participants to rate the influence of sustainability factors within their decision-making processes on a scale from 1 to 10. The findings in 2021 revealed an average score of 3.9, with a median score of 3, translating to sustainability factors influencing approximately 39% of the investment evaluation model. This outcome highlighted that, despite the presence of hurdles in applying ESG criteria and mixed perceptions of its added value, a substantial fraction of investment choices were being made with considerations that transcend strictly financial criteria. In a surprising turn, the average score declined to 3 by 2024, indicating a slight retreat in the enthusiasm for integrating ESG factors, reflecting a nuanced shift in the prioritization of sustainability within investment decisions.

When dividing the scores into the operation type subgroups (Fig. 9), in line with the assumption, the results show that banks and asset management companies currently put the most effort on the ESG factor inclusion, while PE/VC funds and early-stage funds are slightly below. Particularly high the result is for banks, which means that already now there are companies in the Baltic countries, which most likely cannot obtain bank financing due to the non-financial factors.

All in all, the study confirms that financier's impact on the ESG adoption in the Baltic countries is already present and likely to grow in its strength over the forthcoming years. Overall, this study not only confirms the tangible impact financiers currently have on the integration of ESG practices in the Baltic countries but also forecasts a substantial amplification of this influence in the foreseeable future. This trend indicates a paradigm shift where ESG considerations are becoming

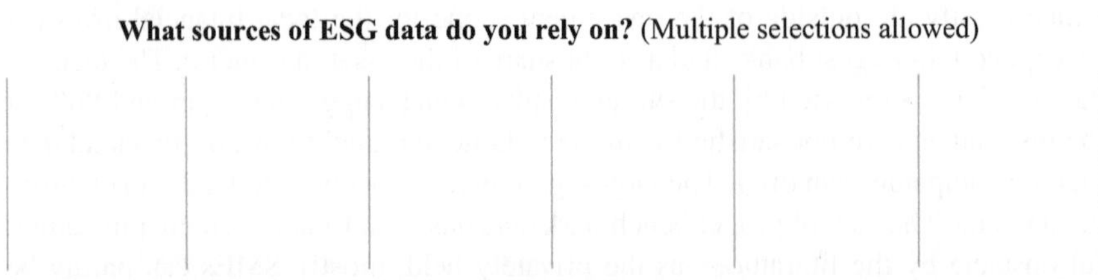

Fig. 8 What sources of ESG data do you rely on? (Multiple selections allowed). Created by the authors

Fig. 9 Average weight ESG factors play in investment evaluation measured in scale from 1 to 10. Created by the author based on survey results

increasingly central to financial decision-making. As awareness and regulatory frameworks around sustainability continue to evolve, it is anticipated that the role of financiers in promoting ESG adoption will become even more pivotal.

Transitioning from the broader impact of financiers on ESG adoption in the Baltic region, we narrow our focus to a specific segment of the financial market: asset management within the realm of pension funds.

6 Baltic Pension Fund Managers Survey

ESG investing within the pension fund management area in Baltic countries like in other CEE countries is still developing following the general global trend. Despite pension management is in the development stage, the overall asset under management exceeds 10 bn EUR in Baltics and for the large number of participants it is regarded as their largest wealth pile. Given social role of the pension funds and their huge size, there is a huge pressure for the pension fund manager to integrate ESG principles into investing processes.

ESG principles appear on the radar of all pension management stakeholders we interviewed: customers, financial regulators, and fund managers. By conducting expert interviews and employing the analytic hierarchy process, we aim to evaluate which of the key stakeholder groups are the readiest to introduce or to demand (in case of customers) ESG layer to be present within the provided investment services. The evaluation of the readiness and awareness is being performed by assessing the opinion on four dimensions: investment universe, investment strategy, overall perception as well as supervision. We conducted interviews with experts from Estonia, Latvia, and Lithuania covering the dimension of fund managers, investment product end-users via their representatives, and financial market regulators.

Most of the experts admit the polarization of ESG knowledge among the Baltic financial industry players and emphasize that there is certain share of clients, who is truly interested that their funds are invested in companies complying with highest ESG standards. At the same time, the performance usually comes first, assigning ESG principles of secondary importance.

To assess the readiness to implement sustainability concepts for investments of the Baltic pension funds, the authors conducted expert interviews and used the analytic hierarchy process for processing answers. Analytic hierarchy process (AHP) is considered to be a common tool for structured decision-making. Decision-making has become a mathematical science today. It involves many criteria and sub-criteria used to rank the alternatives of a decision. Data are collected from experts or decision-makers corresponding to the hierarchic structure, in the pairwise comparison of alternatives on a qualitative scale. The Fig. 10 presents an overview of assessment criteria and alternatives. Assessment criteria include investment universe (C1), investment process C2), awareness and perception by customers (C3), and supervision and transition policies (C4). Alternatives represent key stakeholder groups—fund managers (A1), customers (A2) as well as financial watchdogs (A3).

The criteria have been chosen based on previous research in the area. Importance of the investment universe (C1) topic was emphasized by Yu and Luu (2021) as well as other authors. Awareness and perception (C3) received decent attention by Sauders (2020), Zumente and Bistrova (2021), Index Industry Association (2022), and SEB (2021). Finally, supervision (C4) was studied by Barnett Waddingham (2022), Karadima (2022).

In total, eight experts were selected within the research (see Table 2). Choice of experts was balanced to the extent it was possible in terms of countries of their origin and stakeholders groups they represent. In particular, there are three fund managers' representatives, three customer representatives and ultimately two representatives of the Baltic financial watchdogs. All of them were considered to possess sufficient knowledge on ESG and Baltic pension fund investments.

In the course of the analytic hierarchy process, experts were required to rate the comparison either as equal, marginally strong, strong, very strong, or extremely strong. A set of pairwise comparison matrices (with response scores ranging from −9 to 9 whereas 0 is excluded from the scale) was constructed in accordance with the overall hierarchy of the process. In order to ensure robustness of the pairwise comparisons, consistency checks are conducted by calculating consistency ratios for each iteration of the AHP comparison.

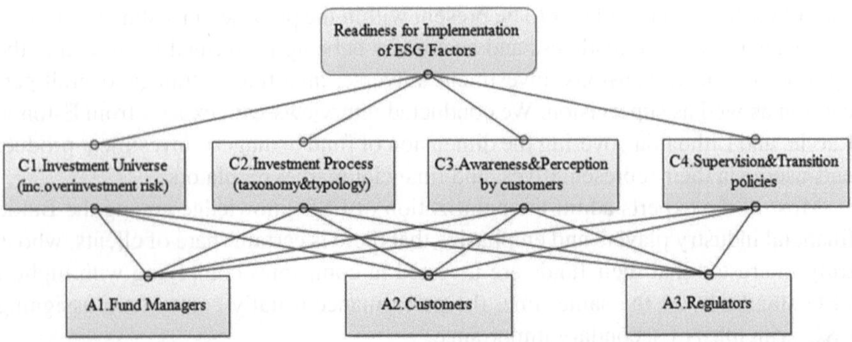

Fig. 10 AHP for ESG introduction readiness by key stakeholders' assessment. Source: Compiled by the authors

Table 2 Stakeholder assessments according to AHP (Source: compiled by the authors)

	A1. Fund managers	A2. Customers	A3. Regulators	Total
C1. Investment universe	0.08	0.04	0.14	0.25
C2. Investment process	0.09	0.03	0.14	0.25
C3. Awareness & perception	0.08	0.05	0.08	0.21
C4. Supervision	0.12	0.03	0.15	0.30
Total	0.36	0.15	0.51	1.00

The stakeholder assessment of the alternatives for readiness of implementation of ESG principles in the Baltic pension fund management is presented under Table 2.

The experts equally assessed readiness of the investment universe (C1) and the investment process (C2). Overall, these two criteria earned a healthy score of 0.25 each. However, the leading criterion proved to be supervision (C4), which was granted the biggest score of 0.30. Customer awareness and perception (C4) somewhat lagged behind the rest of criteria with 0.21 point. Nonetheless it is worth mentioning that the overall distribution of the AHP criteria weights is rather even and does not exhibit any extreme cases. This is an important interim conclusion in terms of readiness of ESG implementation for the chosen criteria.

The assessment of the alternatives (defined as key stakeholder groups in the given research) revealed a somewhat less even score distribution. The leading weight of 0.51 was earned by the financial regulators (A3), followed by the fund managers (A1), which received a score of 0.36. The stakeholder group of customers (A2) enjoyed the least weight of 0.15. It comprises less than a third of the regulators' score. Thus, the overall readiness of customers to welcome ESG-compliant investments of Baltic pension funds can be seen as alarmingly low. Customers (A2) demonstrated the lowest readiness scores in understanding of the investment universe (C2) and supervision (C4) while the highest score was assigned to the awareness and perception (C3), followed by the investment universe (C1). That constitutes an extra educational effort, which is needed to raise financial literacy of customers in the described areas. We discovered that fund managers (A2) are most ready in terms of supervision (C4), while the smallest extent of their readiness was witnessed in the areas of investment universe (C1) and awareness and perception by customers (C3).

The regulators (A3) were surprisingly found to be nearly equally ready in terms of the investment universe (C1), investment process (C2), and supervision (C4). However, a considerable weakness was discovered with regard to the awareness and perception by customers (C3). This, as admitted by the experts, is a result of the low financial literacy level, which creates lack of demand for ESG-tilted products and poorly stimulates the supply from asset management industry. Consistently, there is an obvious polarization among clients as certain minority is extremely aware of ESG and is quite indifferent to their impact in terms of the investments. Several experts have also stressed out ESG knowledge polarization among the fund managers, who are on their way to introduce ESG investing feeling pressure from the regulation. However, the main obstacle for fund managers to deep-dive into adopting of ESG principles was lack of data, which coincided with the responses of global asset managers (BNP Paribas, 2021).

Interviewed experts quite often mentioned a chicken and egg problem with regard to ESG introduction, whether these should be customers, regulators, or fund managers who are taking the lead in the ESG investing. Though, it was discussed that it all should start with awareness, which drives regulations in turn. Regulations and supervision were admitted being also in the starting phase though often it might be flawed, while regulators can lack experience. Nevertheless, it remains one of the driving forces in the current conditions imposing a new rule and addressing the consequences of non-complying, while also educating society and promoting ESG investing among pension fund managers and their clients.

AHP assessment indicated an important role of the regulator as well as its very high level of readiness. The role of the regulator is being also emphasized in the international studies, indicating that the regulatory body is being the top mover of the ESG trend globally (Karadima, 2022; Barnett Waddingham, 2022). However, in the developed financial markets client demand (Sauders, 2020) and organizational reputation (BNP Paribas, 2021) are being mentioned as main drivers to adopt ESG investing.

7 Concluding Remarks

The global trend toward the adoption of ESG principles in investing is being recognized and gradually embraced in the Baltics as well. However, it is often not driven by a desire for additional value creation but rather by global trends, pressure, and regulations. The major barriers to the full implementation of an ESG framework tend to be the lack of trustworthy data reported by companies and a deficit of knowledge in implementation. Additionally, clients are not overly demanding when it comes to responsible investing, largely due to a lack of understanding regarding the importance of ESG factors and the impact on the society and environment exerted by responsible strategies.

Asset managers, followed by private equity firms and banks, are noted to be the most attentive to the ESG dimension when allocating capital. A reality check reveals that expectations generally align with the actual situation. On average, ESG factor evaluations score 3.8 out of a maximum of 10 points for banks, 3.3 for asset managers, and 2.4 for private equity/venture capital firms.

The quality of ESG reporting among Baltic companies suggests room for improvement compared to the global average, particularly concerning the environmental pillar. However, it should be acknowledged that environmental reporting is substantially improving, as is governance reporting, driven by scrutiny from regulators and market participants. The social pillar, while deserving more attention in reports compared to the environment, has not shown significant improvement in corporate reporting.

The recognition of the importance of sustainability operations among Central and Eastern European and Baltic companies is evident through mission statement analysis. Over nearly a decade, mission statements have undergone significant changes, shifting focus from profitability and market position to sustainability, people, and innovation. This reflects an adaptation to global changes in priorities and companies' intentions to achieve profitability sustainably.

References

Aguinis, H., & Glavas, A. (2012). What we know and don't know about corporate social responsibility. *Journal of Management*. https://doi.org/10.1177/0149206311436079

Amel-Zadeh, A., & Serafeim, G. (2017). *Why and how investors use ESG information: Evidence from a global survey*. Harvard Business School Working Paper. Retrieved from https://ssrn.com/abstract=2925310

Analoui, F., & Karami, A. (2002). CEOs and development of the meaningful mission statement. *Corporate Governance, 2*, 13–20. https://doi.org/10.1108/14720700210440044

Bakar, A. B. S. A., Ghazali, N. A. B. M., & Ahmad, M. B. (2019). Sustainability reporting and board diversity in Malaysia. *International Journal of Academic Research in Business and Social Sciences, 9*(2), 1044–1067. https://doi.org/10.6007/IJARBSS/v9-i2/5663

Barnett Waddingham. (2022). *Sustainable investor insight survey*. Retrieved from https://view.barnett-waddingham.co.uk/sustainable-investor/p/1

Barth, M. E., Beaver, W. H., & Landsman, W. R. (2001). The relevance of the value relevance literature for financial accounting standard setting: Another view. *Journal of Accounting and Economics, 31*(1–3), 77–104.

Bartkus, B., Glassman, M., & McAfee, B. (2006). Mission statement quality and financial performance. *European Management Journal, 24*(1), 86–94. https://doi.org/10.1016/j.emj.2005.12.010

Berg, F., Koelbel, J. F., & Rigobon, R. (2019). Aggregate confusion: The divergence of ESG ratings. *MIT Sloan Research Paper, 5822*(19), 1–43. https://doi.org/10.2139/ssrn.3438533

Bistrova, J., & Lace, N. (2012). Quality of corporate governance system and quality of reported earnings: Evidence from CEE companies. *Economics and Management, 17*(1), 55–61. https://doi.org/10.5755/j01.em.17.1.2251

Bizoumi, T., Lazaridis, S., & Stamou, N. (2019). Innovation in stock exchanges: Driving ESG disclosure and performance. *Applied Corporate Finance, 31*(2), 72–79. https://doi.org/10.1111/jacf.12348

Bloomberg. (2021). *ESG 2021 Midyear Outlook*. Retrieved from https://www.bloomberg.com/company/press/esg-assets-rising-to-50-trillion-will-reshape-140-5-trillion-of-global-aum-by-2025-finds-bloomberg-intelligence/

BNP Paribas. (2021). *The ESG Global Survey 2021*. Retrieved from https://www.theia.org/sites/default/files/2021-09/The%20ESG%20Global%20Survey%202021.pdf

Bruno, G. (2018). ESG and socially responsible investing: A critical review. *SSRN Electronic Journal*. https://doi.org/10.2139/ssrn.3309650

Capital Group. (2023). *ESG global study 2023*. Retrieved from https://www.capitalgroup.com/advisor/pdf/shareholder/ITGEOT-073-1043294.pdf

Chatterji, A. K., Durand, R., Levine, D. I., & Touboul, S. (2016). Do ratings of firms converge? Implications for managers, investors and strategy researchers: Do ratings of firms converge? *Strategic Management Journal, 37*(8), 1597–1614. https://doi.org/10.1002/smj.2407

Chen, Z., & Xie, G. (2022). ESG disclosure and financial performance: Moderating role of ESG investors. *International Review of Financial Analysis, 83*, 102291. https://doi.org/10.1016/j.irfa.2022.102291

Dorfleitner, G., Halbritter, G., & Nguyen, M. (2015). Measuring the level and risk of corporate responsibility—An empirical comparison of different ESG rating approaches. *Journal of Asset Management*. https://doi.org/10.1057/jam.2015.31

Doyle, T. M. (2018, July). *Ratings that don't rate: The subjective world of ESG ratings agencies* (p. 17).

Dumrose, M., Rink, S., & Eckert, J. (2022). Disaggregating confusion? The EU taxonomy and its relation to ESG rating. *Finance Research Letters, 48*, 102928. https://doi.org/10.1016/j.frl.2022.102928

Duygulu, E., Ozeren, E., Işıldar, P., & Appolloni, A. (2016). The sustainable strategy for small and medium sized enterprises: The relationship between Mission statements and performance. *Sustainability, 8*, 16. https://doi.org/10.3390/su8070698

Eccles, R. G., Stroehle, J., & Lee, L.-E. (2019). The social origins of ESG? An analysis of Innovest and KLD. *Organization & Environment*, 1–36.

Eurosif. (2016). *European SRI study*. Retrieved from http://www.eurosif.org/wp-content/uploads/2016/11/SRI-study-2016-HR.pdf#page=52

Friede, G., Busch, T., & Bassen, A. (2015). ESG and financial performance: Aggregated evidence from more than 2000 empirical studies. *Journal of Sustainable Finance and Investment, 5*(4), 210–233. https://doi.org/10.1080/20430795.2015.1118917

Friedman, F. (1970, September 13). A Friedman doctrine. *New York Times Magazine*. Retrieved from https://www.nytimes.com/1970/09/13/archives/a-friedman-doctrine-the-social-responsibility-of-business-is-to.html

Index Industry Association. (2022). *IIA global ESG study*. Retrieved from https://www.indexindustry.org/

Invesco. (2022). *Invesco 2022 Stewardship Report*. Retrieved from https://invesco.com/content/dam/invesco/corporate/en/pdfs/reports/2022_ESG_Investment_Stewardship_Report.pdf

Invest Europe. (2021). *Sustainable finance*. Retrieved from https://www.investeurope.eu/policy/key-policy-areas/sustainable-finance/

Ioannou, I., & Serafeim, G. (2012). What drives corporate social performance? The role of nation-level institutions. *Journal of International Business Studies, 43*(9), 834–864. https://doi.org/10.1057/jibs.2012.26

Jensen, M. (2002). Value maximization, stakeholder theory, and the corporate objective function. *Business Ethics, 12*, 235–256.

Karadima, S. (2022). *FDI drivers in 2022: ESG [investment monitor]*. Retrieved from https://www.investmentmonitor.ai/features/fdi-drivers-in-2022-esg-sustainability/

Kotsantonis, S., & Serafeim, G. (2019). Four things no one will tell you about ESG data. *Journal of Applied Corporate Finance, 31*(2), 50–58. https://doi.org/10.1111/jacf.12346

Matisone, A., & Lāce, N. (2017). *Venture capital in Latvia*. Retrieved from https://www.researchgate.net/profile/Valdis_Avotins/publication/320842368_The_analysis_of_business_start-up_factors/links/59fd8118458515d07068b10d/The-analysis-of-business-start-up-factors.pdf#page=101

Moir, L., Kennerley, M., & Ferguson, D. (2007). Measuring the business case: Linking stakeholder and shareholder value. *Corporate Governance, 7*(4), 388–400.

Morrison & Foerster LLP. (2020). *Sustainable finance disclosure regulation (SFDR): What to expect?* Retrieved from https://www.lexology.com/library/detail.aspx?g=56ca382e-e1a4-4090-9c19-a5bdba7dfd6b

OECD. (2020). *Sustainable and resilient finance*. OECD. https://doi.org/10.1787/eb61fd29-en

PwC. (2021). *PwC's 2021 annual corporate directors survey*. Retrieved from https://www.pwc.com/us/en/about-us/newsroom/press-releases/pwc-launches-acds-2021.html

Roca, L., & Searcy, C. (2012). An analysis of indicators disclosed in corporate sustainability reports. *Journal of Cleaner Production, 20*(1), 103–118. https://doi.org/10.1016/j.jclepro.2011.08.002

Sakuma Keck, K., & Hensman, M. (2013). A motivation puzzle: Can investors change corporate behavior by conforming to ESG pressures? In S. Young & S. Gates (Eds.), *Institutional investors' power to change corporate behavior: International perspectives (Critical studies on corporate responsibility, governance and sustainability)* (Vol. 5, pp. 367–393). Emerald Group. https://doi.org/10.1108/S2043-

Sauders, M. (2020). *ISS ESG survey*. Retrieved from https://www.issgovernance.com/iss-esg-survey-results-highlight-increased-asset-manager-focus-on-social-issues-in-light-of-covid-19-pandemic/

Schramade, W. (2016). Integrating ESG into valuation models and investment decisions: The value-driver adjustment approach. *Journal of Sustainable Finance & Investment, 6*(2), 95–111.

SEB. (2021). *Significant pension*. Retrieved from https://www.seb.lv/en/significant-pension

UN Global Compact. (2020). *Communicate the value of sustainability to investors*. Retrieved from https://www.unglobalcompact.org/take-action/action/value-driver-model

van Duuren, E., Plantinga, A., & Scholtens, B. (2016). ESG integration and the investment management process: Fundamental investing reinvented. *Journal of Business Ethics*. https://doi.org/10.1007/s10551-015-2610-8

Yu, E. P., & Luu, B. V. (2021). International variations in ESG disclosure—Do cross-listed companies care more? *International Review of Financial Analysis, 75*, 101731. https://doi.org/10.1016/j.irfa.2021.101731

Zumente, I., & Bistrova, J. (2021). Do Baltic investors care about environmental, social and governance (ESG)? *Entrepreneurship and Sustainability Issues, 8*(4), 349–362. https://doi.org/10.9770/jesi.2021.8.4(20)

Zumente, I., & Lāce, N. (2021). ESG rating—Necessity for the investor or the company? *Sustainability, 13*(16), 8940. https://doi.org/10.3390/su13168940

Ilze Zumente, PhD, originating from Riga, Latvia, has an extensive academic and professional background. She holds a master's degree in business administration from the University of Cologne, Germany, and a PhD degree from Riga Technical University. She has a professional background in finance, including roles in M&A, private equity, and strategy.

Julija Bistrova, PhD, CFA, is Assoc. Professor at Riga Technical University. She is also the head of an asset management company. Her research interests include ESG and sustainability, earnings plausibility, stock market investments, and role of innovations. In 2014, she received a doctoral degree in economic sciences from Riga Technical University.

Nataļja Lāce is a professor at Riga Technical University (RTU), Latvia. She graduated from RTU (former Riga Polytechnic Institute), Faculty of Engineering Economics, in 1982 with a diploma of engineer-economist. Her doctoral thesis (1990) focused on alternative choices in engineering decision-making. Her pedagogic activities encompass bachelor's, master's, and doctoral programs: lecturing, supervising, and reviewing bachelor's, master's, and PhD theses. Natalja Lace is the leader of the academic research group "Finance" and Economics and the director of the master's program "Business Finance" at RTU, Faculty of Engineering Economics and Management. She is on the editorial boards of several academic journals and an expert on the Latvian Council of Science. She is an author, co-author, and editor of more than 200 scientific papers and books. She is involved in executing research projects sponsored by the Latvian Government, the Scientific Council of the Republic of Latvia, and the European Commission. Her research interests are focused on critical success factors of small and medium-sized enterprises and innovation as well as financial aspects of business.

Ilja Arefjevs, PhD, is a seasoned finance area executive and a fintech co-founder. He has been working for both the leading financial groups of the Baltic region and innovative financial technology companies. While being mainly a practitioner, Ilja Arefjevs is teaching several finance-related courses for the double degree programs of BA School of Business and Finance and SBS Swiss Business School.

Practical Implementation of the Idea of Sustainable Development: The Perspective of Polish Companies

Ewa Mazur-Wierzbicka and Anna Cierniak-Emerych

1 Introduction

The subject matter of sustainable development is particularly important in the time of changes that are happening in global economies and also due to the observed climate change and its adverse effect on the functioning of societies. Activities taken up in individual economies to create conditions that facilitate implementation of the 17 SD goals (SDGs) adopted in 2015 are crucial from the perspective of implementation of sustainable development. The goals gain increasingly greater popularity not only at the level of economies or economy sectors, but also within individual enterprises. They have an ever greater effect on, for example, business models or how investment decisions are made.

Given the above, the main goal of this chapter is to assess the understanding of sustainable development and to identify how it is practised in Polish enterprises.

The main goal was assigned the following detailed goals:

- Shedding light on the development of importance of the SD concept in Poland and its legal grounding
- Assessment of the implementation of SDGs in Poland
- The attitude of Polish companies to social and environmental issues
- Identification of benefits of adopting sustainable practices in Polish companies
- Identification of sustainable practices in selected Polish companies

E. Mazur-Wierzbicka
University of Szczecin, Szczecin, Poland

A. Cierniak-Emerych (✉)
Wrocław University of Economics and Business, Wrocław, Poland
e-mail: anna.cierniak-emerych@ue.wroc.pl

S. O. Idowu, S. Vertigans (eds.), *Sustainability in Global Companies*, CSR, Sustainability, Ethics & Governance, https://doi.org/10.1007/978-3-031-77971-8_5

The design on this chapter serves to pursue the main and detailed goals. The first, theoretical, part, presents the SD concept with a particular emphasis on its development in Polish realities and its legal determinants. Part two presents issues of implementation of SD from the perspective of Polish companies on the basis of, i.e., their realization of SDGs. This chapter has been written on the basis of government documents, acts of law, relevant literature and research reports.

2 Sustainable Development: Introduction to the Subject Matter

The concept of sustainable development has become one of the key paradigms of today's social, economic and environmental development. In the context of increasingly complex global challenges, such as climate change, poverty or social inequalities, sustainable development is a framework of thought and action that strives to ensure a balance between the needs of today and the possibilities of future generations. It is in this context that the concept of sustainable development was first discussed extensively in the Brundtland Report (1987). It was defined as "development that meets the needs of the present without compromising the ability of future generations to meet their own needs". This definition highlights a balance between the three main areas: social, economic and environmental. This means that sustainable development requires that simultaneous needs of economic development, social development and environmental protection are given due consideration (cf. Velazquez et al., 2011; Byrch et al., 2007).

The implementation of sustainable development, therefore, has numerous implications for the society, economy and natural environment. In the social sphere it means a fight with social inequalities through improvement of the quality of life, ensuring equality of access to education, health care and other basic services and through proportion of public participation and gender equality. In the economic sphere, it pertains the need to transition to a more sustainable model of economic growth, which includes social and environmental aspects through, i.e., promoting innovation, creating new jobs, improving efficiency of using resources and promoting ethical and transparent business. In the sphere of the natural environment, on the other hand, it means supporting initiatives aimed at protecting the natural environment and biodiversity, reduction of greenhouse gas emissions and adaptation to climate change, reducing exploitation of natural resources and promotion of sustainable use and protection of water, soil and forest resources (cf. Mazur-Wierzbicka & Swiatkiewicz, 2023).

Therefore, referring to each of these dimensions, one may conclude that from the economic perspective sustainable development may be understood as, e.g., a model of development that allows achievement of economic growth with a minimal impact on the natural environment and with maintaining social and economic balance. From the perspective of the social dimension, sustainable development may be considered as development that ensures fair access to resources, decent live conditions and an opportunity for development for all members of society, without discrimination and exclusion, whereas from the perspective of the environmental protection dimension, sustainable development may be defined as development based on a harmonious interaction between the economy and the natural environment, which ensures effective use of natural resources and minimization of the negative impact on the environment.

Given the above, it may be stated that understanding sustainable development reflects the complexity and m**ultifaceted character** of this concept, taking into account the economic, social and environmental dimensions (Kolk et al., 2017; Arora & Mishra, 2019; Hirons, 2020). A strive to achieve a balance between them so as to ensure lasting and sustainable development of society and natural environment is key.

3 Sustainable Development Determinants: Poland's Perspective

In the face of increasingly obvious socio-economic and environmental challenges, the sustainable development concept has become of the fundamental approaches taken into account in environmental (ecological) and socio-economic policies of states, including Poland. The beginnings of an interest in the sustainable development concept trace back to the 1990s. Under international obligations (especially in the context of Poland's accession to the European Union) and social pressure, the Polish government began to take into account SD principles in its documents, policies and development strategies (Mazur-Wierzbicka, 2012).

The dynamic progress of sustainable development in Poland took place in 2000–2010. A number of actions were initiated in this period in development plans to integrate social and economic aspects and aspects relating to the natural environment. Many strategic documents were written, such as the national Low Emission Economy Plan or the Sustainable Development Strategy, which specified the goals and directions of action in the area of sustainable development.

The last decade has seen Poland's even greater involvement in promoting sustainable development. Increasingly more focus is given to intensifying green economy, improving energy efficiency, protection of natural resources and promoting corporate social responsibility.

Implementation of the principles of sustainable development requires concrete legislation that regulates actions of public administration, the private sector and civil society. There are a number of legal acts in Poland that play a key role in promoting and implementing the SD concept. They include both, general legislation acts that refer to the questions of sustainable development and more specialized laws that regulate specific areas of operation.

Key legislative acts at the national level include:

1. The Constitution of the Republic of Poland (1997) that is a basic statutory document that includes instruction on, i.e., environmental protection and sustainable development.
2. Sectoral statutes, such as these following major acts:
   - Act of 27 April 2001 Environmental Law, which is a key legislative act addressing sustainable development in Poland. It specifies rights and obligations in environmental protection, including issues pertaining to waste management, nature conservation, protection of ambient air and waters and natural resources management. It provides a basis for many other environment-related legislative acts and regulations in Poland.
   - Act of 20 July 2017b on renewable energy sources, which is a key legislative act that specifies principles of supporting producers of energy from renewable sources and that regulates questions of the RES sector.
   - Act of 11 September 2015 on used electrical and electronic equipment, which introduces laws relating to electrical and electronic waste management, promotes recycling and reuse, which helps reduce the negative impact on the natural environment. It is an especially important legislative act in the context of circular economy and effective resource management, which are essential elements of the sustainable development concept.
   - Act of 21 December 2012 on underground waters and protection against contamination, which provides a legal basis for actions aimed at protection of underground and ground waters, ensuring their sustainable use and protection against contamination. Protection of water resources is a key element of sustainable development, especially in the context of climate change and threats to the quality of waters.
   - Act of 20 May 2016 on energy efficiency that promotes energy savings, improvement of energy efficiency and reduction of greenhouse gas emissions through regulating actions in the energy sector.
3. Relevant strategies and programmes include: government strategies and programmes, such as the National Low-Emission Economy Plan or the Responsible Development Strategy (Resolution no 8 of the Council of Ministers, 2017a). They are key tools in promoting SD and outlining priorities for action.

Moreover, as a Member of the European Union, Poland is obliged to implement EU directives and regulations on sustainable development. Recently, the European Union has intensified its actions for environmental protection and SD promotion by adopting, i.e., the European Green Deal (European Commission, 2019), which is now extensively debated in EU countries, including Poland.

We need to point to the signing of international agreements that focus on sustainable development. One of the most crucial documents that has been drafted recently (2015) is the 2030 Agenda for Sustainable Development that presents 17 sustainable development goals (SDGs) (UN, 2015).

Legislative acts that pertain to the SD concept in Poland are key tools of implementing the policy of environmental protection and promotion of sustainable development. The agreements made in these legislative acts reflect Poland's commitments arising from international agreements and conventions and are a basis for the activity of public administration, the private sector and civil society towards the achievement of lasting and sustainable development. Further provision and adjusting of legislative acts to the changing socio-economic and environmental determinants are key for effective promotion of SD in Poland.

4 Presence of the Sustainable Development Concept in the Activity of Polish Enterprises

Today's companies increasingly realize that they must take into account social and environmental issues in their activity. Polish enterprises are no exception here. The actions undertaken by them result from internal and external determinants alike.

When it comes to the former, we need to point to increased social and ecological awareness of the governing institutions and their sense of responsibility for actions taken towards both internal and external stakeholders and an increased awareness and knowledge of the significance of sustainable development and its impact on long-term profitability of activity and on reinforcing their competitive position. As a consequence, companies that plan to rely on guidelines of the SD concept in their activity increasingly intertwine its principles in their business strategies and take actions to balance economic, social and environmental interests.

When it comes to external determinants, one needs to point to the trend, a major interest recently, of increased social and ecological awareness among consumers, suppliers and investors alike. It is in response to this that companies increasingly implement CSR and SD strategies.

Sustainable development has recently undoubtedly become an attribute of competitiveness of Polish enterprises. However, it is difficult to imagine the possibility of implementing it without greater engagement of the public sector and without developing adequate action plans, beginning with governmental principles. In order to be able to strive to implement principles of sustainable development, it is necessary to adopt specific foundations, plans and visions, both for the macro- and microeconomic level (the level of individual organizations). An essential role here goes to the 2050 sustainable development vision for Polish business, which identifies key significance of six selected areas (Table 1).

Despite the passage of time, the adopted principles for Polish business in the context of fitting in the implementation of the sustainable development concept are still valid. They were additionally strengthened by the implementation of sustainable development goals by Polish entities. It needs to be noted that SD goals have been very well received by Polish organizations and have been reflected in their basic (operational) and strategic activity. Polish companies declare their readiness to undertake sustainable development actions and believe that achievement of SDGs until 2030 is possible.

Table 1 Vision 2050. The New Agenda for Business in Poland

Area	Social capital	Human capital	Infrastructure	Natural resources .	Energy	State and institutions quality
Important and urgent	Creation of solutions which initiate and stimulate cooperation	Change of life-style and values	Strategic thinking about infrastructure as a tool for sustainable development	Building an attitude of co-responsibility for the consequences of natural resources scarcity	Common strategic decisions concerning energy sources	Change of thinking of roles and tasks of the state and the entrepreneurs
Vision 2050	Competitive and innovative Polish economy develops based on social capital	Optimum number of workers qualified adequately to the market needs	New, safe, environmentally friendly and accessible infrastructure	Resources used in sustainable way throughout their life cycle	Diverse and safe energy for a reasonable price, non-generating social or environmental costs	Foresee able and clear attitude to entrepreneurs based on cooperation
Vision 2050. The New Agenda for Business in Poland.						

Source: Vision 2050. The New Agenda for Business in Poland. Executive Summary, May 2012, Warsaw, p. 9.—PODAJE LINK https://odpowiedzialnybiznes.pl/wp-content/uploads/2014/03/Vision-2050_summary.pdf

When it comes to measuring the implementation of SD goals by Polish organizations, an adequate tool was created in 2016—"Impact Barometer", which was the first Polish set of SDG indicators for business. It is composed of 30 indicators (1 indicator from Goal 3; 5 indicators from Goal 4; 6 indicators from Goal 5; 8 indicators from Goal 8; 3 indicators from Goal 9 and 7 indicators from Goal 12) (Table 2).

Each Polish company may use this tool free of charge. It allows measurement of the impact of the business on the implementation of the 2030 Agenda in Poland. It is a set of indicators tailored to the specific characteristics of business, including instructions on how to make calculations, developed as part of the 17 SD goals, with substantive support from Statistics Poland (GUS) and cooperation of experts from the world of business and science. It allows Polish companies to independently assess their contribution to sustainable development. One of the main goals of this tool is to encourage companies to include SD goals in their business strategies and to undertake ambitious, effective actions that foster their implementation. This tool allows companies to diagnose where at the ladder of implementation of SDGs they are, which areas have been managed successfully and which need improvement. Jointly collected results will allow creation of an image of the impact of Polish business on SDGs and a diagnosis of progress in implementing priority SDGs and will show areas that need specific actions.

If one looks at the involvement of Polish companies in the implementation of SDGs, good practices of Polish companies in the area of corporate social responsibility focus on 7 core subjects identified as part of the ISO 26000 standard, that is

Table 2 SDGs important for Polish business in the context of the impact Barometer

SDGs	Important for the Polish business
Goal 3: Good health and well-being	Innovative solutions and technologies supporting patients in diagnosis and treatment processes. Creating products and services that promote a healthy and active lifestyle. Taking care of employee safety and mental health; supporting their health-promoting activities and habits. Promoting health and safety issues throughout the supply chain (most road fatalities occur during work). Developing public–private partnerships and increasing access to the health care system. Partnership cooperation of companies for prevention and health education. Reducing the harmful impact of some burdensome industries on human health and the natural environment.
Goal 4: Ensure high-quality education for all and promotion of lifelong learning	Cost-effective education products and services, in particular digital (e-learning, m-learning, b-learning) ones, that remove barriers in access and improve quality of teaching. Fostering employees' lifelong learning and development of skills in new technologies. Promotion of and investment in STEM education (i.e. Science, Technology, Engineering, Mathematics), with a particular focus on girls and women. Supporting and initiating new educational activities for local communities (e.g. as part of life-long learning), with a particular focus on disfavoured groups. Education in and various incentives for sustainable lifestyle.

(continued)

Table 2 (continued)

SDGs	Important for the Polish business
Goal 5: Achieve gender equality and empower all women and girls	Increasing the level of women's employment in managerial positions. Ensuring fair and equal pay and additional performances for work of equal value. Appreciating and supporting the parent's care over children or other dependents by providing services, resources and information. Policy of "zero tolerance" to all forms of abuse at work, including verbal and physical abuse, and counteracting sexual harassment. Developing business relations with companies run by women (including small enterprises and women-entrepreneurs). New technologies (including ICT) that strengthen the role of women and their digital competences necessary in the future. Empowering women in the entire value chain and elimination of practices that are detrimental to them. Satisfaction of social and environmental needs and teaching new client groups—using the knowledge and experience of companies whose business strategies are closely related to Goal 5.

(continued)

Table 2 (continued)

SDGs	Important for the Polish business
Goal 8: Promote stable, sustained and inclusive economic growth, full and productive employment and decent work for all	Improving economic efficiency through diversification, technological modernization and innovations, and also through a focus on high value-added and labour-intensive sectors. Sustainable production, including increasing efficiency of use of natural resources and reduction of the negative environmental impact in the production process. Implementation of the strategy of sustainable supply chain. Developing sustainable tourism that creates workplaces and promotes local culture and products. Supporting micro-, small- and medium-sized enterprises and including them in sustainable supply chains. Running skills development programmes throughout the supply chain. Supporting the culture of entrepreneurship and investing in mentoring of beginning entrepreneurs. Facilitating access to banking, insurance and financial services for all. Creating decent workplaces and safe working environment. Combating mobbing and harassment in a workplace. Elimination of unfair practices in recruitment and employment, especially against groups vulnerable to abuse (e.g. migrant workers, persons with disabilities); creation of innovative forms of social protection of such employees. Equalling remuneration of men and women and excluded groups for work of equal value Creating opportunities for work and for the development of qualifications for young people. Effective mechanisms of detecting child labour and forced labour in global supply chains and implementation of remedies if violations are detected.
Goal 9: Build resilient infrastructure, promote inclusive and sustainable industrialization and foster innovation	Developing sustainable high-quality and high value-added industry, also by improving efficiency of the use of resources and application of clean and environmentally friendly production technologies and processes. Investment in research and development, supporting research and development centres. Modernization of and investment in new infrastructure—More sustainable, resilient, people- and environment-friendly. Engaging local communities, especially vulnerable groups, in consultation processes on changes in infrastructure. Promotion of innovation and openness to stakeholders' ideas. Sustainable management of company projects and initiatives, including standard setting. Supporting micro-, small and medium-sized enterprises, including industrial enterprises, by including them in value chains and facilitating their access to financing. Enhancing access to information and communication technologies and ensuring accessible and universal Internet access.

(continued)

Table 2 (continued)

SDGs	Important for the Polish business
Goal 12: Ensure sustainable consumption and production patterns	Implementation of the idea of circular economy throughout the value chain by, e.g., organizing a circular supply chain, recovery and recycling of raw materials, extending product life, delivering products as shared goods and services. Implementation of innovations and innovative business models for sustainable consumption and production. Product ecodesign—Introduction of an additional criterion to the design stage that allows identification of environmental factors associated with the product (including those relating to its recycling and utilization) in the early stage of its development. More effective use of raw materials and departure from fossil fuels, replacing primary raw materials with recovered or recycled raw materials Reduction of the impact of production processes on the environment and on the accessibility of natural resources Limiting the number of produced waste and its responsible management. Implementation of sustainable development practices, from "green office" to changes in the company's value chain. Measuring the environmental and social footprint throughout the product life cycle and activity and publishing information on it in regular reports Education of stakeholders in sustainable development, sustainable consumption and counteracting wastage. Developing sustainable tourism that creates workplaces and promotes local culture and products.

Source: Author's own compilation on the basis of: https://kampania17celow.pl/agenda-2030/

organizational governance, human rights, labour practices (relations with employers and counter-parties), environment (climate change mitigation and adaptation), consumer issues (fair practices, building consumer awareness), community involvement and development (introduction of public dialogue that includes participation of community organizations in the process of planning and implementing projects with consideration to social needs in education, culture, health, development and access to technologies), fair operating practices (fair competition, preventing corruption, promoting reliability in supply chains and respecting ownership rights).

It is especially noticeable in reports published by one of the organizations that strongly support actions relating to sustainable development, corporate social responsibility, ESG in the education, information and mentoring angle—Responsible Business Forum (RBF). These reports are published annually as "Responsible business in Poland. Good practices". They include, i.e., a compilation of initiatives implemented by companies under the adopted SDGs that are accommodated under the core subjects of the ISO 26000 standard. They are the most comprehensive review of ESG, CSR and SD actions taken by organizations in Poland.[1] In the latest report for 2023 (RBF, 2024) (it was the 22nd edition of the annual RBF report), it was mainly large enterprises (75 percent) that offered their practices. 16 percent of submissions came from medium-sized companies, while it was 7 percent and 2 percent for small and micro enterprises, respectively. The trend in the core subjects of the practices and initiatives reported has been similar for a few years now. Most actions presented in latest reports pertained to community involvement (in 2023, it was 418 good practices), labour practices (249 good practices in 2023) and natural environment (210 good practices in 2023). The least robust areas are those relating to human rights (54 good practices in 2023), consumer issues (46 practices n 2023), fair operating practices (37 good practices in 2023) and organizational governance (32 practices in 2023). The practices presented in the report are also described in the context of contribution to the implementation of SD goals. They most often involve actions concerning Goal 3: Good health and well-being (221 good practices in 2023), Goal 12: Responsible consumption and production (218 practices in 2023) and Goal 4: Quality education (214 good practices in 2023). In turn, the greatest growth in popularity was observed for good practices that support implementation of Goal 12 (Responsible consumption and production), e.g., in the development of circular economy, and Goal 17 (Partnerships for the goals). Those mentioned most often include Goal 2: Zero hunger (13 good practices in 2023), Goal 16: Peace, justice and strong institutions (12 good practices in 2023) and Goal 14: Life below water (3 good practices in 2013) (RBF, 2024).

A novelty in the 2024 Report was the possibility to assign a practice to specific SD reporting standards (ESRS—European Sustainability Reporting Standards) intended to ensure that European companies have a uniform method for reporting ESG information. The most numerous group were actions associated with ESRS S1—Employees (207 good practices), followed by ESRS S3—Affected communities (163 good practices) and ESRS E1—Climate change (84 good practices) (RBF, 2024). It is also worth adding that 73 percent of companies that took part in the 2023 review declared that they are subject to reporting under the Corporate Sustainability Reporting Directive (CSRD).

The subject of non-financial reporting is slowly, yet increasingly, taken up by Polish companies. It may be concluded that the formal requirement encourages reflections on the impact of business on local communities and the environment and contributes to implementation of corporate governance. Capital investors are increasingly interested in investing in companies that transparently communicate non-financial information. Despite the growing awareness of Polish entrepreneurs on the importance of ESG factors, they are still perceived in the categories of challenges rather than business opportunities. This is, among other things, down to

[1] It should be noted that there are more organizations in Poland undertaking ESG, CSR and sustainable development activities under the adopted SD objectives. The aforementioned Reports provide a cross-sectional view of companies and are not a compilation of all organizations undertaking SD goals.

companies lacking structured knowledge in this area, few experts in this field, the cost of employing them, or a lack of IT tools that facilitate reporting.

However, we need to point out that there are organizations in Poland that support business in the implementation of sustainable development-related principles (cf. Cierniak-Emerych et al., 2021). One of the first such organizations in Poland that understood the idea of SDGs was the already-mentioned Responsible Business Forum. Polish business is also supported by an organization called The 17 Goals Campaign (Pl. Kampania 17 Celów), which mobilizes stakeholders to take up joint actions for the implementation of sustainable development goals and taking advantage of business opportunities coming from the 2030 Agenda. Since 2017, when the Campaign was initiated, it has organized more than 50 events in which more than 7 thousand entities have taken part. The National Chamber of Commerce (KIG) also acts to promote sustainable development and corporate social responsibility. For years it has been carrying out many educational projects directed to members of regional and industrial chambers of commerce. It has also launched an education programme "ESG School" under which it teaches how a company may get involved in achieving the 17 SD goals. KIG also awards most socially-sensitive companies under the "Fair Play Company" programme (www.fairplay.pl).

The analysis of RBF reports shows that sustainable business practices are becoming increasingly popular in Poland, both among large corporations and smaller enterprises (RBF, 2016–2024). Striving for a balance between economic, social and environmental aspects does not only bring benefits for the natural environment and society, but may also lead to increased competitiveness and long-term business success (Table 3).

It is also worth looking at Polish companies in the context of implementation of SD goals for priorities they set for themselves and challenges that stand before them. An analysis of RBF reports for 2016–2024 has shown that priority areas for Polish companies are those pertaining to: climate action (Goal 13), consumption and production (Goal 12), building resilient infrastructure, promotion of sustainable industrialization and fostering innovation (Goal 9). The perception of the question of energy efficiency, accommodated under Goal 7, has changed since 2022. Increased energy prices have encouraged a reflection on the need for energy savings. Given the above, Polish companies began to invest in their own generation capacities at their premises and also in intelligent energy management systems. Goal 5 and Goal 10 are also important core subjects for Polish companies because the global climate crisis, armed conflicts and economic difficulties contribute to deepening inequalities not only within individual states, but also between regions and countries. Due to the armed conflict in Ukraine, companies operating in the territory of Poland and also those that have their branches in Ukraine have become involved in humanitarian initiatives by giving donations, collecting necessary equipment, food and clothing to give them to those in need. Therefore, it may be assumed that implementation of Goal 1 in the current circumstances (armed conflict) is also very important for Polish companies.

The SD goals set by Polish organizations are part of the specificity of their activities. As with any goals, these goals are initially declarative, while the approach to

Table 3 Identification of major benefits of adopting sustainable practices by Polish companies looking at economic and social aspects alike

Major benefits	Description
Reduction of operating costs	Adopting sustainable practices may lead to a significant reduction of operating costs for Polish companies. Actions such as improvement of energy efficiency, reduction of the use of raw materials and water and optimization of production processes may contribute to a reduction in the use of resources and related costs. Companies that successfully reduce their operating costs and generate cash savings may thus, in the long run, increase the company's profitability and competitiveness in the market.
Improvement of image and increased customers' trust	Consumers increasingly point to social and environmental aspects of companies' activity while making purchasing decisions. Companies that get involved in pro-social and pro-green actions may gain the trust of customers and build a positive brand image, which translates into customer loyalty and increased sales. Companies that get involved in pro-social and green actions build a positive image in the eyes of customers, employees, investors and local communities. Improved image may contribute to increasing customers' trust and company's reliability in the market.
Improvement of relationships with stakeholders	Companies that get involved in pro-social and green actions may gain recognition and support of their stakeholders, which translates into stability and lasting economic activity. Involvement of companies in social and green initiatives may contribute to increased engagement of their employees, to acquisition of new customers and to gaining approval of the local community.
Obtaining and retaining talents	Companies that get involved in sustainable practices are becoming attractive employers for potential employees. Young generations increasingly care for social and environmental values, which is why companies that promote such values and offer attractive working conditions may retain their existing employees and attract new talents easier.
Fostering economic and social growth	Investment in employee education and development, promoting diversity and equality in a workplace and supporting local communities may contribute to creating workplaces, increased social engagement and reduction of social inequalities.

Source: Authors' own compilation

them and their implementation depends on the level of (social, environmental) awareness of the organization, its maturity of functioning, and its responsibility for its stakeholders. It is not always possible to realize the set goals. This is conditioned by many determinants from both the internal and external environment of the organization. When it comes to holding organizations "accountable" for their goals, it is important that organizations present concrete actions, facts—this will exclude being suspected of being declarative and only taking "timely" initiatives without taking concrete action. Particularly important in this area is the organization's long-term position in the market—its credibility. The mere declarativeness of an organization in terms of adopting SR goals (but not only) not supported by actions does not build a credible, long-term position of the organization—on the contrary, they may be perceived as those using SD issues for marketing activities, which may consequently result in a negative perception among their stakeholders.

Due to, in particular, legislative changes, Polish companies face a number of challenges in the area of sustainable development. They primarily concern the obligation to report non-financial actions that in the coming years will be imposed on companies under the Corporate Sustainability Reporting Directive (SCRD) or the Corporate Sustainability Due Diligence Directive (CSDD). The next challenge concerns the planned amendment of the Unfair Commercial Practices Directive, which stipulates that traders who label their products "eco" will have to base this information on specific data. The EU Ecodesign Directive (now 2009/125/EC) is also being amended now. It will allow customers to make informed choices when it comes to products' circularity. Polish companies that operate on the basis of sustainable development goals will have to put in even more care in the coming years to whether their counter-parties, customers, sub-contractors, etc., are guided by the principles of sustainable development in their actions and will have to observe new ESG sustainable development standards.

5 Conclusions

The concept and essence of sustainable development are key elements of today's political debate and policy. Sustainable development is based on a balance between economic, social and environmental aspects. It has crucial practical implications for companies (Schaltegger, 2018), society and the natural environment (Gunawan et al., 2020). Despite the challenges, the perspectives for the future are promising because more and more companies and societies note the benefits of sustainable development and take actions in this realm.

Legislative acts that pertain to the concept of sustainable development in Poland are key tools for implementing the policy of environmental protection and promotion of sustainable development. Provisions laid down in them reflect Poland's commitments arising from international agreements and conventions and are a basis for the activity of public administration, the private sector and civil society towards the achievement of lasting and sustainable development. Further development and adjusting legislative acts to the changing socio-economic and environmental determinants are key for effective promotion of sustainable development in Poland.

Observing the behaviour of Polish companies, it is worth noting that the implementation of the sustainable development concept in their actions and initiatives is becoming increasingly advanced and comprehensive. Implementation of CSR programmes, pro-environment investment and the development of social and ecological awareness are key elements of business strategies of many companies. Despite the challenges, sustainable development perspectives in Poland are promising. Continuation of actions towards sustainable development is key to building a lasting and prospering economy and society.

References

Arora, N. K., & Mishra, I. (2019). United Nations sustainable development goals 2030 and environmental sustainability: Race against time. *Environmental Sustainability, 2*(4), 339–342.

Byrch, C., Kearins, K., Milne, M., & Morgan, R. (2007). Sustainable what? A cognitive approach to understanding sustainable development. *Qualitative Research in Accounting & Management, 4*(1), 26–52. https://doi.org/10.1108/11766090710732497

Cierniak-Emerych, A., Mazur-Wierzbicka, E., & Rojek-Nowosielska, M. (2021). Corporate social responsibility in Poland. In S. O. Idowu (Ed.), *Current global practices of corporate social responsibility* (CSR, sustainability, ethics & governance). Springer. https://doi.org/10.1007/978-3-030-68386-3_13

European Commission. (2019, December 11). *Communication from the Commission to the European Parliament, the European Council, the Council, the European Economic and Social Committee and the Committee of the Regions the European Green Deal, COM(2019) 640 final.*

European Union. (2009). *Directive 2009/125/EC of the European Parliament and of the Council of 21 October 2009 establishing a framework for the setting of ecodesign requirements for energy-related products (recast).*

Forum Odpowiedzialnego Biznesu. (2024). *Raport "Odpowiedzialny biznes w Polsce 2023. Dobre praktyki".*

Gunawan, J., Permatasari, P., & Tilt, C. (2020). Sustainable development goal disclosures: Do they support responsible consumption and production? *Journal of Cleaner Production, 246*, 118989. https://doi.org/10.1016/j.jclepro.2019.118989

Hirons, M. (2020). How the sustainable development goals risk undermining efforts to address environmental and social issues in the small-scale mining sector. *Environmental Science & Policy, 114*, 321–328. https://doi.org/10.1016/j.envsci.2020.08.022

ISAP. (1997). *Konstytucja Rzeczypospolitej Polskiej z dnia 2 kwietnia 1997 r.*

ISAP. (2001). *Ustawa z dnia 27 kwietnia 2001 r. Prawo ochrony środowiska.*

ISAP. (2012). *Ustawa z dnia 21 grudnia 2012 r. o wodach podziemnych i ochronie przed zanieczyszczeniem.*

ISAP. (2015). *Ustawa z dnia 11 września 2015 r. o zużytym sprzęcie elektrycznym i elektronicznym.*

ISAP. (2016). *Ustawa z dnia 20 maja 2016 roku o Efektywności Energetycznej.*

ISAP. (2017a). *Uchwała nr 8 Rady Ministrów z dnia 14 lutego 2017 r. w sprawie przyjęcia Strategii na rzecz Odpowiedzialnego Rozwoju do roku 2020 (z perspektywą do 2030 r.).*

ISAP. (2017b). *Ustawa z dnia 20 lipca 2017 r. o odnawialnych źródłach energii.*

Kolk, A., Kourula, A., & Pisani, N. (2017). Multinational enterprises and the sustainable development goals: What do we know and how to proceed? *Transnational Corporations, 24*, 9–32. https://doi.org/10.18356/6f5fab5e-en

Mazur-Wierzbicka, E. (2012). *Ochrona środowiska a integracja europejska. Doświadczenia polskie.* Difin.

Mazur-Wierzbicka, E., & Swiatkiewicz, O. (2023). Sustainable development and corporate social responsibility. In *Organizing sustainable development.* Routledge. https://doi.org/10.4324/9781003379409-9

Organizacja Narodów Zjednoczonych. (2015). *Przekształcamy nasz świat: Agenda na rzecz zrównoważonego rozwoju do roku 2030 (A/RES/70/1), 25 września 2015.*

Orłowski, W. M. (2012). *2050 Wizja zrównoważonego rozwoju dla polskiego biznesu, Raport-streszczenie, maj 2012.*

Schaltegger, S. (2018). Linking environmental management accounting: A reflection on (missing) links to sustainability and planetary boundaries. *Social and Environmental Accountability Journal, 38*(1), 19–29. https://doi.org/10.1080/0969160X.2017.139535

Velazquez, L. E., Esquer-Peralta, J., Munguıa, N. E., & Moure-Eraso, R. (2011). Sustainable learning organizations. *The Learning Organization, 18*(1), 36–44. https://doi.org/10.1108/09696471111095984

World Commission on Environment and Development (Brundtland Report). (1987). Retrieved from http://www.ace.mmu.ac.uk/eae/Sustainability/Older/Brundtland_Report.html

Ewa Mazur-Wierzbicka is Professor at the University of Szczecin, Faculty of Economics and Management, Institute of Human Capital Management. She specializes mainly in the field of corporate social responsibility (with emphasis on diversity management, ethical and equality-related actions), sustainable development, and human capital management. She also deals with the issues of soft competences. She is an expert in the field of corporate social responsibility (CSR)—external expert of the Responsible Business Forum; University of Szczecin's plenipotentiary to the Technical Committee no. 305 for Social Responsibility operating at the Polish Committee for Standardization; advisor of the Polish Agency for Enterprise Development in terms of corporate social responsibility; Initiator and Chair of the Cycle of Seminars and Conferences titled "Corporate social responsibility—the management and economy perspective" associating both researchers and business practitioners; and expert of the EIGE's (the European Institute for Gender Equality) Experts' Forum (mandate period of 01.12.2018–30.11.2021 (IV term), 01.12.2021–30.11.2024 (V term)). She is a participant of Polish and international research projects; reviewer of research projects of the National Science Centre; member of program councils, scientific committees, and organizing committees; panelist of Polish and international conferences devoted to corporate social responsibility (also addressing equal treatment and diversity management issues) and human capital management; and business consultant. She cooperates with practitioners in the field of management, training companies, and institutions, i.e., Polish Entrepreneurs Foundation, and is author and co-author of training programs, numerous scientific studies (including those addressing equal treatment and diversity management issues), analyses, research projects, and expert opinions for business.

Anna Cierniak-Emerych is Professor in the Department of Labour, Capital and Innovation, Wroclaw University of Economics and Business (Poland); Head of Department of Labour, Capital and Innovation; Dean for student's affairs, Faculty of Business, Wroclaw University of Economics and Business; and Director of postgraduate studies "Occupational Safety and Health." The author of more than 180 publications in both Polish and English, among others, monograph entitled "The participation of employees in the management of potential operating company," ed. Wroclaw University of Economics, Wrocław 2012, and co-author of the monograph "Labour potential management and the satisfaction of employee interests," ed. Wroclaw University of Economics, Wrocław 2022. Anna Cierniak-Emerych's research interests relate mainly to the human and his work in the company, particularly corporate social responsibility (CSR), development of tangible and intangible working conditions (health and safety), employee participation, greater flexibility in employment, human resources management, organizational culture, satisfaction of employees, the interests of employees and employers in the enterprise and their respect, and contemporary management concepts (e.g., lean management and TQM).

Agriculture Sustainability: Strategies of the Olive Oil in Portugal

Rute Abreu, Ermelinda Oliveira, and Francisco Tomé

1 Introduction

Agriculture sustainability is essential to human life, and olive oil strategies are a prime focus to research of how sustainability good practices can be implemented to answer to needs and expectations of the citizen. To address these (Darnhofer, 2010; WFCS, 2023) many olive oil producers in Portugal have implemented sustainable farming practices. Some of these practices focus on reducing the environmental impact of olive oil production by decreasing the use of water, fertilisers, and pesticides and, even, the increasing use of the solar energy sources to obtain water (Figueiredo, 2011).

Furthermore, olive oil production has been a traditional product of the Portuguese food and markets for centuries and without no doubt that this market has an extended tradition based on the olive production and, more recently, the industry is growing as an essential economic sector (Mâcedo et al., 2010). However, due to market changes, concerns about the environmental impact of olive oil production have intensified, while financial performance in this market has become increasingly scrutinized.

The research starts to focus on the main strategy—assess the external context—that it will be used by Portuguese olive oil producers to promote sustainability. So, the use of strengths, weakness, opportunities, and risks (SWOR) analyses allows to

R. Abreu (✉)
Instituto Politécnico da Guarda, CICF-IPCA, CISeD-IPV, CITUR, Guarda, Portugal
e-mail: ra@ipg.pt; ermelindaol@ipg.pt; tome@ipg.pt

E. Oliveira · F. Tomé
Instituto Politécnico da Guarda, Guarda, Portugal
e-mail: ermelindaol@ipg.pt; tome@ipg.pt

S. O. Idowu, S. Vertigans (eds.), *Sustainability in Global Companies*, CSR, Sustainability, Ethics & Governance, https://doi.org/10.1007/978-3-031-77971-8_6

combine relevant policy objectives of agriculture sustainability with comprehensive analysis of stakeholder's perspectives (Xie et al., 2019: 1).

Methodologically, this research focuses, on the one hand, on the literature review to contextualise the olive oil sector, in general, and its sustainability in Portugal and to connect those approaches with spatial, ecological, socio-cultural, technological, and economic dimensions. On the other hand, it promotes the research of the SWOR analysis applied to olive oil in Portugal. Also, the research is supported on the Portuguese statistical database published by Official Bodies that enable to understand the challenges of agriculture sustainability in rural areas.

The position of the olive oil sector in Portugal demonstrates that sustainable agriculture practices can be implemented successfully in traditional industries, such as olive oil production (Gray, 2013). By adopting these practices, olive oil producers in Portugal are not only reducing their environmental impact but also improving the quality of their products, increasing the new cooking tendencies, and protecting the well-being of consumers (Pires, 2005).

This paper focus on agriculture sustainability and deals with socio-economic concerns (Graaf and Eppink, 1999), because the authors aim to propose strategies that will help stakeholders to promote olive oil in Portugal. Furthermore, the motivation is proposed by the authors through the research that it will promote the high value of farming and the accountability of the olive and olive oil production.

Olive oil is one of Portugal's main agricultural products, occupying a prominent position in international trade (GPPAP, 2020). The olive oil sector traces its origins to the Arabic term meaning "olive juice," reflecting the enduring ancestral influence of Arab culture on olive cultivation (Santo, 2022). Indeed, the olive tree is an ancient traditional crop in the Mediterranean Basin (Fraga et al., 2020). But there has been a significant growth in the quantity and the quality of olive oil produced in Portugal (INE, 2023).

Portugal is the eighth largest producer of olive oil in the world (INE, 2023) and it is relevant are the area of production, where *Alentejo* region is the main producer, accounting for around 75% of Portugal's total olive oil production. Other prominent regions are *Trás-os-Montes*, *Beira Interior*, and *Ribatejo* (INE, 2023). Olive oil production in Portugal is a traditional agriculture activity and it is a very important economic activity as Table 1 shows.

Table 1 illustrates the remarkable growth in olive oil production in Portugal, which increased by 379.26% between 1995 and 2021. This strong increase is subdivided by 474.92% on the industrial press production, 205.34% on the cooperative press production, and 107.40% on the particular press production (i.e. small farmers). Thus, it confirms the change of the dimension trend of the olive oil production.

This demonstrates that the only way for farmers to survive is through sustainable agriculture (Cazimoglu, 2017), but why this is important?

The Brundtland Report of 1987 (UN, 1986) defined sustainable development as meeting the needs of the present without compromising the ability of future generations to meet their own needs. In agriculture, the Food and Agriculture Organization (FAO) further elaborated on this concept, emphasising practices that are environmentally non-degrading, technically appropriate, economically viable, and socially acceptable (FAO, 1991).

Table 1 Olive oil production in Portugal, 1995–2021

	Years	1995	2000	2005	2010	2015	2016	2017	2018	2019	2020	2021
	Total	477,728	249,433	318,174	686,832	1,190,524	757,373	1,470,352	1,094,433	1,540,630	1,070,620	2,289,549
Press type	Particular	33,361	11,063	14,650	32,582	84,729	52,229	72,302	45,914	90,925	41,259	69,192
	Cooperative	124,039	96,134	109,156	240,292	236,145	142,401	268,380	202,790	240,415	169,738	378,743
	Industrial	320,328	142,235	194,368	413,958	869,650	562,744	1,129,670	845,729	1,209,290	859,623	1,841,615

Source: INE (2023)

These new strategies underscore the importance of aligning agricultural practices with societal priorities, particularly concerns for the future of agriculture (Delonge et al., 2020; Dinis et al., 2015). By focusing on sustainability, agricultural development aims to ensure that current practices do not deplete natural resources or harm the environment, thereby preserving these resources for future generations (Pruss, 2019). Additionally, sustainable agriculture seeks to employ techniques that are technically feasible and economically viable, ensuring that farming practices can be maintained over the long term without causing undue financial burden (Dixon et al., 2001).

Moreover, by emphasising social acceptability, sustainable agriculture acknowledges the importance of considering the well-being of communities and stakeholders involved in agricultural activities (Monaghan et al., 2017). This includes addressing issues such as equity, social justice, and cultural preservation, ensuring that agricultural development benefits society as a whole (Mairech et al., 2020).

In summary, these new strategies for agricultural development justify their importance by addressing societal priorities and aligning with the principles of sustainability outlined in the Brundtland Report. By adopting practices that meet environmental, technical, economic, and social criteria, sustainable agriculture seeks to promote long-term viability and resilience in the agricultural sector while safeguarding the needs of both present and future generations (Richardson, 1975; Pretty, 1995).

As Schaller (1993: 90) argues "the idea of sustainability, at least with regards to agriculture and natural resources, is not new". Thus, it is possible to identify the first research explaining the definition of agriculture sustainability that has been published, in 1798, by Malthus with "An Essay on the Principle of Population". Indeed, Oberč and Arroyo Schnell research (2020: 5) argue that agriculture needs to focus on two societal priorities, such as: "preserving the environment and providing safe and healthy food for all". Thus, the authors agree that the guiding principles of sustainable agriculture promoted by Feher and Beke-Lisanyi (2013: 9) are "stewardship (the importance of environmental protection), and economic justice (profitability and social impacts)". In this sense, in Table 2, it is presented the average per farm in thousands of euro of several indicators that shows the economic situation of specialised farms or olive oil producers, between 2017 till 2021.

Table 2 Economic situation of olive oil producers, in Portugal, 2017–2021

Year	Balance current subsidies and taxes (SE600)	Gross farm income (GFI) (SE410)	Farm net value added (FNVA) (SE 415)	Family farm income (FFI) (SE420)	Farm net value added/AWU (FNVA/AWU) (SE425)	Farm net income/FWU (FFI/FWU) (SE430)
2017	7.6	28.4	24.8	19.4	23.0	26.3
2018	4.0	26.9	24.5	18.4	25.5	31.2
2019	3.1	30.8	28.2	23.1	30.4	35.3
2020	3.0	29.9	27.6	21.1	27.4	32.4
2021	3.3	37.6	35.1	27.2	31.5	39.3

Source: EC (2023a)

One of the more important conclusions in Table 2 is related with decrease of subsidies and taxes on the Current Balance from 7.9 thousand of euro in 2017 to 3.3 thousand of euro in 2021. But the farm net value added has increased from 24.8 thousand of euro in 2017 to 35.1 thousand of euro in 2021. The need for investments in technology and infrastructure, as well as improvements in environmental management and sustainability of production (Silveira et al., 2018) offers opportunities to promote the accountability of producers has main concern due to the harmful consequences.

Porter's strategies will help farmers position themselves to gain sustainable competitive advantages and therefore succeed in the marketplace (Porter, 1986; Santos, 2022). This research will discuss strategic planning tools, namely on the contextual analysis on the agriculture sector and the SWOR analysis (Abreu, 2023), to propose strategies that will create value for production in a more sustainable and environmentally friendly sector.

This paper is organised into six sections. The first one is the summary of the research. The second is the introduction with presentation of the aim of the research—Agriculture and Olive Oil—and its sustainability. The third section is presented the methodology used on this research. The fourth section presents theoretical analysis related with the agriculture sustainability. The fifth section exposes the strategies of Olive Oil in Portugal. The sixth section presents conclusions, with limitations and future developments.

2 Methodology

The methodology of the research will be developed based on two complementary analyses. The first is the theoretical analysis through the literature review of the contribution to the desired agriculture sustainability conjoin with Porter's (1986) and Ansoff's (1975) strategies, which will allow to make suggestions to stakeholders involved in this sector.

Moreover, through the Portuguese context, the empirical analysis will assess the olive oil sector through a qualitative (Krippendorff, 2018) and quantitative analysis (Hair et al., 2022). Indeed, it is also an exploratory and deductive analysis (Blaxter et al., 1996) that will systematise a proposal of strong (S) and weak (W) factors, opportunities (O), and risks (R) known by the SWOR analysis (Abreu, 2023), both for the olive oil as a whole and for organic production.

This research is the first that plans the effectiveness of the strategies in the framework of the sustainability agriculture which can be subjective because strategies are presented as final goals and should be added a more widely vision related with stakeholder impact due to changes on perspectives and criteria of values, norms, beliefs, and attitudes of each one.

As a result, the authors know that agriculture sustainability is an area that "has no easy problems and no easy solutions" (Agranoff, 2013: 23). The authors aim to answer the research question recognising the relevance of indicators that illustrate trends (Herrick, 2016), examine difficulties on the markets (Bangsa and Schlegelmilch, 2020) to evaluate benefits (Erraach et al., 2013) that must be derived by literature (Béné et al., 2019). The most important step is to define the research question to be analyzed:

How can agriculture sustainability promote strategies of the olive oil in Portugal?

The methodology helps researchers to answer to the research question. Indeed, the authors conducted the empirical analysis that it will provide a clear identification of weakness, strengths, factors, opportunities, and risks. This analysis aims to corroborate findings from surveys, interviews, observations, and document analysis of different stakeholders of the agriculture sustainability (Siebrecht, 2020). Also, the authors through the literature review interpret findings to better understand the complexity of this research question. The objective is to produce generalisations based on the Portuguese experience that allow to justify these strategies that will help stakeholders to influence the transfer of knowledge to the society (Partidario et al., 2009).

3 Literature Review

Due to the extensive use of lands and resources, changes in consumption patterns (to fast food), and urbanisation, the citizen, farmers, and agri-food producers show concerns about the future, especially related with sustainable food systems (UNEP, FAO, and UNDP, 2023). This means that this research focuses on the olive oil sector because it requires specific strategies to address these problems in a sustainable manner (FAO, 2023).

Therefore, agriculture sustainability must be accountable, and to be understood, it requires a clear definition of the concept of sustainable agriculture (Velten et al., 2015). In the literature, Schaller (1993: 89–90) presents several concepts and defined that "it is generally regarded as an alternative to modern industrialized, or conventional agriculture, an agriculture described as highly specialized and capital intensive, heavily dependent upon synthetic chemicals and other off-farm inputs". This definition demands a critical analysis to deal with its impacts.

Indeed, the authors agree with Larrinaga and Garcia-Torea (2022: 5), when they claim that "the laws of physics make complete recycling virtually impossible and require paying attention to global limits". So, promoting knowledge that could transfer to the society will allow stakeholders to be more sustainable. This evidence is presented on several studies, such as: Lowrance et al. (1986), Dunlap et al. (1992), and Hansen (1996) that argue about consequences that may be unnoticed, difficult or remain unrelated to their source of problem.

At least, the agricultural sustainability and food self-sufficiency are currently on the agenda in the EU (EC, 2023b, 2023c). Olive grove levelling involves the implementation of practices that ensure sustainable olive production and the preservation of the environment, allowing cultivated soils and ecosystems around the olive grove to be protected and conserved and reducing the ecological footprint (Sarvade et al., 2019).

In this way, henceforth agricultural production must increasingly be carried out in an environmentally responsible manner, with practices aimed at conserving natural resources, avoiding contamination of soil, water, and air. This includes the proper use of fertilisers and pesticides, erosion control, biodiversity conservation, and the implementation of strategies for the efficient use of water and energy. In addition, the agricultural sustainability of olive groves must also consider socio-economic aspects, ensuring food security, improving the quality of life of rural communities, and promoting the sustainable development of the region and creating local wealth.

4 Strategies of the Olive Oil in Portugal

To understand the strategies of the olive oil sector in Portugal, it is essential to identify the stakeholders that legitimize its development. The definition of Freedman (1984: 86) of stakeholder is "any group or individual who can affect or is affected by the achievement of the organization's objectives". Indeed, Bonnafous-Boucher

Table 3 Stakeholder's identification of the Olive Oil Sector in Portugal

Stakeholders	Details
Farmers	Small, medium, and large olive farmers
Producers	Small, medium, and large producers who process and package olive oil
Competitors	Companies or entrepreneurs competing in the olive or olive oil market that differentiate themselves in price, quality, products, marketing, distribution, and sustainability reporting policies.
Employees	Employees, consultants, and service providers in the field of olive farming and olive oil production
Family of farmers and producers	Family of small, medium, and large olive farmers and producers who process and pack olive oil
Local community	Inhabitants of the area of influence of olive farmers and olive oil producers
Retailers	Small, medium, and large retailers who buy olives from farmers or olive oil from producers to sell to distributors or consumers
Distributors	Small, medium, and large intermediaries in the supply chain for both olives and olive oil
Final consumers or customers	People who buy and consume olive oil
European Union	Entities responsible in the European Union for regulating the olive and olive oil sector
Government	Entities responsible for regulating the olive and olive oil sector in Portugal (Ministry of Agriculture and Ministry of Economy)

(Continued)

Table 3 Continued

Stakeholders	Details
Public authorities	Entities responsible for ensuring food safety (international: FAO—Fund and Agriculture Organisation and in Portugal: DGAV—Directorate General for Food and Veterinary)
Supervisory bodies	Entities supervising the olive oil sector (in Portugal: ASAE-*Autoridade de Segurança Alimentar e Económica*), ensuring that quality standards are met and protecting consumers from unfair commercial practices.
Olive oil producers' associations and co-operatives	Associations and co-operatives of a private nature, but in the public interest, to promote and defend the interests of olive oil producers.
Environmental and biodiversity associations	Environmentalist associations, of a private nature but in the public interest, which are dedicated to assessing the impact of the olive tree on biodiversity and the sustainability of natural resources.
Trade unions	Organisation formed by workers who unite to protect and promote their interests and rights in the workplace
Tourists and visitors	People from different countries and Portugal who promote the olive and olive oil sector as Portugal's agricultural, cultural, and gastronomic heritage
Education and training organisations	Universities, Polytechnic Institutes, Other Entities, and Vocational Training Companies
Research organisations	Research Units of Polytechnic Institutes and Universities

Source: Our elaborations

and Rendtorff (2016: 7) present several concepts of stakeholders to "establish its strategies by ensuring that they are consonant with societal expectations".

By doing so, Allen et al. (1991: 34) argue that "researchers, educators, policy-makers, and activists have initiated sustainable agriculture programs and efforts the world over". In this line, the authors propose the Olive Oil stakeholder's identification. The authors agree with Chakhovick and Virtanen (2023) that defend the stakeholder perspectives have often been presented as the source of sustainability.

According to NP EN ISO 9000:2015 (IPQ, 2015), stakeholders are entities relevant to provide significant risk to the sustainability of the organisation if their needs and expectations are not met. In this sense, and to carry out the strategy planning, the results that are necessary to provide value to these relevant stakeholders to reduce this risk must be defined, so that they attract, capture, and retain the support of the relevant stakeholders on whom their success depends (IPQ, 2015). Thus, Table 3 presents the olive oil stakeholder's identification.

Stakeholders in the olive oil sector in Portugal include olive producers, olive oil processors, distributors, retailers, exporters, consumers, government and regulatory organisations, producer associations and local communities. These stakeholders are interconnected and have complex interactions with each other (Duarte, 2005; Duarte et al., 2008). For example, farmers provide raw material (olives) to processors or mills to produce olive oil, who in turn sell the final product to distributors and retailers. Consumers buy the oil at the point of sale and often make choices based on brand, quality, and price. Local communities are affected by the sector's activity, which can generate jobs and economic development in rural regions.

Final customers and consumers generally prefer an olive oil that differentiates itself and are even willing to pay a little more for the satisfaction of that difference (Contini et al., 2016; Latino et al., 2022). The market for luxury olive oils in limited quantities and editions aimed at niche markets with high purchasing power may be suitable for small producers (Stasi et al., 2018).

Government, public bodies, and inspection organisations oversee compliance with regulations and quality standards and provide policies, incentives, and guidance for the sector. Olive oil producers' associations and cooperatives play an important role in defending the interests of producers and promoting the olive oil sector in Portugal and internationally (IOL, 2019).

Tourists and visitors see olives and olive oil production as a tourist attraction for Portugal, generating wealth for the inland region and promoting the country's cultural and gastronomic heritage. The families of farmers and producers are affected by the change in consumption patterns, so a drop in olive oil consumption is almost certain, which will continue over the next few years (Karanikolas et al., 2021).

Although stakeholders may have different objectives, such as more regulation by enforcement or regulatory bodies, more environmental sustainability or health promotion by Environmental Associations, everyone stands to gain from the success of the olive oil sector, and it is important that they work together in favour of improving the sector. Co-operation and collaboration between stakeholders and all research and government support can lead to strengthening the economy, environmental sustainability, and the provision of quality products to consumers.

According to NP 4469-1:2008 (IPQ, 2008), the olive oil sector could classify stakeholders according to the following criteria:

- Linkage—Those to whom the olive oil sector has, or may have, legal, economic, environmental, financial, or operational obligations, between others.
- Influence—Those to whom the olive oil sector can or do influence and the ability to achieve its objectives, regardless of whether they are hindering or facilitating the performance of the business.
- Proximity—Those with whom olive oil sector interacts most, including internally, with whom it has long-standing relationships, on whom it depends on day-to-day operations, and whose housing is close to the company's premises.
- Dependency—Those who are directly or indirectly dependent on olive oil's activities and products, in economic or financial terms, or in terms of regional or local infrastructure and fulfilment of basic needs.
- Representation—Those who, through legal provisions, statutes, customs, values, beliefs, or culture of the olive oil sector can legitimately claim and represent other individuals.

Table 4 categorises the Olive Oil stakeholders based on the criteria of linkage, influence, proximity, dependency, and representation explained before. Stakeholders are presented in descending order of significance, i.e. from the most significant to the least significant to the olive oil sector.

Table 4 Stakeholder's classification of the Olive Oil Sector in Portugal

	Classification		Criteria				
Stakeholders	Internal	External	Linkage	Influence	Proximity	Dependency	Representation
Farmers	●		●	●	●	●	●
Producers	●		●	●	●	●	●
Competitors		●	●	●	●		
Employees		●	●			●	●
Family of farmers and producers		●	●		●	●	●
Local community		●		●	●		
Retailers		●				●	
Distributors		●		●		●	
Final consumers or customers		●	●		●	●	●
European Union		●	●	●	●	●	
Government		●	●	●	●	●	
Public Authorities		●	●	●	●	●	
Supervisory bodies		●	●	●	●	●	
Olive oil producers' associations and co-operatives		●		●		●	
Environmental and biodiversity associations		●		●		●	
Trade Unions		●		●		●	●
Tourists and visitors		●		●		●	
Education and training organisations		●		●	●	●	
Research organisations		●	●		●	●	

Source: Own elaborations

In Table 4 are presented several stakeholders and their level of significance on each criterion. In relation with customers or consumers can be direct (when they themselves turn to the company that sells olives or produces olive oil) or indirect (when there is a retailer or distributor that liaises between them).

The olive farmers and olive oil producers were categorised as an external stakeholder and with the criteria of influence, proximity, dependence, and representation. Indeed, proximity refers to the degree of closeness or direct interaction between customers or consumers and the olive oil sector. In this context, it encompasses both physical and relational closeness.

Physically, customers or consumers may have direct access to olive oil producers through farmer's markets, local retailers, or online platforms. They may visit olive groves, participate in tasting events, or engage in agritourism experiences, fostering direct connections with producers and the product itself. This physical closeness facilitates transparency, trust, and a sense of community between customers and the olive oil sector.

Relationally, proximity reflects the ongoing engagement and communication channels maintained between customers or consumers and the olive oil sector. Whether through feedback mechanisms, customer support channels, or marketing initiatives, maintaining regular and meaningful interactions fosters a sense of partnership and responsiveness to customer needs and preferences (Kotler and Armstrong, 2020).

Ultimately, considering the proximity criterion acknowledges the importance of fostering close relationships and ongoing engagement between customers or consumers and the olive oil sector, enhancing trust, loyalty, and mutual understanding for sustainable business relationships.

The Supervisory Bodies were categorised as external stakeholders as they directly affect the activity of the olive oil sector. In addition, they were also considered to be influenced by the criteria of linkage, influence, proximity, and dependence. In the context of the criteria of linkage and influence, it is crucial to recognise the multifaceted obligations that the olive oil sector must uphold with olive farmers and producers. These obligations stem from legal, economic, and operational imperatives, all of which significantly impact the functioning of the sector, such as:

Legal Obligations: The olive oil sector is bound by a myriad of legal requirements that govern its operations. These encompass regulations related to working conditions, occupational safety, health standards, and, most importantly, food safety. Compliance with these legal obligations is not only mandatory but also critical for maintaining the integrity and reputation of the sector. Adhering to stringent food safety standards ensures that olive oil products meet the highest quality and safety benchmarks, safeguarding consumer health and trust.

Economic Obligations: Beyond legal mandates, the olive oil sector is subject to various economic obligations concerning fair labour practices, sustainable production methods, and equitable economic partnerships with olive farmers and producers. Ensuring fair compensation for farmers, promoting environmentally responsible farming practices, and fostering inclusive economic growth are integral to the sector's long-term sustainability and resilience.

Operational Obligations: Operational obligations encompass the day-to-day responsibilities and commitments that the olive oil sector must fulfil to ensure smooth and efficient functioning. This includes maintaining transparent communication channels with farmers and producers, providing necessary support and resources, and implementing effective quality control measures throughout the production process. By prioritising operational excellence, the sector can uphold its reputation for reliability, consistency, and product excellence.

By recognising and addressing these legal, economic, and operational obligations, the olive oil sector can strengthen its linkages with olive farmers and producers while exerting a positive influence on their practices. This collaborative approach not only enhances compliance with regulatory requirements but also fosters a culture of responsibility, accountability, and mutual benefit within the olive oil supply chain.

In Portugal, several public authorities are involved, with the Food and Economic Safety Authority specifically responsible for inspections. Stakeholders in the sustainability of the sector and in compliance with the legislation in force may indicate: the General Directorate of Food and Veterinary, the General Directorate of Agriculture and Rural Development, the Regional Directorates of Agriculture, the Office of Planning, Policies and General Administration, the European Commission, and the Ministry of Agriculture and the Sea. In compliance with the legislation in force: all olive oils fit for consumption must be packaged, labelled, with the presentation and advertising legally required for marketing in the European Union. Capsules of olive oil for catering must be tamper-proof (EC, 2023a). According to Casa do Azeite (2020) and the GPPAG (2020), the legislation for labelling must follow the laws: Ordinance no 24/2005; Decree-Law n.° 76/2010; Regulation (EEC) n.° 2568/1991; Regulation 28 (EEC) n.° 1234/2007 and Regulation (EEC) n.° 29/2012.

Strategies are determinant to understand the increase, the gap, and the significant decrease in olive oil sector. For example, at the production level, impact factors have introduced significant pressure on olive oil prices within the agricultural sector. Regarding olive oil consumption in Portugal, it has been stable in recent years, with a slight downward trend. This phenomenon is not exclusive to Portugal, as it is the case in all the main European producing countries. This trend was only counteracted in 2020, when there was an increase in olive oil consumption, which is attributed to a greater increase in consumption at home, due to the lockdown and the fact that people are more concerned with healthy eating.

A reduction in olive oil consumption is expected, due to the increase in the price of olive oil at source and the increase in all operating costs (water, electricity, labour, among others), which sooner or later will have to be reflected in the final price of olive oil for the final consumer. In this way, the sector faces a serious situation, where 70% of olive groves worldwide are not profitable, i.e. more than 8 million hectares of olive groves in the world produce olive oil above the market price, and this situation has become structural, not circumstantial (WB, 2023).

The Annual Growth Rate (AGR) of olive oil production in Portugal exhibited a notable evolution over the periods 2004–2014 and 2014–2021. During the former period, the AGR stood at 3%, while in the latter period, it surged to an impressive 19%. Notably, the AGR for the finest quality olive oil, extra virgin olive oil with an acidity of less than 0.8%, reached an astonishing 25%. By 2022, the production of extra virgin olive oil constituted over 90% of the total output, a significant increase from the 55% share it held in 2004 (INE, 2023).

This exponential growth can be attributed to various factors, foremost among them being the expansion of intensive and super-intensive olive groves. Additionally, investments in olive production within the *Alqueva* irrigation perimeter have played a pivotal role in stimulating industry growth (INE, 2023).

Another example is substantial increase in the production in recent years is also due to the installation of mills with more modern technology, allowing greater production in less space and with fewer manual processes. The existence of these groves is diminishing, and they are now more than half of those in operation than in 2004 (INE, 2023). In short, the authors can say that in Portugal the sector has been characterised in recent years by greater production of olives and olive oil (because of the irrigation made possible by the *Alqueva* area and intensive olive groves) and by greater mechanisation at various stages of the value chain.

In recent years, Portugal has managed to create a significant niche in the olive oil sector, the *Alentejo* region, where *Alqueva* is located (dam with 4.150 million m^3). This region, located in the south of mainland Portugal, has become a global reference for modern, sustainable, and efficient olive groves in the last decade. The *Alentejo* region, which in 2004 produced 32% of the national olive oil, will produce 80% in 2021 in an olive grove area that is only half of that existing in the country. In this region, between 2004 and 2021, and because of investment in intensive olive groves, the cultivated area grew by 33% but olive oil production grew by more than 1000%. This is a surprising increase in productivity, quite different from other regions of the country and clearly visible from the moment when intensive olive groves begin to yield.

Another example occurs in the year 2021, favourable weather conditions throughout the 2021 olive oil season cycle, especially during flowering and fruit set, combined with the increased importance of intensive irrigated olive groves contributed to the highest ever production, at around 2.29 million hectolitres of olive oil (+48.6% than in 2019), the second-best record since 1915 (INE, 2023). In general, the olive oil produced was of good quality, with low acidity and good organoleptic characteristics.

Another example is the year 2022, which was among the most challenging in recent history for Portugal's olive oil sector. In terms of production, there was a drop in olive production, negatively impacted by several climatic and phytosanitary problems, with great repercussions on both the quantity of olives and their quality. We can highlight the high temperatures at the time of flowering in spring, which greatly impaired the setting of the fruit, and the severe drought during most of the production cycle, which particularly affected rainfed olive groves. In addition, the significant increase in all factors of production (energy, fuel, fertilisers, etc.), resulting from the war in Ukraine, as well as the shortage of labour has posed additional difficulties for olive growers. As a result of all these factors, it is estimated that the drop in production in Portugal in the current 2022/2023 campaign will be in the order of 30–40%, and in some traditional olive grove regions these losses will certainly be higher. In terms of the other European producing countries, such as Spain or Italy, the scenario was quite similar (INE, 2023).

This analysis allows the authors to present a Strength (S), Weakness (W), Opportunity (O), and Risk (R) analysis, with the acronym of SWOR, because it is a strategic tool supported on a "squared matrix based on past-experience, present-practice, and future-expectations associated with the internal situation strengths and weaknesses and with the external environment opportunities and risks to assess a company, a product or service, an idea, a partnership, an investment opportunity, or a new decision-making process" (Abreu, 2023: 2).

This tool proves to be very useful in the search for competitive strategies because it combines the internal and external scenario of olive oil sector to understand the current and future situation of it, allowing effective and solid strategies to be created for the future. The discussion obtained from the SWOT analysis can be used to face changes in the business environment in a more efficient and effectiveness perspective, always seeking the maximum positive results for the Portuguese olive oil sector.

The SWOT analysis applied to the Olive Oil Sector aims to systematise the strategies and propose solutions to the challenges they face, which can be used as opportunities for the stakeholders of the sector to boost sales and monetise their business and own activities. However, always coordinated with values and ethical principles as Abreu (2023) argues "at last, the SWOR analysis incorporates the strategic plan made with assessment of the mission, the vision, and values of the organisation to create value to the future".

Table 5 presents the strategy formulation based on the cross-checking of internal factors (weaknesses and strengths) with external factors (risks and opportunities), which consists of SWOT analysis. As a food commodity, it is subject to compliance with national and community legislation: labelling, safety and hygiene standards, waste, controlled designations of origin, marketing, protection of geographical indications and designations of origin of agricultural products, foodstuffs, and organic production.

Furthermore, Table 5 presents the relevant weakness by the olive oil sector, such as: market price, distribution channels, sustainable management capacity, skilled labour, and production costs. For example, in this period of high inflation rate and price increases are currently expected to have a substitution effect on the demand for less healthy fats, affecting the demand for olive oil. Recent years have seen investment in the technological modernisation of mills and processing and transformation systems, which has strengthened external competitiveness. Social networks and the emergence of online commerce and digital communication now make it possible to reach consumers more effectively and quickly.

The most relevant strongness of the olive oil sector are traditional production, value of the private label, agriculture 5.0, organic production and management quality, environmental and safety certification. For example, the management quality, environmental and safety certification demands that the production must be classified in different categories of olive oils, according to quality parameters related to

Table 5 SWOR analysis of the Olive Oil Sector in Portugal

SWOR		Weakness					Strengths				
		Market price	Distribution channel	Sustainable management capability	Skilled labour	Production costs	Traditional Production	Value of private label	Agriculture 5.0	Organic production	Management Quality, Environmental and safety systems
Risks	Climate change (with extreme events)	+	–	+	+	+	–	+	+	+	+
	Diseases and pests	–	+	+	+	+	+		+	+	+
	Competing products	+	+	+	+	+	+	–	–	+	+
	Bargaining power of distribution and resale	+	+	+	+	+	–	+	+	+	–
	DOP demarcated areas	+	+	+	+	+	–	+	+	–	+
Opportunities	Olive oil consumption	+	–		–		–		+	+	+
	Regulation at the market	–	–	–	–	–	–	+		+	+
	Potential of nature tourism and healthy food	+	–	–	+	+	–	–	+	–	+
	Research and development	+	+	+	+	+	+	+	+	+	+
	Environmental regulation	+	–		–	–	–	+	–	–	–
	(–) negative interaction: Potential risk or missed opportunity										
	(+) positive interaction: Risk mitigated, or opportunity realised										

Source: Own elaborations

their physicochemical characteristics, such as acidity content, peroxide index, fatty acid content and sterol composition and their organoleptic (sensory) characteristics, such as fruitiness and the absence of organoleptic defects (EC, 2020).

The risks that must be mitigated by the olive oil sector are climate change (with extreme events), diseases and pests, competing products, bargaining power of distribution and resale, and Denominação de Origem Protegida (Protected Designation of Origin—DOP) demarcated areas. For example, the DOP demarcated areas classify the olive oil marketed under a specific category, its characteristics must comply with limits set by the European Union rules for each category (EC, 2020). The top category is extra virgin olive oil and acidity does not need to be specified, as it is guaranteed to be below 0.8%, fruity and without organoleptic defects. Next on the quality scale is virgin olive oil, with acidity between 2% and 0.8%. This is an olive oil that is likely to have minor sensory defects, but at a very low level and is excellent for use in cooking, particularly in marinades, roasts, or soups. In terms of conservation, olive oil is photosensitive and should therefore be kept away from light, and after a year of storage it may have already lost the virtues that were initially recognised (Melo, 2019).

The opportunities that could be leveraged are olive oil consumption, regulation at the market, potential of the nature tourism and healthy food, research and development, and environmental regulation. For example, the regulation at the market could be EU marketing standards that aim to ensure that the market is supplied with agricultural products of harmonised, security food and quality, to meet consumer expectations, to facilitate trade and to ensure a level playing field for EU producers. EU olive oil legislation defines the different categories of olive oils and olive-pomace oils, as well as the methods of analysis applicable by national control authorities and lays down rules for labelling and packaging.

The support scheme for the conversion and restructuring of vineyards also allows producers to receive financial incentives for olive oil production if collaborations between olive oil producers and winegrowers are realised. In addition, olive oil producers in Portugal can also apply for funding through the Portugal 2030 programme applications, a programme that aims to encourage investment in various sectors of the national economy and scientific research. Olive oil production can also generate tax deductions, such as the possibility of deducting part of the expenses with investments in agricultural activities. In addition to the above, there are also regional incentives, such as some municipalities that offer specific incentives for the development of olive oil production, such as tax exemptions, technical support, among others.

Along with the evolution in production, there has been an increase in consumption in the national market, justified by the health benefits of the Mediterranean diet. In fact, the value of this food is recognised by the medical community as reducing the levels of bad cholesterol and the prevalence of cardiovascular diseases, partly due to the presence of monounsaturated fatty acids. In addition, "vitamin E has an antioxidant function on artery walls, thus helping to prevent the development of cardiovascular diseases" (Santo, 2022: 25).

The olive oil sector shows the negative image in relation to the modern and the intensive olive groves, knowing the high internal and external competition and bargaining power of the large distribution and retail companies. Furthermore, olive grove pests, olive tree diseases, and insect attacks, combined with the costs of meeting environmental certification requirements, are significant challenges, compounded by the impact of climate change (with increasingly drier and hotter years). There has been an increase in production in Mediterranean basin countries and other regions with favorable soil and climate conditions, such as California, Argentina, Chile, South Africa, and Australia, where production costs are lower.

The formulation of strategies stems from the crossing of Internal Factors (Strengths and Weaknesses) with External Factors (Opportunities and Threats), present in the previous tables of the SWOT analysis. From there, and in a dynamic analysis among the group of researchers involved in this research, suggestions were found that include identifying their distinctive competences vis-à-vis the competition that distinguish and differentiate them. The crossing of the strengths and weaknesses, with the opportunities and threats, basically results in the following strategies and future action plans, explained in the following points.

Olive oil producers seek olives from efficient olive groves, and the productive competitiveness of these groves and this perspective will increase the value chain and it would be highly advisable to promote the differentiation of olive oils, through organic, bioregenerative, biodynamic, and living olive groves. In this way, the olive grower obtains an additional margin on his oils, improving the net income from them through an increase in added value.

Instituto Politécnico da Guarda invests in research and innovation to find solutions to the problems faced by the olive oil sector in various areas, such as economy, accounting, sustainability, informatics, technology, engineering, medicine, environment, culture, and tourism, among others, to create new ideas, improvements in processes and products, and new scientific discoveries. This innovation can lead to the creation of new companies and increase Portugal's competitiveness in the global market with the support for entrepreneurship tendencies.

At least, it is important to promote high education for different students, such as young people and adults, by offering higher education programmes in strategic areas to the sector, and preparing students to become highly qualified professionals, able to contribute to the development of the sector. This research serves as a strong example of promoting international collaboration by fostering partnerships among public entities, civil society, businesses, academics, and students.

5 Conclusions

This research answers to the research question of how agriculture sustainability can promote strategies to olive oil sector. After all, the authors show that the olive oil sector, in Portugal, has a great potential for growth and development in Portugal, Europe and Global markets, due to the following strategies.

In the context of economic strategy, the assessment of olive oil sector is of extreme importance for Portuguese economy and its social development, ensuring the generation of employment and income, providing farmers a better quality of life. Additionally, this sector holds significant economic importance for Portugal's economy, as olive oil is one of its principal agricultural export products, contributing to employment and income generation across various regions.

In the context of market strategy, Portuguese olive oil is globally recognized as one of the leading products exported by the country. Furthermore, Portuguese olive oil is internationally recognised for its quality and unique taste, which boosts foreign trade and provides great competitiveness in the market.

In context of tourist strategy, the olive oil sector contributes to the preservation of the environment by promoting sustainable agriculture and the conservation of the Portugal's natural heritage as well as promoting the Portugal's cultural and gastronomic identity in the world.

In the context of corporate governance strategy, the olive oil sector faces risks and weakness, but it is fundamental to intensify the opportunities and the strengths that guarantee a promising future for olive groves in Portugal, due to the creation of differential value that stands out from the competition.

In the context to achieve sustainability, the olive oil sector is fundamental at political, legal, economic, socio-cultural, environmental, ethical, and governance positive impacts because it is a guarantee of a well-defined strategy for farmers which allows them to be resilient with aggressive methods to diminish the existent problems.

In the context to promote accountability, management practices are needed that will leverage the efficient use of olive trees as a natural resource. These are initiatives that generate long-term benefits for society as a whole and contribute to building a fairer, more responsible, and sustainable world.

In the context of higher education strategy, the olive oil sector requires Higher Education Institutions to offer distinct academic levels (undergraduate, master's, and PhD) across diverse scientific areas. Beyond agriculture, there is an urgent need for education in accounting and accountability to equip farmers with the skills to prioritize sustainability alongside profitability.

In the context of the competitive world strategy, stakeholders must understand the value chain analysis and it was possible to propose long-term factors, that it will integrate decision-making process to focus on fight the unsustainability agriculture (exacerbated productions, abandonment of traditional agriculture practices and territories, historic territory trend in land use and poverty, and degradation of small cities).

This paper promotes a new design for several strategies to be adopted by stakeholders of the olive oil sector in Portugal, but it is the authors' conviction that sustainability agriculture is the present to determine the best of our future.

The limitation of this research is the demanding task of collecting data from the political, legal, economic, socio-cultural, environmental, ethical, and governance impacts of the olive oil sector. These information's are dispersed, presented by distinct public bodies, and need to evaluate the accuracy of these sources.

The future developments of this research will assess the olive oil strategies proposed on this research. This assessment, through a stakeholder's survey, will identify reasons, estimate the level of knowledge, and discuss the perceived intention to adopt them in the management of agriculture firms.

Acknowledgements The authors would like to thank David Crowther, PhD, and José-Ángel Pérez-López, PhD, of Seville University (Spain) for their comments. This research has been supported by national funds through the Instituto Politécnico da Guarda. Special acknowledgment is extended to Professor Antonella Tamborrini and Alessandro Leone of the Università degli Studi di Bari Aldo Moro for her invaluable support and the coordination of the Project Horizon Europa Seeds ECOnomia circolare, Sostenibilità e profili di evoluzione Normativa nella produzione dell'Olio extravergine Di Oliva - (ECOSNODO).

References

Abreu, R. (2023). SWOR analysis. In S. Idowu, R. Schmidpeter, N. Capaldi, L. Zu, M. Del Baldo, & R. Abreu (Eds.), *Encyclopaedia of sustainable management*. Springer. https://doi.org/10.1007/978-3-030-02006-4_871-1

Agranoff, R. (2013). *Managing within networks: Adding value to public organizations*. Georgetown University Press.

Allen, P., van Dusen, D., Lundy, J., & Gliessman, S. (1991). Integrating social, environmental, and economic issues in sustainable agriculture. *American Journal of Alternative Agriculture, 6*(1), 34–39.

Ansoff, H. I. (1975). Managing strategic surprise by response to weak signals. *California Management Review, 18*(2), 21–33.

Bangsa, A. B., & Schlegelmilch, B. B. (2020). Linking sustainable product attributes and consumer decision-making: Insights from a systematic review. *Journal of Cleaner Production, 245*, 118902.

Béné, C., Oosterveer, P., Lamotte, L., Brouwer, I. D., de Haan, S., Prager, S. D., & Khoury, C. K. (2019). When food systems meet sustainability—Current narratives and implications for actions. *World Development, 113*, 116–130.

Blaxter, L., Hughes, C., & Tight, M. (1996). *How to research*. Open University Press.

Bonnafous-Boucher, M., & Rendtorff, J. D. (2016). *Stakeholder theory: A model for strategic management*. Springer Nature.

Casa do Azeite – Associação do Azeite de Portugal. (2020). *Site of the Portuguese olive oil association*. Casa do Azeite. Retrieved from www.casadoazeite.pt

Cazimoglu, I. (2017). *Sustainable agriculture: Why and how*. The Science Innovation Union. Retrieved from http://science-union.org/articlelist/2017/3/22/sustainableagriculture-why-and-how.html

Chakhovick, T., & Virtanen, T. (2023). Accountability for sustainability—An institutional entrepreneur as the representative of future stakeholders. *Critical Perspectives on Accounting, 91*, 102399. https://doi.org/10.1016/j.cpa.2021.102399

Contini, C., Boncinelli, F., Casini, L., Pagnotta, G., Romano, C., & Scozzafava, G. (2016). Why do we buy traditional foods? *Journal of Food Products Marketing, 22*, 643–657.

Darnhofer, I. W. (2010). Strategies of family farms to strengthen their resilience. *Environmental Policy Governance, 20*, 212–222.

Delonge, M., Robbins, T., Basche, A., & Haynes-Maslow, L. (2020). The state of sustainable agriculture and agroecology research and impacts. *Journal of Agriculture, Food Systems, and Community Development, 9*, 1–26.

Dinis, I., Ortolani, L., Bocci, R., & Brites, C. (2015). Organic agriculture values and practices in Portugal and Italy. *Agricultural Systems, 136*, 39–45. https://doi.org/10.1016/j.agsy.2015.01.007

Dixon, J. A., Gibbon, D. P., & Gulliver, A. (2001). *Farming systems and poverty: Improving farmers' livelihoods in a changing world.* FAO - The World Bank.

Duarte, F. (2005). *Olivero - Project communication no. 7 - Overview of olive oil marketing and processing and olive sector organisation. The future of olive plantation systems on sloping and mountainous land; scenarios for production and natural resource conservation.* Retrieved from http://www.olivero.info

Duarte, F., Jones, N., & Fleskens, L. (2008). Traditional olive orchards on sloping land: Sustainability or abandonment? *Journal of Environmental Management, 89*(2), 86–98. https://doi.org/10.1016/j.jenvman.2007.05.024

Dunlap, R. E., Beus, C. E., Howell, R. E., & Waud, J. (1992). What is sustainable agriculture? An empirical examination of faculty and farmer definitions. *Journal of Sustainable Agriculture, 3*, 5–39.

Erraach, Y., Sayadi, S., & Parra-López, C. (2013). Olive oil origin preferences: Incidence of socio-economic variables and lifestyle. In J. Vilar Hernández (Ed.), *The olive oil producing sector: A multidisciplinary study* (pp. 473–494). Centro Internacional de Excelencia para Aceite de Oliva–GEA Westfalia Separator Ibérica.

European Commission (EC). (2020). *Olive oil.* Retrieved from https://agriculture.ec.europa.eu/system/files/2020-03/factsheet-olive-oil_en_0.pdf

European Commission (EC). (2023a). *Farm economy focus by sector—Farms specialised in olives in the EU.* EC. Retrieved from https://agridata.ec.europa.eu/

European Commission (EC). (2023b, July 14). *Short-term outlook report: European farmers' resilience continues to be challenged.* Retrieved from https://agriculture.ec.europa.eu/news/short-term-outlook-report-european-farmers-resilience-continues-be-challenged-2023-07-14_en

European Commission (EC). (2023c, July 5). *European Green Deal: more sustainable use of plant and soil natural resources.* Retrieved from https://ec.europa.eu/commission/presscorner/detail/en/IP_23_3565

Feher, I., & Beke-Lisanyi, J. (2013). Approaches to sustainability in the agricultural policy. *Economics and Rural Development, 9*(2), 7–15.

Figueiredo, E. (2011). *O Rural Plural: Olhar o Presente, Imaginar o Futuro.* 100 Luz.

Food and Agriculture Organisation of the United Nations (FAO). (1991). *Sustainable agricultural resources development.* Retrieved from http://www.nationsencyclopedia.com/United-Nations-RelatedAgencies/The-Food-and-Agriculture-Organizationof-the-United-Nations-FAO-sustainableagricultural-resources-development.html

Food and Agriculture Organisation of the United Nations (FAO). 2023), *Statistical database of agricultural resources.* Retrieved from https://www.nationsencyclopedia.com/United-Nations-RelatedAgencies/

Fraga, H., Moriondo, M., Leolini, L., & Santos, J. A. (2020). Mediterranean olive orchards under climate change: A review of future impacts and adaptation strategies. *Agronomy, 11*(1). https://doi.org/10.3390/agronomy11010056

Freedman, R. E. (1984). *Strategic management: A stakeholder approach.* Pitman.

Gabinete de Planeamento, Politicas e Administração Geral (GPPAG). (2020). *Analise Setorial do Azeite.* GPP. Retrieved from https://sima.gpp.pt/images/PEPAC/Consultaalargada16Nov2020_revisao/Indice_analissectorial_Azeite.pdf

Graaf, J., & Eppink, L. A. A. F. (1999). Olive oil production and soil conservation in southern Spain, in relation to EU subsidy policies. *Land Use Policy, 16*(4), 259–267. https://doi.org/10.1016/S0264-8377(99)00022-8

Gray, L. C. (2013). *Introduction to agricultural economics.* Read Books.

Hair, J. F., Babin, B. J., Anderson, R. E., & Black, W. C. (2022). *Multivariate data analysis.* Cengage Learning.

Hansen, J. W. (1996). Is agricultural sustainability a useful concept? *Agricultural Systems, 50*, 117–143.

Herrick, J. (2016). *Unlocking the potential of land resources: Evaluation systems, strategies and tools for sustainability; A report of the working group on land and soils of the international resource panel.* United Nations Environment Programme.

Instituto Nacional de Estatística (INE). (2023). *Estatísticas Agrícolas*. INE.
Instituto Português da Qualidade (IPQ). (2008). *Norma NP 4469-1:2008 – Sistema de Gestão da Responsabilidade Social Parte1: Requisitos e linhas de orientação para a sua utilização*. IPQ.
Instituto Português da Qualidade (IPQ). (2015). *Norma Portuguesa NP EN ISO 9000:2015 – Sistema de Gestão da Qualidade: Fundamentos e vocabulário*. IPQ.
International Olive Council (IOL). (2019). *Olive Oil production by country*. Retrieved from https://www.internationaloliveoil.org/estaticos/view/131-world-olive-oil-figures/
Karanikolas, P., Martizez-Gomez, V., Galli, F., Posperi, P., Hernández, P. A., Arnalte-Mur, L., Rivera, M., Goussios, G., Fastelli, L., Oikonomopoulou, E., & Fonseca, A. (2021). Food system integration of olive-oil-producing small farms in southern Europe. *Global Food Security, 28*, 100499.
Kotler, P., & Armstrong, G. (2020). *Principles of marketing*. Pearson.
Krippendorff, K. (2018). *Content analysis: An introduction to its methodology*. Sage.
Larrinaga, C., & Garcia-Torea, N. (2022). An ecological critique of accounting: The circular economy and COVID-19. *Critical Perspectives on Accounting, 82*, 102320. https://doi.org/10.1016/j.cpa.2021.102320
Latino, M. E., Devitis, B. D., Corallo, A., Viscecchia, R., & Bimbo, F. (2022). Consumer acceptance and preference for olive oil attributes—A review. *Food, 11*(23), 3805.
Lowrance, R., Hendrix, P. F., & Odum, E. P. (1986). A hierarchical approach to sustainable agriculture. *American Journal of Alternative Agriculture, 1*(4), 169–173.
Mâcedo, S., Nóbrega, K. C., Queiroz, J. V., & Whebber, P. (2010). Planejamento e Gestão Estratégica: Um Estudo sobre Adoção e Práticas em Indústrias do Rio Grande do Norte. *Revista Eletrônica Sistemas & Gestão, 7*(1), 58–75.
Mairech, H., López-Bernal, A., Moriondo, M., Dibari, C., Regni, L., Proietti, P., Villalobos, F. J., & Testi, L. (2020). Is new olive farming sustainable? A spatial comparison of productive and environmental performances between traditional and new olive orchards with the model OliveCan. *Agricultural Systems, 181*, 102816. https://doi.org/10.1016/j.agsy.2020.102816
Malthus, T. (1798). An essay on the principle of population. Politzer Zsigmond es Fia Sorozat. Nemzetgazdasagi Irok Tara.
Melo, F. (2019). Azeite: o trunfo mais discreto de Portugal. https://ocio.dn.pt/sucesso/azeite-otrunfo-mais-discretode-portugal/21249/.
Monaghan, K., Swisher, M., Koenig, R. L., & Rodriguez, J. C. (2017). Education for sustainable agriculture: A typology of the role of teaching farms in achieving learning goals and objectives. *Environmental Education Research, 23*, 749–772.
Oberč, B. P., & Arroyo Schnell, A. (2020). *Approaches to sustainable agriculture: Exploring the pathways towards the future of farming*. IUCN, European Regional Office. https://doi.org/10.2305/IUCN.CH.2020.07.en
Partidario, M. R., Sheate, W. R., Bina, O., Byron, H., & Augusto, B. (2009). Sustainability assessment for agriculture scenarios in Europe's mountain areas: Lessons from six study areas. *Environment Management, 43*, 144–165.
Pires, D. (2005). *Estudo de Valorização Comercial do Azeite de Trás-Os-Montes e Alto Douro*. Instituto Politécnico de Bragança.
Porter, M. E. (1986). *Competition in global industries*. Harvard Business Press.
Pretty, J. N. (1995). Participatory learning for sustainable agriculture. *World Development, 23*, 1247–1263.
Prus, P. (2019). The role of higher education in promoting sustainable agriculture. In L. Tauginienė (Ed.), *Corporate social responsibility and business ethics in the central and Eastern Europe* (pp. 99–119). Nomos Verlagsgesellschaft.
Richardson, E. W. (1975). Growth centers, rural development, and national urban policy: A defense. In J. Friedmann & W. Alonso (Eds.), *Regional policy: Readings in theory and practice* (pp. 97–132). MIT Press.
Santo, A. C. X. B. S. (2022), *A hora de dar voz à marca "Azeite de Portugal": Plano de comunicação*. Dissertação de mestrado. ISCTE - Instituto Universitário de Lisboa. Retrieved from http://hdl.handle.net/10071/26413

Sarvade, S., Upadhyay, V. B., Kumar, M., & Imran Khan, M. (2019). Soil and water conservation techniques for sustainable agriculture. In M. K. Jhariya, A. Banerjee, R. S. Meena, & D. K. Yadav (Eds.), *Sustainable agriculture, forest, and environmental management* (pp. 133–188). Springer.

Schaller, N. (1993). The concept of agricultural sustainability. *Agriculture, Ecosystems, and Environment, 46*, 89–97.

Siebrecht, N. (2020). Sustainable agriculture and its implementation gap—Overcoming obstacles to implementation. *Sustainability, 12*(9), 3853.

Silveira, A., Muñoz-Rojas, J., Pinto-Correira, T., Guimarães, M. H., Ferrão, J., & Schmidt, L. (2018). The sustainability of agricultural intensification in the early 21st century: Insights from the olive oil production in Alentejo (southern Portugal). In A. Delicado, N. Domingos, & L. de Sousa (Eds.), *Changing societies: Legacies and challenges* (The diverse worlds of sustainability) (Vol. 3, pp. 247–275). Imprensa de Ciências Sociais.

Stasi, A., Diotallevi, F., Marchini, A., & Nardone, G. (2018). Italian extra-virgin olive oil: Impact on demand on being market leaders, private labels or small producers. *Review of Economics and Finance, 13*(3), 39–54.

UNEP, FAO and UNDP. (2023). *Rethinking our food systems: A guide for multi-stakeholder collaboration*. Nairobi, Rome and New York. https://doi.org/10.4060/cc6325en

United Nations (UN). (1986). *Report of the world commission on environment and development: Our common future*. Brundtland Report.

Velten, S., Leventon, J., Jager, N., & Newig, J. (2015). What is sustainable agriculture? A systematic review. *Sustainability, 7*, 7833–7865.

WFCS. (2023). *Good agricultural practices (GAP)*. World Food Safety Council. Retrieved from http://www.wfsccouncil.org/gap.html

World Bank (WB). (2023). *The World Bank, world development indicators*. The World Bank Group.

Xie, H., Huang, Y., Chen, Q., Zhang, Y., & Wu, Q. (2019). Prospects for agricultural sustainable intensification: A review of research. *Land, 8*(11), 1–27.

Rute Abreu is a Coordinator Professor with Accounting Habilitation at the Polytechnic Institute of Guarda, Portugal. Rute teaches financial accounting, auditing, quality, environment and safety standards, sustainability, and social responsibility management. Rute has an Industrial Engineering Master's degree from the Universidade Nova de Lisboa, Portugal, and a PhD Degree in Accounting and Finance from the Universidad de Salamanca, Spain. She is the Accounting Degree Director. Rute is an active and a long-time member of the Social Responsibility Research Network and the Global Corporate Governance Institute (GCGI). Rute researches at the Center for Research in Accounting and Taxation of the Polytechnic Institute of Cávado and Ave (CICF-IPCA), Digital Services Research Center of the Polytechnic Institute of Viseu (CISeD-IPV), and CITUR, Portugal where she researches on CSR, corporate governance, accounting and firm valuation and many of her papers have been published in several journals. Also, she actively engages in the double-blind peer-review process for several academic journals and has acted as Guest Editor of *The Social Responsibility Journal and Accounting and Management Review* of Portuguese Chartered Accountants. She regularly participates in scientific activities related to the organisation of international conferences and meetings in different parts of the world. Additionally, Rute promotes research projects related to Accounting Education, Tax Citizenship, Management Quality Systems in Social Economy, and Performance Analysis at Beauty Firms that transfer to the society knowledge.

Ermelinda Oliveira is a Adjunct Professor at the Department of Economics and Business of Escola Superior de Tecnologia e Gestão at the Polytechnic Institute of Guarda, Portugal, where she teaches Financial Accounting to the Management and Accounting degree. She holds a PhD in Economics from the University of Beira Interior, Portugal; Economics Master's degree from the University of Minho, Portugal; and a Bachelor's degree in Economics from the University of Coimbra, Portugal. Her main research interests are in Economics, Social Responsibility and Hospitality. Additionally, for a few years, Ermelinda has been a member of the research project: Local Housing Strategy in the municipality of Manteigas, Portugal.

Francisco Tomé is an Adjunct Professor at the Department of Economics and Business of Escola Superior de Tecnologia e Gestão at the Polytechnic Institute of Guarda, Portugal, where he teaches Economics to the Management and Accounting degree. He holds Economics Master from the University of Portugal and a Degree in Economics from the University of Coimbra, Portugal. His main research interests are in Economics, Social Responsibility and Hospitality.

The EU Taxonomy for Sustainable Activities: What It Really Implies for Companies

Magnus Frostenson

1 Introduction

The EU Taxonomy Regulation (2020/852, 'the taxonomy') is part of the EU action plan for financing sustainable growth. On 12 July 2020, the taxonomy went into force (EU, 2020). The main idea behind it is to direct flows of capital to sustainable sectors of the economy. In the context of the EU 'Green Deal', the taxonomy is one (of several) recent governance initiatives supposed to promote sustainable investments and the transition to a greener economy (Berg & Olsson, 2023; Papari et al., 2024).

The taxonomy involves a change of policy. But the taxonomy is not a system of rules detailing how companies should invest or how they should perform their business. Rather, it is a *transparency tool* aimed at classifying and visualising economic activities and whether these are sustainable or not. This is done through connecting the operations of the companies to their contributions to the environmental objectives of the EU, their capacity not to cause harm and their correspondence with established social standards. As a transparency tool, the taxonomy is not a requirement or regulation about how to perform business. It does not formulate 'dos and don'ts'. Rather, it pertains to self-assessment in the sense of describing an understanding of business activities in relation to environmental objectives, potential harm, and correspondence with established standards.

The transparency aspect entails that investors and other stakeholders should be better placed to take decisions relating to sustainability. In the long run, that is also assumed to affect companies, non-financial and financial ones as defined by the EU, into allocating resources to sustainable activities and alternatives. Through the

M. Frostenson (✉)
School of Business, Society and Engineering, Mälardalen University, Västerås, Sweden
e-mail: Magnus.Frostenson@mdu.se

S. O. Idowu, S. Vertigans (eds.), *Sustainability in Global Companies*, CSR, Sustainability, Ethics & Governance, https://doi.org/10.1007/978-3-031-77971-8_7

taxonomy, the entire chain of investment can be visualised and highlighted with regard to sustainability, which may impact, for example, investment behaviour and capital allocation. Another assumption is that the taxonomy will affect the costs of debt and equity and lower the cost of capital of companies with sustainable operations. As argued by Köppl-Turyna and Schwarzbauer (2022, p. 252): 'The scope of application means that the taxonomy will be applicable to both new projects, for which companies might seek financing (e.g., public grants) and involving operational expenses (e.g. issuance of debt or equity to cover day-to-day operations). /…/ (T)his measure is expected to affect both the costs of debt and equity, in which entities with a higher share of revenue from taxonomy-aligned activities will entail a lower cost of capital'.

The taxonomy primarily targets larger companies, but it may very well be the case that it has trickle-down effects also to smaller companies (cf. O'Reilly et al., 2023). Projecting into the future, some even suggest 'colossal' impact also on small and medium-sized companies in terms of existential business risks and unprecedented opportunities (Sancak, 2023). Reporting procedures prescribed by the taxonomy, to take one example, may spread to smaller companies. And despite its aim to reallocate capital, its effects are not limited to the financial sector and its instruments but may very well impact non-financial companies in a significant way. How you do business and in which way may be affected. However, according to accounting practitioners (O'Reilly et al., 2023), implementation costs may be substantial, which means that small and medium-sized companies may only align with the taxonomy if they are coerced to do so.

In other words, the taxonomy rests on ambitions and expectations. Now, a couple of years after it went into force, one may start to observe its empirical consequences, at least in a rudimentary way. Basically, companies have begun to relate to the taxonomy. If one considers the consequences of the taxonomy, one may go beyond obvious and prescribed ones such as increased and extended demands on sustainability reporting by larger firms and information on their activities' taxonomy eligibility and alignment (cf. O'Reilly et al., 2023). As noted by, for example, Tettamanzi et al. (2023), the actual consequences of the taxonomy remain to be studied and understood, in addition to the predictive discussions about, for example, governance, financial, and market implications of the taxonomy (e.g. de Oliveira Neves, 2022). Especially, that is the case since these consequences have been relatively obscure in the sense that the general expectations have been foreseeable but the more direct effects on stakeholders not equally clear. Norang et al. (2023) corroborate the picture that stakeholders are unsure about which consequences that the taxonomy implies. Consequences, that is, understood in a broad sense, as either what the companies tend to do or what the taxonomy will achieve in terms of, for example, reallocating capital and improving environmental performance.

The consequences of the taxonomy are not necessarily limited to specific companies or activities but affect business more generally. One could also ask whether these consequences impact companies in deeper ways and how. The purpose of the chapter is to discuss the consequences of the EU Taxonomy Regulation for companies and their sustainability work. Doing so sheds light on the consequences of the taxonomy and the transformative power of policy and regulation on corporate activities and operations.

The chapter is an analytically oriented discussion rather than an extensive empirical investigation. It uses examples to point to the consequences of the taxonomy from a micro-level perspective, showing what happens within and among companies that must adapt to the taxonomy. In the following, these consequences will be discussed and understood in a more general or categorical sense, under the headings of self-assessment, accounting concerns, and business transformations. That is, companies in general will be affected by the taxonomy in terms of actively describing and judging one's own business activities, accounting for them, and, possibly, repositioning and changing how they perform business. Before going into that discussion, however, a closer description of the taxonomy follows.

2 The EU Taxonomy Regulation

As part of the EU action plan of financing sustainable growth, the EU taxonomy regulation is a potential game-changer for both financial and non-financial companies in Europe. In short, it requires companies to identify, assess, and describe (economic) activities in terms of sustainability. In fact, it standardises specific understandings of sustainability and requires companies to relate their business to it. The taxonomy contains a classification system, where classification relates to whether the economic activities of companies are sustainable or not, according to certain definitions.

The taxonomy builds on and is oriented towards the environmental objectives of the EU. These are:

- Climate change mitigation
- Climate change adaptation
- Sustainable use and protection of water and marine resources
- Transition to a circular economy
- Pollution prevention and control and
- Protection and restoration of biodiversity and ecosystems

In the taxonomy, economic activities are related to these objectives in the sense that companies are supposed to assess and describe whether the activities performed by the company contribute to them. This is done based on four conditions that economic activities should meet. If they do so, they are aligned with the taxonomy. The conditions are:

- Making a substantial contribution to at least one environmental objective
- Doing no significant harm to any other environmental objective
- Complying with minimum social safeguards
- Complying with the applicable technical screening criteria

Notably, the technical screening criteria referred to are developed over time in the delegated acts. In the technical screening criteria, environmental performance requirements are stipulated ensuring a substantial contribution to the environmental objective that it refers to and, analogously, that it does no significant harm to other environmental objectives.

The EU originally prioritised economic activities that are the most relevant ones in relation to the first two environmental objectives (climate change mitigation and climate change adaptation). The first objectives were focused in the delegated act on climate issued in late 2021 (EU, 2021a). Since then, further delegated acts have been issued. One act, covering the remaining four objectives was released in late 2023, and in effect from January 1st, 2024 (EU, 2023a).

The taxonomy is aimed at sectors and activities most relevant for reductions in greenhouse gas emissions as well as for improving climate resilience. Other sectors and activities may be covered in the future, but already from the outset, and by covering these sectors, the EU pinpoints activities responsible for some 80% of all greenhouse gas emissions in the EU (Schütze & Stede, 2020). Activities are targeted in specific sectors, such as transportation, building, energy, forestry, agriculture, and manufacturing, for which technical screening criteria are developed. Other activities enabling the transformation of these sectors are also included. The choice of these sectors reflects their high contribution to CO2 emissions. Notably, even though the taxonomy targets EU-based companies, in principle larger ones with more than 500 employees and market actors providing financial products within the EU, it is likely to impact businesses located outside Europe. For example, non-EU investors or financial undertakings that offer products on the EU market are subject to regulation (the Sustainable Finance Disclosure Regulation, SFDR), requiring alignment with the EU taxonomy.

The taxonomy follows a specific timeline and concerns more companies over time. That means, for example, that the reporting requirements extend to further companies and the specific issues to be reported multiply. The first reporting according to the taxonomy took place in 2022, first including companies that fall under the scope of the CSRD (the Corporate Sustainability Reporting Directive). Reporting first included *eligibility* (whether the activities of the companies were covered by the taxonomy). Later, *alignment* was supposed to be reported (whether the eligible activities of companies comply with the criteria included in the delegated acts of the EU). Reporting, however, was open also to other companies already from the outset.

The EU has made it clear that the taxonomy is not a rating system or a form of input as to what one should invest in (EU, n.d.). Nor does it imply any hints as to financial performance relating to the classifications. Furthermore, what is not 'green' is not necessarily 'brown', according to the EU. As mentioned, the taxonomy is a classification system aiming at helping investors to make well-informed decisions. Transparency, facilitating transition and technology neutrality have been important lodestars. One should also bear in mind that the taxonomy is a regulation under constant development, reflecting, for example, technological and policy developments. The EU, however, accentuates several advantages with the taxonomy. The common definition of environmentally sustainable activities is one thing.

But it is also instrumental in scaling up sustainable investment in the EU, as well as providing investors with more security and protecting them from greenwashing. In general, the idea is that the taxonomy should be a tool to make companies more climate-friendly and mitigate the fragmentation of markets stemming from unclear understandings of sustainability.

The reception of the taxonomy has been mixed. Researchers have pointed to both possibilities and disadvantages. To Ahlström and Sjåfjell (2022, p. 24), the taxonomy 'is welcomed, as it represents a significant shift regarding the integration of sustainability in financial market and business law, which has for far too long been disregarded in important areas'. Others have voiced criticism, both against the process of developing the taxonomy and the entire regulative approach that it implies. Kooths (2023), for example, attacks it and brings up several issues. The taxonomy is claimed to stand in opposition to market principles and requires substantial resources. It also hampers, according to critics, entrepreneurial initiatives and does not belong to the most efficient policy instruments (as it is opposed to market forces). This causes allocation problems, and in addition, hidden political agendas and lobbyism may prevail. As Kooths (2023, p. 248) argues, the taxonomy 'absorbs high-skilled labor that is in high demand elsewhere in the economy and creates a massive extra bureaucratic burden without any expectable benefit for the total economy'. Such criticism, thus, points to both the risk of impaired productivity and to the costs of increased bureaucracy that the taxonomy implies. In a more practical sense, substantial financial resources must be devoted to understanding and dealing with the taxonomy.

From an opposite perspective, criticism has been voiced against the taxonomy for not challenging the market-based approach and profit motives of our contemporary economic system. For such critics, for example, Knapp et al. (2024), sustainable outcomes tend to stand in opposition to economic growth and the profit motive of actors. In addition, the taxonomy has been subject to lobbyism by actors with interests in, for example, the finance sector and the fossil fuel industry, which has influenced the taxonomy to the extent that it does not ban 'brown' activities. It has also been suggested (e.g. Och, 2020) that the taxonomy does not go far enough, but fails to cover and frame less and non-sustainable economic activities.

Positive voices, on the other hand, include Pacces (2021), who relates the taxonomy to investor concerns and corporate governance. The taxonomy, Pacces (2021) argues, aligns investors' and beneficiaries' interests and fosters inclusion of sustainability into corporate governance concerns. The taxonomy reduces potential greenwashing through standardisation. Beneficiaries are also informed about the sustainability character of the investors and portfolio companies and can act according to preferences, and the taxonomy fosters sound competition among investors for sustainability-oriented beneficiaries. From a more practical perspective, Ahlström and Sjåfjell (2022) point to the difficulties of identifying and measuring sustainability aspects and to consider these in, for example, investment decisions. Even though the taxonomy is helpful in that respect, it is not an exemplary standardising tool for solving the problems. Others observe empirical consequences as (side) effects of the taxonomy. Dumrose et al. (2022), for example, suggest that ESG (environmental, social, governance) ratings made by rating firms may converge because of the guidance and standardisation aspects of the taxonomy.

Whatever the pros and cons of the taxonomy, it is in force and companies must take action and adapt to it. In the following, the consequences of the taxonomy will be discussed from a corporate perspective. To put it in another way, what happens within and in companies as a consequence of the taxonomy? In the remainder of the chapter, such consequences will be discussed as micro-level phenomena. Macro-phenomena, for example, capital allocation, will in principle not be the focus of the discussion.

3 Increased Self-assessment

Self-assessment according to the taxonomy implies a form of constructivist approach in response to regulation (Frostenson, 2023). Companies are to assess and explain eligibility and alignment with the taxonomy. In doing that, companies are supposed to produce accounts building on reflections on their activities. In the taxonomy and its delegated acts, the issue of self-assessment can be exemplified by the emphasis on determination. As expressed in EU (2023a, p. 8), relating to the screening criteria of the objective of circular economy:

> The technical screening criteria for determining the conditions under which an economic activity qualifies as contributing substantially to the transition to a circular economy and for determining whether that economic activity causes no significant harm to any of the other environmental objectives /.../.

As with the other objectives, the task of companies is about vetting and *assessing* one's own's business, in order to understand both taxonomy eligibility and alignment.

An example from the Norwegian oil and gas company Equinor is telling. The company suggests that it performed EU Taxonomy alignment evaluations in 2022, and it did so for the activities that were considered significant in its portfolio of eligible activities. The testing was focused on both producing assets and assets under development, including current investments related to wind power electricity generation and photovoltaic technology. Full alignment with the taxonomy was not possible for some new investments in storage of electricity, but the company expects these to contribute positively to alignment KPI's in the future. Further, the company argues (Equinor, 2023, p. 253):

> All tested eligible activities passed the substantial contribution criteria. When assessing compliance with the Do No Significant Harm criteria the following interpretations and judgments were applied:
>
> - Climate change adaption—relevant climate related hazards have been assessed based on a risk assessment.
> - Circular economy—durability and recyclability have been assessed where feasible.
> - For DNSH criteria that reflect legal requirements under EU regulations, the technical screening criteria are considered met when the operations are conducted within

> normal, lawful operations, comply with emission permits, environmental impact assessments have been performed and necessary action have been taken when required. Based on the alignment testing performed the tested assets and associated activities are aligned with the technical screening criteria by year-end 2022.

Further, the company states that the assessment process was organised in various ways. For example, regarding the level at which assessment is made of substantial contribution and compliance with technical screening, this is done individually for the identified economic activities or, in some cases, at the level of the entire economic activity, an 'operating segment' or the group as a whole. When it comes to not doing significant harm, assessment primarily reflects legal requirements under EU regulation. The company does this to prevent investment processes from being too narrowly focused on one EU objective while downplaying the others. When it comes to assessing minimum safeguards, Equinor (2023, p. 259) has

> a group wide approach to ensuring compliance with the minimum safeguards. Equinor is committed to respecting human rights in all business processes. To prevent human right violations, Equinor adhere to external standards and defines its own principles and policies.

In other words, there are processes of self-assessment going on in companies. Without taking a stand on the accuracy and reasonableness of these assessments and tests made by Equinor, it is fair to say that they are carried out without detailed information about how and by whom (other than the company). In other words, there is self-assessment but with relatively limited transparency for a reader of the report. That, however, tends to be the case for many other companies. Process descriptions, rather, tend to be of general kind in annual reports. Another example comes from the Danish health giant Novo Nordisk (2023, p. 26):

> Eligibility and alignment. We followed a two-step process to arrive at our present Taxonomy disclosures. Firstly, we screened the Taxonomy rules to create a list of economic activities that could potentially be eligible. The description of each economic activity was assessed against how we perform the economic activity. Then, after applying materiality considerations, we decided to report Taxonomy-eligible CapEx for economic activities 7.1 and 7.2, i.e., regarding the construction of new buildings and the renovation of existing buildings, respectively. Secondly, we evaluated whether we could classify any of our Taxonomy-eligible CapEx as Taxonomy-aligned.

This is another example of very general process descriptions, and the discretionary nature of them is evident. For example, Novo Nordisk only suggests that taxonomy-eligible capital expenditure is 'classified' by the company itself. In this case too, limited information is given as to the very process.

Whatever the case, the activities of determining eligibility and alignment through assessment and testing are central to the taxonomy. All companies are required to perform them, and there is no reason to doubt that this is done. Self-assessment is central, and it does take place. It is not necessarily so, however, that the very process is transparent.

4 Accounting Concerns

The self-assessment that the taxonomy generates is closely related to accounting concerns and growing reporting requirements. Consequences of the taxonomy related to accounting can be widely understood, for example as affecting the role of the auditor (Holhaugen & Kjelsvik, 2022). But more specifically, the taxonomy involves expectations on how to report. What has been found in research is that companies are aware of the challenges that the taxonomy implies, both from a general and a reporting perspective. In a study of 134 Swedish listed firms by Andersson and Arvidsson (2023), the authors found that most of the surveyed firms (85% of the financial firms and 78% of the goods producing) had analysed the effects of the taxonomy on their firms. Other firms were still in the process of doing so, with only some single firm having had no process at all. To the surprise of the authors, preparations for the EU taxonomy seemed to be more advanced compared to planned effects of and adjustment to the CSRD directive on sustainability reporting.

Another study by Hummel and Bauernhofer (2023) focused on taxonomy consequences for Austrian firms in terms of reporting, capital-market issues and actions and outcomes related to the taxonomy. During the first year of reporting, compliance with the taxonomy was important. However, even though firms disclose the required key performance indicators (KPIs), they tend to lack the reporting infrastructure necessary to provide detailed information. Capital-market consequences were to some extent vaguer, but firms emphasised clearly defined KPIs and the important role of, for example, banks to integrate taxonomy-related information in their lending decisions. As for actions, mainly internal discussions on strategy were noted.

Distinctively, however, consequences of the taxonomy tend to be understood very much in terms of accounting and reporting. Reporting requirements were stated at an early stage in 2020. In Article 8 of the taxonomy (EU, 2020), it is stated that

> 1. Any undertaking which is subject to an obligation to publish non-financial information pursuant to Article 19a or Article 29a of Directive 2013/34/EU shall include in its non-financial statement or consolidated non-financial statement information on how and to what extent the undertaking's activities are associated with economic activities that qualify as environmentally sustainable under Articles 3 and 9 of this Regulation.
>
> 2. In particular, non-financial undertakings shall disclose the following:(a) the proportion of their turnover derived from products or services associated with economic activities that qualify as environmentally sustainable under Articles 3 and 9; and (b) the proportion of their capital expenditure and the proportion of their operating expenditure related to assets or processes associated with economic activities that qualify as environmentally sustainable under Articles 3 and 9.

The requirements have been further detailed in the delegated act of 2021 (EU, 2021b). There, the specific content, methodology, and presentation of information to be disclosed by financial and non-financial undertakings were specified. A consequence of the reporting requirements has been that already from the financial year of 2021 and onwards, reporting according to taxonomy requirements has been a reality for European firms, including large public interest entities (PIE) with more than 500 employees. Specifically, such companies are expected to disclose information about how and to what extent their activities are eligible (covered by) and aligned with (meeting the technical criteria of) the taxonomy. Concretely, KPIs are described (EU, 2021b) and related to total revenue, operational expenditure (OpEx), and capital expenditure (CapEx). Simply put, companies are assumed to disclose the proportions of turnover, capital expenditure, and operating expenditures that can be related to environmentally sustainable activities (Krasodomska et al., 2023), and whether these eligible activities contribute positively to the objectives or do not cause harm, and are aligned with minimum standards for social responsibility (Andersen & Gjølberg, 2022). When it comes to large financial entities, such as banks, asset managers, investment firms, insurance and reinsurance companies, expected disclosures reflect the proportions of environmentally sustainable economic activities in the total assets that these entities finance or invest in (Moneva et al., 2023).

What does the reporting look like in a more concrete sense? Examples are manifold. The Finnish-Swedish paper and pulp company Stora Enso mentions the taxonomy 110 times in its annual report of 2022. The company notes, for example (Stora Enso, 2023, p. 45):

> In 2022, Stora Enso continued its EU Taxonomy reporting for investors and analysts. The Company carried out assessments for Taxonomy-eligibility and Taxonomy-alignment during the year. Investor Relations participated in an internal working group developing the EU Taxonomy reporting

A process is described and the concrete results of it are reported. The KPIs that the taxonomy requires are presented to the reader in separate tables relating to turnover, CapEx and OpEx, in line with how these are defined in the taxonomy. An example showing taxonomy reporting in relation to turnover is found in Table 1:

This is just one example of how assessment is made, and figures and information reported. Some conclusions can be drawn. First, following self-assessment, only a limited percentage of the business activities (6.5%) are seen as taxonomy-eligible. That means that only a limited part of the company's business activities is assessed in terms of alignment with the environmental objectives (relating to contribution to their fulfilment), whether they do no harm or live up to minimum safeguards. The reporting aspects of the taxonomy have also been subject to limited assurance by third-party auditors.

The accounting consequences of the taxonomy are obvious. It is not, however, altogether clear that accounting as such has a severe impact on companies apart from the frequently painstaking processes of producing reports and information. Reflections are few as to change as a consequence of the KPIs presented in the reports. One reason for this may be the absence of distinct quality criteria other than the technical screening criteria provided by the EU. That is, even though the EU suggests that companies, for example, should assess whether they contribute to environmental objectives, there is no explicit information or discussion about the quality and nature of a company that scores 'low', in case it does not contribute to

Table 1 Proportion of turnover from products or services associated with taxonomy-aligned economic activities 2022 (adapted from Stora Enso, 2023, p. 133)

EUR million				Substantial contribution criteria						DNSH criteria ('does not significantly harm')									
Economic activities	Code(s)	Absolute turnover EUR	Proportion of turnover %	Climate change mitigation %	Climate change adaptation %	Water and marine resources %	Circular economy %	Pollution %	Biodiversity and ecosystems %	Climate change mitigation y/n	Climate change adaptation y/n	Water and marine resources y/n	Circular economy y/n	Pollution y/n	Biodiversity and ecosystems y/n	Minimum safeguards	Taxonomy-aligned proportion of turnover, year N	Category (enabling activity) E	Category (transitional activity) T
A TAXONOMY-ELIGIBLE ACTIVITIES																			
A.1 Environmentally sustainable activities (taxonomy-aligned)																			
Forest management	1.3	111	0.9%	100%						N/A	y	y	y	y	y	y	100.0%		
Manufacture of energy efficiency equipment for buildings	3.5	595	5.1%	100%						N/A	y	y	y	y	y	y	100.0%	E	
Cogeneration of heat/cool and power from bioenergy	4.20	54	0.5%	100%						N/A	y	y	N/A	y	y	y	96.6%		
Turnover of environmentally sustainable activities (taxonomy-aligned) A.1		760	6.5%														99.7%		

A.2 Taxonomy-eligible but not environmentally sustainable activities (not taxonomy-aligned activities)																			
Cogeneration of heat/cool and power from bioenergy	4.20	2	0.0%																
Turnover of taxonomy-eligible but not environmentally sustainable activities (not taxonomy-aligned activities) (A.2)		2	0.0%																
Total (A.1 + A.2)		762	6.5%																
B. TAXONOMY-NON-ELIGIBLE ACTIVITIES																			
Turnover of taxonomy-non-eligible activities (B)		10,932	93.5%																
Total (A + B)1		11,694	100%																

the objectives. Accounting implies no automatic checkbox or tool for ranking companies or suggesting that they should reassess their business. It may be argued that this is an indirect consequence of accounting, but as it seems, companies have not so far taken this as an impetus to severely reorganise or revamp their business activities.

5 Business Transformations

The question of whether the taxonomy changes business operations, priorities and whether it, as a fact, contributes to positive consequences is an open one. At a general level, for example in relation to trade specialisation, consequences have been highlighted: 'We believe that the introduction of the taxonomy will alter trade patterns and trade specialisation, in spite of the signs that EU members are specialised in industries and economic activities that are favoured by its introduction. The taxonomy is likely to further work not only directly but also through supply chains, as obtaining more attractive funding for companies is indirectly linked to the compliance of the company's suppliers as well' (Köppl-Turyna & Schwarzbauer, 2022, p. 252).

Some scholars relate potential business transformation to the criteria that are formulated in the taxonomy. Schütze et al. (2020), for example, discuss whether the taxonomy, as it is formulated, contributes to meeting the goal of climate-neutrality in 2050. They suggest that the criteria in the taxonomy will lead to climate-neutrality in some economic sectors but not in others. The criteria will, for example, push the automotive sector in the right direction but will be insufficient for the emission-intensive basic materials sector. In another article, Schütze and Stede (2024) point to certain aspects contained in the taxonomy that are likely to affect and incentivise companies in various ways, depending on activities, sectors, and the thresholds of the taxonomy, implying the levels where the activities are assumed to be climate-friendly. One point that they make is that the taxonomy is not incentivising innovative technology as much as it should.

> …the Taxonomy does not currently indicate a path towards climate neutrality for several of the economic activities it covers. While activities labelled 'green' or 'enabling' are assumed to contribute directly to reaching the climate goals, for the sectors in 'transition' technical thresholds need to be met for an activity to be Taxonomy-eligible. /…/ However, typically capital-intensive breakthrough technologies are needed to decarbonise these economic activities. These innovative technologies are currently not incentivised by the Taxonomy, due to its binary structure: Once a company reaches the threshold, there is no further incentive to improve beyond the threshold (Schütze & Stede, 2024, p. 143).

Generally, Schütze and Stede (2020, 2024) point to threshold issues of the taxonomy and, specifically, the use of a single threshold in emission-intensive sectors. The researchers suggest that new investments should have stricter criteria than current corporate activities. And in cases where sectors are not covered by the taxonomy, low-emissions activities should be differentiated from high-emission activities not compatible with a low-carbon future.

Another aspect is that the taxonomy simply does not cover all relevant activities, thereby creating few incentives to transform these. Yet again, one could refer to the Finnish-Swedish company Stora Enso, describing the implications of the EU Taxonomy on the company (Stora Enso, 2023, p. 131):

> The Taxonomy Regulation is a developing regulation and not yet covering all sustainable activities in the market. The forest industry is not at the core of the current legislation and therefore has only few relevant economic activities to report on. From Stora Enso's main products, only wood-based solutions for construction industry are included in the EU Taxonomy through their contribution to buildings' energy efficiency. Other main products, production of pulp, consumer board, containerboard, corrugated packaging and paper, are out of the scope of the EU Taxonomy and therefore the reported Taxonomy-eligible KPIs are low. EU Taxonomy is anticipated to expand to four other environmental objectives during 2023 with the next delegated act, but the amendments are not expected to bring major impact to Stora Enso's Taxonomy eligibility.

Simply put, if your business activities are not eligible, they are not covered by the taxonomy and for that reason hardly affected and transformed by it. As for micro-level company-related changes, it has been claimed by Norang et al. (2023) that the consequences of the taxonomy and what will change is an ongoing topic of discussion. We are still in a state of expectation rather than having full knowledge. Further green investments or reorientation of existing business practices into green ones are options (but not yet facts).

One aspect emphasised in the literature, however, is that the taxonomy incentivises investors and firms to engage in sustainability transformation (Andersson & Arvidsson, 2023). As noted in the study by Hummel and Bauernhofer (2023), implications in the form of discussions on corporate strategy have been noted. However, as for business transformation, decoupling is an issue that is commonly present in relation to regulative reforms. A study by de Gier et al. (2022) shows that the taxonomy is a source of decoupling in the sense that different organisational levels respond differently to the demands of the taxonomy. Ensuring compliance is important at a strategic level whereas a practice-orientated approach can be found on the project level of the large Danish construction firm under study. In the study, it is argued that although a policy tool, the taxonomy is not developed and mature enough to avoid the consequence of decoupling between policy and corporate practice.

It is important to note that the taxonomy or the delegated acts do not contain specific action-guiding principles on how to realign or transform business. The taxonomy is not a management system or a toolkit. What the EU does is to suggest ways of thinking given the assessments made by the companies. For example, the EU has issued a 'user guide' (EU, 2023b) to facilitate dealing with the taxonomy. Strikingly, most of the case studies referred to concern assessment and reporting of business activities in relation to the taxonomy. It is not specifically directed towards action and business transformation. At a more general level, it has some suggestions about, for example, how to align with best practices when adapting to the impacts of climate change. Only general advice is offered, about selecting and implementing suitable adaptation solutions, and setting up control systems.

To understand business transformation, in case it occurs, one must go far beyond the mere EU documents and intentions. Reasonably, the taxonomy contains little that automatically changes business activities. The central issue is the link between self-assessment, reporting, transparency and potential action and adaptation. That, however, suggests a potential role of the taxonomy in sustainable transformation, but probably only a partial one. Other aspects than the taxonomy tend to impact business, other regulations, business priorities or demands of the market.

6 Conclusion

Understanding the consequences of the EU taxonomy requires an understanding at different levels. At the macro-level, it is obvious that it constitutes an important aspect of *policy change* that must be taken into account by states and companies. There are both expectations and hopes for a significant change of the allocation of capital, investment decisions, and the greening of companies in general following the taxonomy. At the micro-level, the focus of this chapter, the more immediate consequences of the taxonomy are making themselves known. In a very direct sense, an effect of the taxonomy has been a need for competence. The taxonomy has brought about an overwhelming need for knowledge, about what the taxonomy means, implies, how one should report, and so on. Advisory services are prospering, as so often in times of new regulation that requires knowledge.

In this contribution, however, three aspects of micro-level consequences have been highlighted, all relating to corporate activities following the taxonomy, namely self-assessment, accounting concerns, and business transformations. As far as one can see at the time of this writing, the EU taxonomy tends to elicit self-assessment and accounting concerns among virtually all companies affected by it (see Andersson & Arvidsson, 2023). But as for the third (potential) consequence, the taxonomy is not necessarily a sharp instrument when it comes to bringing about business transformation. At least one cannot draw such a conclusion based on what the companies themselves claim to do and what the taxonomy requires.

An explanation for the not so obvious transformative power of the taxonomy is possibly its indirect character. Transformation does not come about as a direct issue based on specific requirements of the taxonomy. It is, most likely, a macro phenomenon 'translated' to the micro context of business operations. It may very well be conditioned by, for example, resource allocation decisions by investors or higher costs of capital experienced by the companies. That, in turn, is a long-run consequence that the EU and others are probably hoping and aiming for. Also, the character of the taxonomy, excluding some industries (so far), not relating to social issues, and the discretionary character of it (allowing for self-assessment and classification of activities), implies that certainly not all 'brown' activities are exposed as non-aligned, not fulfilling environmental objectives, or doing significant harm. In other words, the visualisation aspect of the taxonomy is not waterproof. It does not necessarily visualise that which may cause pressure to transform in the longer run—when acted upon by investors and others. It may be that the taxonomy needs to be further developed to gain the transformative power that some want it to have.

The very nature of the taxonomy, with its emphasis on assessment and reporting, tends to lead to direct consequences. But it also carries with it a reasonable portion of 'wishful thinking', to put it a bit bluntly. The suggested logic, more or less explicitly formulated by the EU, is that transparency leads to visualisation, which leads to pressure and transformation. That is a long chain of action that remains to be proven. In any case, the assumption is that transparency on issues related to sustainability will occur and that it will have the intended consequences. The belief, or should one say the hope, is strong. Long-term effects remain to be seen, however.

This contribution of the chapter is, on the one hand, its discussion about the micro-level consequences of the EU taxonomy. On the other hand, it can also be related to the power of policy in general. By prescribing what companies must do, policy and regulation do matter and get visible consequences. But at the same time there is no guarantee that the things that policymakers are hoping for will materialise and that business transformation will occur. Time will tell, however. From a research perspective, further studies should look into the consequences of the taxonomy and other regulation both in a more direct and indirect long-run sense, in particular to capture the transformative aspect of policy.

References

Ahlström, H., & Sjåfjell, B. (2022). Complexity and uncertainty in sustainable finance: An analysis of the EU taxonomy. In T. Cadman & T. Sarker (Eds.), *De Gruyter handbook of sustainable development and finance* (pp. 15–40). Walter de Gruyter GMBH.

Andersen, K. R., & Gjølberg, M. (2022). Bærekraft i selskapsrapporter: Fra floskler og frivillighet til finansielt relevant og lovpålagt rapportering. *Praktisk økonomi & finans, 38*(3), 240–250.

Andersson, F. N. G., & Arvidsson, S. (2023). The EU's sustainable finance platform: A new game plan in the quest for competitive advantage. In P. N. Ghauri, U. Elg, & S. M. Hånell (Eds.), *Creating a sustainable competitive position: Ethical challenges for international firms* (pp. 237–249). Emerald Publishing.

Berg, M., & Olsson, J. (2023). Managing public value conflicts – Institutional strategies and the greening of public pension funds. *Scandinavian Journal of Management, 39*(4), 101301.

de Gier, A. J., Gottlieb, S. C., Koch, C., & Frederiksen, N. (2022). EU taxonomy for sustainable activities: A source of decoupling or a pathway for greening the construction industry? In *Proceedings of the 38th annual ARCOM conference*, pp. 572–581.

de Oliveira Neves, R. (2022). The EU taxonomy regulation and its implications for companies. In P. Câmara & F. Morais (Eds.), *The Palgrave handbook of ESG and corporate governance* (pp. 249–265). Springer.

Dumrose, M., Rink, S., & Eckert, J. (2022). Disaggregating confusion? The EU taxonomy and its relation to ESG rating. *Finance Research Letters, 48*, 102928.

Enso, S. (2023). *Stora Enso's annual report 2022*. Stora Enso.

Equinor. (2023). *2022 integrated annual report*. Equinor.

EU. (2020). *Regulation (EU) 2020/852 of the European Parliament and of the council of 18 June 2020 on the establishment of a framework to facilitate sustainable investment, and amending regulation (EU) 2019/2088*. EU.

EU. (2021a). *Commission delegated regulation (EU) 2021/2139 of 4 June 2021 supplementing regulation (EU) 2020/852 of the European Parliament and of the Council by establishing the technical screening criteria for determining the conditions under which an economic activity qualifies as contributing substantially to climate change mitigation or climate change adaptation and for determining whether that economic activity causes no significant harm to any of the other environmental objectives*. EU.

EU. (2021b). *Commission delegated regulation (EU) 2021/2178 of 6 July 2021 supplementing regulation (EU) 2020/852 of the European Parliament and of the Council by specifying the content and presentation of information to be disclosed by undertakings subject to articles 19a or 29a of directive 2013/34/EU concerning environmentally sustainable economic activities, and specifying the methodology to comply with that disclosure obligation*. EU.

EU. (2023a). *Commission delegated regulation (EU) 2023/2486 of 27 June 2023 supplementing regulation (EU) 2020/852 of the European Parliament and of the Council by establishing the technical screening criteria for determining the conditions under which an economic activity qualifies as contributing substantially to the sustainable use and protection of water and marine resources, to the transition to a circular economy, to pollution prevention and control, or to the protection and restoration of biodiversity and ecosystems and for determining whether that economic activity causes no significant harm to any of the other environmental objectives and amending commission delegated regulation (EU) 2021/2178 as regards specific public disclosures for those economic activities*. EU.

EU. (2023b). *A user guide to navigate the EU taxonomy for sustainable activities*. EU.

EU. (n.d.). *EU taxonomy navigator*. EU. Accessed February 7, 2024, from https://ec.europa.eu/sustainable-finance-taxonomy/

Frostenson, M. (2023). Controlling or constructing business through the sustainable development goals. In S. O. Idowu & L. Zu (Eds.), *The Elgar companion to corporate social responsibility and the sustainable development goals* (pp. 130–141). Edward Elgar.

Holhaugen, E., & Kjelsvik, M. R. (2022). *Taksonomiens påvirkning på store foretaks bærekraftsrapportering* (Master's thesis, University of Agder).

Hummel, K., & Bauernhofer, K. (2023). *Consequences of sustainability reporting mandates: Early evidence on the EU taxonomy regulation*. Available at SSRN 4175157.

Knapp, M., Litofcenko, J., Maringele, S., Rogers, C., Schmid, L., Streinzer, A., & Taschwer, M. (2024). Current policy initiatives on green finance in the EU: The green taxonomy in the global context. In J. Jäger & E. Dziwok (Eds.), *Understanding green finance* (pp. 73–87). Edward Elgar.

Kooths, S. (2023). EU taxonomy: Mission impossible. *The Economists' Voice, 19*(2), 243–249.

Köppl-Turyna, M., & Schwarzbauer, W. (2022). Will the EU taxonomy impact the trade specialisation of European economies? *The Economists' Voice, 19*(2), 251–260.

Krasodomska, J., Zieniuk, P., & Kostrzewska, J. (2023). Reporting on sustainable development goals in the European Union: What drives companies' decisions? *Competitiveness Review, 33*(1), 120–146.

Moneva, J. M., Scarpellini, S., Aranda-Usón, A., & Alvarez Etxeberria, I. (2023). Sustainability reporting in view of the European sustainable finance taxonomy: Is the financial sector ready to disclose circular economy? *Corporate Social Responsibility and Environmental Management, 30*(3), 1336–1347.

Norang, H., Støre-Valen, M., Kvale, N., & Temeljotov-Salaj, A. (2023). Norwegian stakeholder's attitudes towards EU taxonomy. *Facilities, 41*(5/6), 407–433.

Nordisk, N. (2023). *Annual report 2022*. Novo Nordisk A/S.

O'Reilly, S., Gorman, L., Mac An Bhaird, C., & Brennan, N. M. (2023). *Implementing the European Union green taxonomy: Implications for small-and medium-sized enterprises* (pp. 1–26). Accounting Forum.

Och, M. (2020). *Sustainable finance and the EU taxonomy regulation – Hype or Hope?* Jan Ronse Institute for Company & Financial Law Working Paper, (2020/05).

Pacces, A. M. (2021). Will the EU taxonomy regulation foster sustainable corporate governance? *Sustainability, 13*(21), 12316.

Papari, C. A., Toxopeus, H., Polzin, F., Bulkeley, H., & Menguzzo, E. V. (2024). Can the EU taxonomy for sustainable activities help upscale investments into urban nature-based solutions? *Environmental Science & Policy, 151*, 103598.

Sancak, I. E. (2023). *The EU sustainability taxonomy: Will it affect small and medium-sized enterprises?* Available at SSRN https://ssrn.com/abstract=4546899 or https://doi.org/10.2139/ssrn.4546899

Schütze, F., & Stede, J. (2020). *EU sustainable finance taxonomy – What is its role on the road towards climate neutrality? DIW Berlin Discussion Paper No. 1923*. DIW.

Schütze, F., & Stede, J. (2024). The EU sustainable finance taxonomy and its contribution to climate neutrality. *Journal of Sustainable Finance & Investment, 14*(1), 128–160.

Schütze, F., Stede, J., Blauert, M., & Erdmann, K. (2020). EU taxonomy increasing transparency of sustainable investments. *DIW Weekly Report, 10*(51), 485–492.

Tettamanzi, P., Gotti Tedeschi, R., & Murgolo, M. (2023). The European Union (EU) green taxonomy: Codifying sustainability to provide certainty to the markets. *Environment, Development and Sustainability, 26*(11), 27111–27136.

Magnus Frostenson works as a professor of industrial economy and organisation at Mälardalen University, Sweden. He has held positions at Østfold University College, Norway, Örebro University School of Business, Sweden, and at other higher education institutions. His research is focused on business ethics, corporate social responsibility (CSR), and the management and control of professions. Sustainability accounting is also one of his research interests. Magnus has published many journal articles, books, book chapters, and reports.

Impact of Developing Iconic Cultural Centres on Sustainability and Socio-Economic Development: A Case Study of the Turner Contemporary Margate

Adebimpe Adesua Lincoln and Jane Croad

1 Introduction

Stakeholders including, customers, shareholders, local communities, and government institutions are paying more attention to Corporate Social Responsibility (CSR) and sustainability and putting pressure on organizations and international businesses to act in the best interest of society. CSR and sustainability are now accepted as crucial requirements for organizations to allocate vital resources (Yin & Jamali, 2016). Businesses that proactively pursue social responsibility and sustainability initiatives are less vulnerable to regulatory changes imposed both on a national and international level (Bern et al., 2009). The importance of international business in the global development agenda cannot be overemphasized. Burritt et al. (2020) state that international businesses have been criticized as being one of the primary institutions contributing towards 'unsustainability' in terms of impacts on society and the environment—these organizations have been criticized for the negative and often unethical aspects of their business practices. Various academics and researchers have opined that international businesses act against the best interest of the local communities in which they operate. For example, Jamali (2010, p. 183) notes that international businesses do this by 'outsourcing dirty operations, sourcing labour below subsistence pay levels, and fostering poor working conditions while taking advantage of the lax social and environmental standards in foreign countries where they set up shop'. This criticism has resulted in various stakeholder groups

A. Adesua Lincoln (✉)
University of Liverpool, Liverpool, UK

J. Croad
Robert Kennedy College, Zurich, Switzerland

S. O. Idowu, S. Vertigans (eds.), *Sustainability in Global Companies*, CSR, Sustainability, Ethics & Governance, https://doi.org/10.1007/978-3-031-77971-8_8

demanding that international businesses refocus their efforts away from short-term economic gain and towards long-term sustainable activity (Burritt et al., 2020). International businesses are in a unique position to promote inclusivity and contribute to the development of socially sustainable communities. This study therefore seeks to examine CSR and sustainability as a means of bringing improvement to the population's situation and changing the business environment within which international businesses operate. Using the Turner Contemporary Margate as a case study, this research focuses on the impact of developing iconic cultural centres in deprived areas on the social-economic development of the location and the corporate social responsibility of the cultural centres to the stakeholder groups specifically those of the lower social economic group that can encounter cultural and economic disruption. The study seeks to answer the following research question:

1. What are the perceptions of the key stakeholder groups on the impact of the Turner Contemporary?
2. What are the economic and social benefits of the Turner Contemporary to people who live, work, study, or visit Thanet and Kent?
3. Whether there is an association between income and social development factors?
4. What strategies can be adopted to improve the contributions of the Turner Contemporary to the economic and social development of the area?

2 Sustainability and Social Responsibility

The definition of sustainability is a subject of academic debate. The term has been defined to encompass environmental management, corporate responsibility, corporate social responsibility, and corporate sustainability (Bansal & Song, 2017; Burritt et al., 2020). The most prominent definition of sustainable development is that proposed by the Brundtland Commission Report (WCED, 1987) According to the Brundtland Commission Report, sustainable development is one which 'meets the needs of the present generation without compromising the ability of future generation to meet their needs' (WCED, 1987). As it relates to international business organizations, sustainability incorporates three dimensions or pillars—economic (economy), environmental, and social (society). The three pillars of (environment, society, and the economy) were first proposed in the 1987 Brundtland Report. These pillars of sustainable development are seen as interacting with each other at the same level. Sustainable development is built on these three pillars. This is because sustainable development can only be achieved when environmental protection, social equity, and economic profitability coexist without one area overshadowing the others. International business organizations must improve performance along all of these dimensions and, apart from improving their operations and products, they must contribute to the sustainable development of markets and society (e.g. Schaltegger et al., 2018; Burritt et al., 2020).

Environmental Pillar The environmental pillar focuses on environmental responsibility of a business and the method adopted to measure environmental variables—such as air quality, water quality, energy consumption, natural resources, solid and toxic waste, and land use. In recent times of particular importance to international business organizations is the need to reduce their carbon footprints and reduce waste generated through excessive packaging. From a 'cost' perspective, valid arguments are proffered vis-à-vis the link between lessening the amount of material used in packaging and eventual reduction in the overall spending on those materials. Other aspects of concern include a reduction in water consumption and the overall effect of the MNEs' business activities on the environment.

Social Pillar The social pillar focuses on how business activities can be conducted in a socially responsible manner. In assessing the social responsibility of a business, factors such as diversity and inclusivity, social justice, poverty reduction, and access to good quality education and social resources are measured. It is essential that business organizations work together with key stakeholder groups such as employees, customers, suppliers, and the local communities in which they operate in devising apt strategies which not only align with the business core values but are of added benefit to various stakeholders, thus positively enhance society as a whole. Some of the initiatives that international businesses seek to adopt focus on targeting human rights, fair treatment of employees, equality and diversity, and the introduction of community support initiatives such as investment in local projects—fundraising, sponsorship, and scholarships.

Economic Pillar The economic pillar focuses on the economic measurement of the business, as well as profit and shareholder wealth maximization and economic measurements of employment, taxes, income, and expenditures. The economic pillar is not only concerned with aspects of profitability and wealth maximization, but also focuses on existing corporate governance frameworks such as compliance, effective governance, and risk management, and alignment of boards of directors, shareholders, and stakeholders' interests.

In recent times, the three sustainable development pillars have been referred to as 'people, planet and profits'—also known as the Triple Bottom Line (TBL), or 3 principles of sustainability (Elkington, 1994). Figure 1 below shows the interrelationship between the three sustainability pillars and the Triple Bottom Line framework proposed by Elkington.

- *People:* People measures the positive and negative impact a firm's activity has on various stakeholder groups such as employees, families, customers, suppliers, and the local communities in which it operates.
- *Planet:* Planet measures the positive and negative impact a firm's activity has on its natural environment. Some of the environmental issues include climate change, greenhouse gas emissions, reducing its carbon footprint, usage of natural resources, waste management and recycling, life on land, and life below water.

Sustainability Pillar	Elkington Triple Bottom Line
Environmental	**Planet:** air and water quality, energy consumption, natural resources, solid and toxic waste and land use.
Social	**People:** poverty reduction, education, equity and access to social resources, social justice, diversity and inclusivity, health and well-being, quality of life and social capital.
Economic	**Profit:** flow of money, corporate profit accounts and the financial health of the firm as well as economic aspects. Economic aspects include income and expenditures, business diversity and climate factors, taxes and employment.

Fig. 1 Aligning the sustainability pillars with Elkington triple bottom line

- *Profit:* Profit measures the positive and negative impact a firm's activity has on the economy's growth and development; for example, employment generation, innovation, wealth creation, and taxation.

Elkington (2018) explains that 'the TBL was not designed to be just an accounting tool. It was supposed to provoke deeper thinking about capitalism and its future, but many early adopters understood the concept as a balancing act, adopting a trade-off mentality'. Its goal was 'system change—pushing toward the transformation of capitalism. It was never supposed to be just an accounting system. It was originally intended as a genetic code, a triple helix of change for tomorrow's capitalism, with a focus on breakthrough change, disruption, asymmetric growth—with unsustainable sectors actively sidelined, and the scaling of next-generation market solutions'. Accordingly, Elkington (2018) notes that 'the triple bottom line is a sustainability framework that examines a company's social, environment, and economic impact. The original idea was encouraging businesses to track and manage economic—not just financial, social, and environmental value added—or destroyed'. Elkington (2018) states 'Success or failure on sustainability goals cannot be measured only in terms of profit and loss. It must also be measured in terms of the well-being of billions of people and the health of our planet, and the sustainability sector's record in moving the needle on those goals has been decidedly mixed. While there have been successes, our climate, water resources, oceans, forests, soils, and biodiversity are all increasingly threatened. It is time to either step up—or to get out of the way'. Business organizations can use these three pillars to conceptualize their environmental responsibility, thus determine any negative or positive social impacts of their business activities and devise measures to reduce potentially negative activities, whilst increasing activities in areas in which they make positive impacts on the supply chain, exchanges with business partners, and use of renewable energy to positively impact society and the environment in addition to their shareholder wealth maximization agenda. Adopting a sustainable approach to business operations brings potential benefits which include:

A. Driving Internal Innovation
B. Improving Environmental and Supply Risk
C. Attracting and Retaining Employees
D. Building Customer and Brand Loyalty
E. Reducing Production Costs
F. Enhancing Competitive Advantage
G. Setting Industry Trends and Standards

The United Nations Sustainable Development Goals (SDGs) are based on the three sustainability pillars. The United Nations proposes 17 Sustainable Development Goals (SDGs). According to the United Nations (2015), 'the Sustainable Development Goals will stimulate action over the next 15 years in areas of critical importance for humanity and the planet'. The SDGs seek to align various environmental, social, and economic initiatives in putting forward a comprehensive framework. The 17 SDGs include:

- Goal 1. End poverty in all its forms everywhere.
- Goal 2. End hunger, achieve food security and improved nutrition, and promote sustainable agriculture.
- Goal 3. Ensure healthy lives and promote well-being for all at all ages.
- Goal 4. Ensure inclusive and equitable quality education and promote lifelong learning opportunities for all.
- Goal 5. Achieve gender equality and empower all women and girls.
- Goal 6. Ensure availability and sustainable management of water and sanitation for all.
- Goal 7 Ensure access to affordable, reliable, sustainable, and modern energy for all.
- Goal 8. Promote sustained, inclusive, and sustainable economic growth, full and productive employment, and decent work for all.
- Goal 9. Build resilient infrastructure, promote inclusive and sustainable industrialization, and foster innovation.
- Goal 10. Reduce inequality within and among countries.
- Goal 11. Make cities and human settlements inclusive, safe, resilient, and sustainable.
- Goal 12. Ensure sustainable consumption and production patterns.
- Goal 13. Take urgent action to combat climate change and its impacts.
- Goal 14. Conserve and sustainably use the oceans, seas, and marine resources for sustainable development.
- Goal 15. Protect, restore, and promote sustainable use of terrestrial ecosystems, sustainably manage forests, combat desertification, halt and reverse land degradation, and halt biodiversity loss.

- Goal 16. Promote peaceful and inclusive societies for sustainable development, provide access to justice for all, and build effective, accountable, and inclusive institutions at all levels.
- Goal 17. Strengthen the means of implementation and revitalize the Global Partnership for Sustainable Development.

Figure 2 below highlights the interrelationship between the three sustainability pillars and the UN SDGs. This paper focuses on Goals 3, 4, 8, 11, and 16. These goals seek to ensure healthy lives and promote well-being for all at all ages; inclusive and equitable quality education and promote lifelong learning opportunities for all; promote sustained, inclusive, and sustainable economic growth, full and productive employment and decent work for all and make cities and human settlements inclusive, safe, resilient, and sustainable. Finally, promoting peaceful and inclusive societies for sustainable development, providing access to justice for all, and building effective, accountable, and inclusive institutions at all levels.

Bowen in 1953 inquired on what responsibilities businessmen within societies may reasonably be expected to assume (Bowen, 1953). Bowen's query is now part of the global Corporate Social Responsibility (CSR) agenda and has formed an essential aspect of how CSR is defined. 'Corporate Social Responsibility' is a management concept whereby companies integrate social and environmental concerns in their business operations and interactions with their stakeholders. CSR is generally understood as being the way through which a company achieves a balance of economic, environmental, and social imperatives, while at the same time addressing the expectations of shareholders and stakeholders (UNIDO, 2023). An early definition of CSR is that proposed by Davis in 1975. Davis defines CSR as 'seriously considering the impact of the company's actions on society' (Davis, 1975). 'Social responsibility is the obligation of decision makers to take actions which protect and improve the welfare of society along with their own interests' (Davis, 1975). Carroll (1983) defines, 'Corporate social responsibility involves the conduct of a business so that it is economically profitable, law-abiding, ethical and socially supportive. To be socially responsible then means that profitability and obedience to the law are

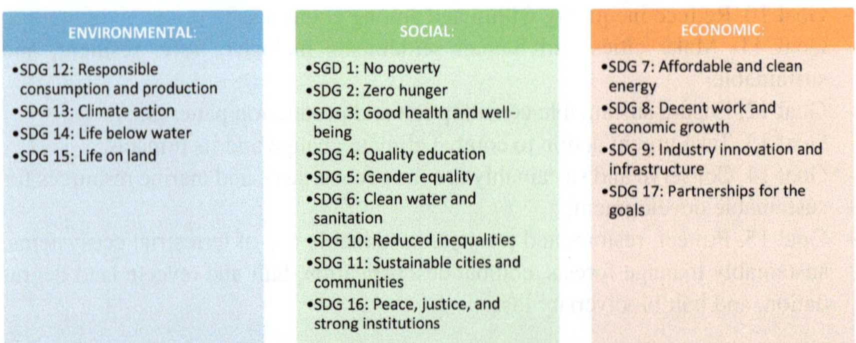

Fig. 2 Interrelationship between the three sustainability pillars and the UN SDGs

foremost conditions when discussing the firm's ethics and the extent to which it supports the society in which it exists with contributions of money, time, and talent' (p. 608). Carroll notes that 'Corporate social responsibility encompasses the economic, legal, ethical, and discretionary (philanthropic) expectations that society has of organizations at a given point in time' (Carroll, 1979, 1991). Hah and Freeman (2014, p. 128) define CSR 'as instances where a firm goes beyond the firm's interests and legal compliance to engage in activities that can advance social good'. Moon (2002) states that CSR 'is only one of several terms in currency designed to capture the practices and norms of new business-society relations. There are contending names, concepts or appellations for corporate social responsibility' (p. 3). According to Visser (2006), 'Carroll's CSR Pyramid is probably the most well-known model of CSR'. Carroll proposes a pyramid that sets out four different obligations or responsibilities that society expects of businesses; these are economic, legal, ethical, and philanthropic responsibilities (see Fig. 3 below).

According to Carroll (1979), the four responsibilities 'are simply to remind us that motives or actions can be categorised as primarily one or another of these four kinds' (p. 500). The hierarchical categorization of these responsibilities 'suggests what might be termed their fundamental role in the evolution of importance…the history of business suggests an early emphasis on the economic and then legal aspects and a later concern for the ethical and discretionary aspects' (p. 500). Carroll (1991) states that the 'responsibilities have always existed to some extent, but it has only been in recent years that ethical and philanthropic functions have taken a significant place' (p. 40). However, 'the more economically oriented a firm is, the less emphasis it places on ethical legal, and discretionary issues' (461). Carroll argues that the responsibilities are not mutually exclusive, but rather 'helps the manager to see that the different types of obligations are in a constant tension with one another' (p. 42). He however stipulates dependence as a rationale for the four responsibilities—beginning with the basic building block notion that economic performance undergirds all else (p. 42).

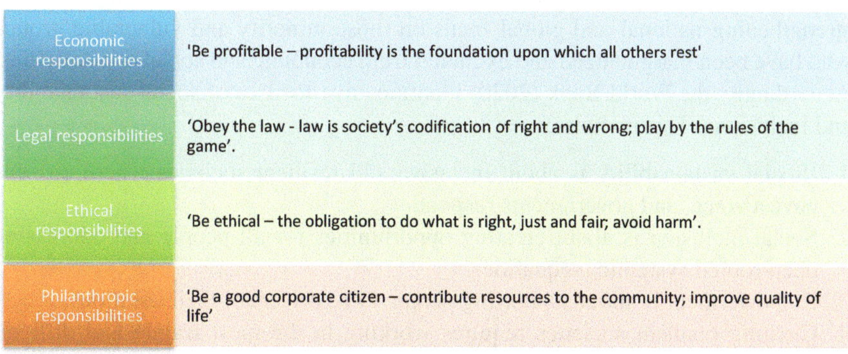

Fig. 3 Carroll's corporate social responsibility pyramid

3 International Businesses' Social Responsibility and Sustainability

Sustainability is now couched to encompass environmental and social elements. There is also a drive to achieve an alignment between financial, social, and environmental elements. Of particular importance is the importance of inclusivity in the social development agenda both on a national and international level. For example, the UN Sustainable Development Goals (SDGs) identify the importance of any sustainability framework incorporating an inclusive approach. The twin concepts of sustainability and inclusivity are interconnected. Social sustainability frameworks require an inclusive approach with the development process focusing on the need to ensure people are at the forefront of any agenda. Various sustainability frameworks in proffering strategies to build cohesive and resilient societies recognize that effective social sustainability expands opportunities for the present generation as well as future generations. Thus, promotes accountability and ensures that institutions are accessible embrace social cohesion and an inclusive approach that seeks to ensure the enhancement and empowerment of vulnerable groups within society as well as those from poor and deprived communities, thereby critical for poverty alleviation and shared prosperity and fostering more resilient and peaceful communities. According to the World Bank (2023), 'global development is at a crossroads. Despite positive efforts made in recent times, there are a growing number of barriers and challenges which threaten to reverse development gains thereby jeopardizing the prospects for continued progress. A large segment of society (For example, women, ethnic minorities, people with disability, the very young and the old) are at risk of exclusion from services and participation in cultural and political spaces. The World Bank (2023) identified the following challenges:

- Rising inequality
- Erosion of trust within societies to climate change
- Pandemics and increasing conflict across the globe.

According to the World Bank (2020), social sustainability and inclusion reflect the importance assigned to addressing inherent challenges to the development and strengthening national and global focus on those minority and vulnerable groups who have been marginalized and excluded from economic and social opportunities. Accordingly, the World Bank (2020) identifies five focuses of social sustainability and inclusions. This includes:

1. 'Social sustainability is about inclusive and resilient societies where citizens have a voice, and governments respond.
2. Social inclusion is about creating opportunities for all people and addressing deep-rooted systemic inequalities.
3. Empowerment is about supporting people to be drivers of their solutions.
4. Creating resilient societies requires working in the most fragile and difficult environments.
5. The Environmental and Social Framework is an integral part of social sustainability' (World Bank, 2020).

From the aforementioned, it is clear that international businesses are proficiently placed to improve local conditions, for example, 'employment and community programs and the transfer of cutting-edge technologies and organizational best practice that might otherwise be unavailable' (e.g. Málovics et al., 2008; Burritt et al., 2020). In addition, activities of international businesses in the communities and societies in which they operate should be tailored to encourage people within the communities to work together to overcome societal challenges, ensure scares resources are allocated in a way that is perceived by all stakeholders to be legitimate and equitable, so that citizens can thrive and prosper. There are also benefits to be gained by international businesses which include cost-saving benefits, competitive advantage, employee satisfaction, and morale, product/service or market innovation, business model and process innovation, effective risk management, and enhanced stakeholder relations (Bern et al., 2009). Most international businesses including those involved in the 'creative arts' embrace CSR and sustainability initiatives due to moral convictions, which can result in several benefits and important social change.

International Art and Culture centres are often identified as exclusive based on the necessity for a level of knowledge and appreciation deemed associated with internationally recognized art (Abbing, 2022). Various academics identify the important role that culture and art play in urbanization strategies. There are however doubts as to the effectiveness of such strategies, and their social, cultural, and sustainable impact (Pratt, 2011; Mathews, 2014; Ward, 2018). This study, therefore, seeks to explore the impact of developing iconic cultural centres in deprived areas on the socio-economic development of the location and the corporate social responsibility of the cultural centres to the stakeholder groups specifically those of the lower socio-economic group that can encounter cultural and economic disruption using the Turner Contemporary as a case study. The Turner Contemporary is an art gallery in Margate, Kent, England, built using public funds. Turner Contemporary is a leading UK contemporary art gallery established in 2011 to regenerate the town of Margate. The art gallery was founded to celebrate art, culture, and artistic talent. With world-class exhibitions on display, the gallery is a centre of excellence that seeks to promote an inclusive and sustainable approach to 'creative learning among children, young people, and families' whilst also establishing the Gallery as an Internationally acclaimed centre for art (Ward, 2018). In addition, Turner Contemporary has made significant contributions to the economic growth of the local economy—Since its opening, the art gallery has generated over £80 million for the Kent economy and has engaged in diverse activities and world-class 'pioneering learning programmes that champion children's leadership through the arts'. The initiatives adopted by the gallery foster interconnectedness with 'thousands of people from the local community, transforming lives and driving real positive impact through the arts'. Of particular importance is its drive for 'regeneration' and 'work with children and young people'. Turner Contemporary prides itself 'as the catalyst for the regeneration of Margate and East Kent' (Turner Contemporary, 2024). Ward (2018) asserts that the Margate culture-led regeneration policy and development of the flagship Turner Contemporary gallery can be better understood in relation to those contemporary discourses around culture, the creative industries, the knowledge economy, and creative cities. The Turner

Contemporary Art Gallery 'is a driver for changing perceptions and experiences of place that utilizes, and trades upon distinctive local characteristics' (Ward, 2018). While Turner Contemporary epitomizes cultural urban policy intervention, such cultural developments have been criticized for their tendency to deliver short-term benefits and thus fail to create long-term value for the local communities and the economy in which they operate (Comunian & Mould, 2014). These cultural centres are devised as part of a strategic drive geared towards improving the image of the city relying on the configuration of symbols, signs, and available local products to produce a specific representation of space. According to Lefebvre (1991), this strategy is not without its complexities, the development of an 'urban product in most cases results in the concealment of labour, the domination and exploitation on which it is founded'.

The economic development of a location through the attraction of people with higher expendable income who can contribute to the location's economic development has been recognized for its socio-economic benefits for decades, on the premise highlighted by Florida (2003) that attracting middle-class skilled people to a location will drive the improvement of the location socially and economically. The attraction of the 'Creative Class' results in rising house prices and increased facilities such as cafes and restaurants, consequently there is increasing employment, creating a virtuous cycle. This was the premise on which much public investment has been made, including the development of Turner Contemporary and other Iconic Cultural developments that attract skilled people to the location. The negative impacts on some groups; for example, the people renting in a location that is the centre of gentrification experience loss of housing and rent increases (Schuerman, 2019.) Gentrification impacts communities and changes a location's landscape with the potential to exacerbate inequalities, e.g. racial and economic—it significantly impacts the demographics resulting in displacement, cultural changes, and economic transformation. The losers and problems with gentrification have been acknowledged and more recently evaluated (Lee & Newman, 2021), the complications of driving economic development with all the visible advantages that this brings whilst avoiding the negative impact of gentrification are complicated and often ignored (Michener & Wong, 2018).

4 Methodology

Data for the study was collected using a questionnaire survey in Margate. Questionnaire surveys were carried out with a wide cross-section of people who have a strong connection to Thanet either through work and residence or those who frequently visit. The questionnaires were completed in the Canteen of the Queen Elizabeth the Queen Mother's Hospital Margate. This allowed a cross-section of patients, staff, visitors, and carers to be recruited as participants on a convenience sampling basis. The questionnaire survey included a mix of closed and open-ended questions as well as a five-point Likert scale question format. The sampling

Table 1 Reliability statistics

Case processing summary			
		N	%
Cases	Valid	221	100.0
	Excluded[a]	0	0.0
	Total	221	100.0

Reliability statistics	
Cronbach's alpha	No. of items
0.943	43

[a]Listwise deletion based on all variables in the procedure

strategy adopted involved the use of convenient and purposive sampling to identify the study participants. A total of 221 participants took part in the study. The Cronbach's Alpha test was used to statistically test the internal consistency of various parts of the questionnaire (Bryman & Bell, 2018). The Cronbach Alpha coefficient in the questionnaire is 0.94, and as the internal reliability value is above 0.80 this is considered to be a satisfactory level of internal consistency (see Table 1 below).

The data from the questionnaire was analysed using the Statistical Package for Social Sciences (SPSS). Descriptive and analytical statistical methods were carried out in analysing the results obtained from the questionnaire. Where relevant, verbatim quotes from the open-ended responses extracted from the questionnaire survey were used in supporting the discussion of findings thus aiding a better understanding of the developing themes. Descriptive statistics were based mainly on frequency, mean values, and standard deviation to examine the differences and similarities between the responses provided. In addition, a non-parametric statistical analysis technique was used. The non-parametric analysis makes no assumptions vis-à-vis the distribution of scores in the population. Non-parametric measures of association were carried out using cross-tabulations to evaluate the responses obtained from the participants to assess the significant strengths of the relationship between the variables. Responses obtained on social development goals (SDG) were cross-tabulated with the income of the participants to identify if a significant association exists. Statistical tests were also employed to test the reliability of the results, confirm the consistency of answers, and identify associations. The Chi-Square test of independence was carried out to identify significant associations between the two independent variables of income and SDG factors. The Chi-Square Test of Independence hypotheses is as follows:

1. H0: The two variables are independent.
2. H1: The two variables are not independent—they are associated.

The study's null hypothesis is that income and perception of SDG factors are associated (H1). If the *p*-value (0.001) of the Chi-Square test is less than 0.05, the null hypothesis is accepted.

5 Analysis and Discussion of Findings

The analysis of findings is split into two main parts. The first part provides descriptive statistics in the form of percentages on demographic information of the participants who took part in the study. The second section involves the use of cross-tabulations, mean value, and Chi-Square in presenting the results vis a vis socio-economic development and sustainability. The result obtained on the demographic information of the respondents shows varied characteristics of the participants. As shown in Table 2 below, the majority of the participants stated that they are female representing 55.7%, and 44.3% are male. The average participant is between 36–49 and 50–64 years old. A high percentage stated that they were either single (23.1%) or empty nesters (24.9%). The finding also shows a high proportion of participants who stated that they were part of a 'couple' with older children living at home (representing 21.3%). Concerning income, the majority of the participants representing 36.7% earn less than £20,000, followed by those who stated that they earn £20,000–£40,000 representing 35.3%. 22.2% of the participants in the sample stated that they earn between £41,000 and £100,000. 5.9% of the participants stated that they earn over £100,000. The findings show varied job functions with the majority of the participants stating that they were unskilled (39.0%). Many of the participants were professionals representing 30.7%. 13.0% stated that they were retired. While 8.6% stated that they were unemployed. The finding also shows varied responses vis-à-vis academic qualifications. The majority stated that they had no qualifications (38.9%). This reflects a significant link between the high level of unemployed and unskilled people with low levels of qualification and a high percentage of people earn either less than £20,000 or between £20,000 and £40,000 (a high percentage of the population earn lower than the average wage—36.7% earn less than £20,000, followed by those who stated that they earn £20,000–£40,000, representing 35.3%). This is below the average wage of £42,000 per annum in the South East of the UK (Corlett et al., 2022). The findings vis-à-vis the low-skill, high employment, and lower-income categories highlighted in our result find support in empirical research work carried out by Beatty et al. (2008) who suggest that coastal towns share specific characteristics ranging from lower income, high unemployment rate, lower levels of achievement, low skills, and high levels of deprivation. The findings support the notion of 'out-migration' in which the highly skilled and those with high levels of education tend to migrate away from coastal towns due to the high level of unemployment and high-skilled employment. 'In-migration' into such locations is usually high for those who are not economically active; for example, those who have retired or nearing retirement age (Beatty et al., 2014). This is what the council is trying to reverse by developing the Turner Contemporary and attracting skilled and highly paid people to move to Margate. A high majority of the participants either 'live and work in Thanet' (40.6%) or 'live in Thanet' (37.6%). Other participants stated that they 'live and work outside Thanet' (9.5%) or 'work in Thanet' (9.0%). 2.8% stated that they 'study in Thanet' (2.8%) or 'visiting Thanet' (0.5%). When asked how frequently they visited the Turner Contemporary,

Table 2 Demographic information of the participants

Entrepreneur characteristics	Variables	Percentages
Gender	Male	44.3
	Female	55.7
Age	18–25	14.5
	26–35	17.6
	36–49	25.3
	50–64	25.3
	65–75	12.2
	75+	5.0
Family status	Single	23.1
	Couple no children	15.8
	Couple young children	11.8
	Older children living at home	21.3
	Empty nesters	24.9
	Single children	2.7
	Couple under 6 and older children	0.5
Job function	Professional	30.7
	Business owners	5.9
	Unskilled	39.0
	Unemployed	8.6
	Retired	13.0
	Student	2.8
Academic qualification	Bachelors	15.8
	Masters/postgraduate/doctorate	13.1
	Professional qualifications	16.3
	HND	15.4
	Secondary	0.5
	No qualification	38.9
Income	Less than £20,000	36.7
	£20,000–£40,000	35.3
	£41,000–£100,000	22.2
	Over £100,000	5.9
Location	I live in Thanet	37.6
	I work in Thanet	9.0
	I live and work in Thanet	40.6
	I am visiting Thanet	0.5
	I study in Thanet	2.8
	I live and work outside Thanet	9.5
Visit to Turner Contemporary	Never	39.4
	Once	17.6
	A few times	35.7
	Frequently	5.9
	I am a patron	1.4

the majority stated that they had never visited (representing 39.4%), 35.7% stated that they had visited a few times, and 17.6% stated that they had visited once. 5.9% stated that they visited frequently and 1.4 stated that they were patrons.

6 Perceptions of the participants on the Impact of the Turner Contemporary

The responses obtained from the open-ended questions show mostly positive responses:

'When they have things going on it draws people to the area...kids' workshop and the art community it has brought to Margate' (Male Participant).

'Artwork and community use, the location, the diversity and adventurous nature of the exhibits' (Male Participant).

'The number of people it has brought and the positive effect it has had on the marketplace' (Male Participant).

'It probably brings wealth, status, and prosperity to Margate. A very colourful and always interesting tourist attraction. The opportunity to view all concepts of art, modern and classic. Ever-changing exhibitions keep it fresh and encourage more visitors' (Female Participant).

'Interesting and educational. The exhibition brings people to the area who are interested—It is free. Attracts people from other places to visit. Constantly being in the national and local press bringing more tourism into the area' (Male Participant).

'It is an important cultural centre it helps to regenerate the local economy and puts Thanet on the visitor map' (Male Participant).

'Great to introduce children to art. Drawing in outside visitors. A great place for days out and art. An eclectic collection, pleasant environment—no pressures to follow a set route' (Female Participant).

'Rejuvenation to the area. A sightseeing opportunity to draw people to Margate. Also promotes the cultural and artistic atmosphere of Margate' (Male Participant).

'Wide range of interests. Bringing culture and arts to the young and impressionable who might not have a chance to benefit otherwise' (Male Participant).

'It has put Margate on the map and help employment. It fills an empty space on the seafront. Greater awareness of the area, and access to exhibits for local people' (Male Participant).

'Educational...Contemporary art...great quality of exhibitions, frequency of exhibition, seeing artists at work, encouragement of local artists/children, an excellent venue for the exhibition' (Male Participant).

'It is good to see art without having to travel too far. Bringing trade and culture to Thanet' (Male Participant).

'The art—encouraging children to take part in arts and crafts. Brings an art culture to the local community. I am a primary school teacher so I feel I can help children to experience art and think outside the box with regard to themes and projects' (Female Participant).

'See art that you would otherwise have to go to London to see—it is more accessible. It has benefitted Margate but mainly that people are from London and abroad' (Female Participant).

The verbatim excerpts above from the questionnaire open-ended responses support arguments posited by Ward (2018) wherein he asserts that 'regeneration policy has effectively utilised local characteristics to recreate Margate as an artful space and has stimulated a local milieu of artistic and cultural activity'.

7 Perception of the Participants on Sustainability and Social Responsibility

This section aims to evaluate the perception of the participants on the activities of Turner Contemporary vis-à-vis against the FIVE SDG goals below:

- Goal 3: Ensure healthy lives and promote well-being for all at all ages.
- Goal 4: Ensure inclusive and equitable quality education and promote lifelong learning opportunities for all.
- Goal 8: Sustained, inclusive, and sustainable economic growth, full and productive employment, and decent work for all.
- Goal 11: Make cities and human settlements inclusive, safe, resilient, and sustainable.
- Goal 16. Promote peaceful and inclusive societies for sustainable development, provide access to justice for all, and build effective, accountable, and inclusive institutions at all levels.

Table 3 illustrates how the FIVE SDG goals were categorized in the questionnaire.

The findings presented in Table 4 below show mixed reactions vis-à-vis the impact of the Turner Contemporary on the SDG factors. A high Mean value can be seen in the following statements 'improved image of the area' with a high mean value of 4.0226; 'increased tourism' with a mean value of 3.9276; 'developed a creative feeling' with a mean value of 3.8371; followed by 'more investment in the area and increase in property values in the area' with a mean value of 3.8054 and 3.8597, respectively. The following statements were also seen to have a high mean value 'more interests in art-based education' (mean value of 3.7964); 'urban renewal' (mean value of 3.7873); 'greater sense of pride' (mean value of 3.7602); enhanced opportunity for enjoyment (mean value 3.7511) and contributed to community social value (mean value 3.7421). The findings vis-à-vis the effectiveness of the Turner Contemporary in improving the image of the area, attracting visitors and

Table 3 Social development goals categories proposed in the questionnaire

Economic growth, entrepreneurship, and employment SDG 8	Inclusive, safe, resilient, and sustainable city SDG 11	Education and skills development SDG 4	Health and well-being SDG 3	Inclusive society SDG 16
Economic improvement	Urban renewal	More interest in art-based education	Greater health benefits	Contributed to community social value
Development of local businesses	Increased facilities	Better opportunities for young people	Positive impact on people's health	More social inclusion
Employment opportunities	Improved image of the area	Development of new skills in people	People are less anxious	Enhanced social contact and interaction
Increased entrepreneurial start-up.	Positive image of the area in the media	Developed a creative feeling	Improved mental health	Social issues which improve life
Increase in property values in the area	Improved safety	More interests in art-based education	Improved motivation and energy	
People feel more affluent	More positive attitude		Positive mood	
More investment in the area	Greater sense of pride		Added well-being.	
Increased prosperity	More confidence			
Attraction and retention of skilled people	Enhanced opportunity for enjoyment			
	Increased tourism			
	Nicer place to live			

Table 4 Frequency and mean value of socio-economic development and sustainability

SDG factors	Strongly agree	Agree	Don't know	Disagree	Strongly disagree	Mean value
Economic improvement	28.1	16.7	27.1	23.5	4.5	3.4027
Increased tourism	31.7	42.5	17.6	3.2	5.0	3.9276
Development of local businesses	28.5	24.4	18.6	23.5	5.0	3.4796
Employment opportunities	24.0	20.8	23.5	25.8	5.9	3.3122
Increased prosperity	24.4	22.6	21.7	25.8	5.4	3.3484
Nicer place to live	24.9	24.0	18.6	27.1	5.4	3.3575
Urban renewal	26.7	40.7	21.3	7.2	4.1	3.7873
Increased facilities	22.6	28.1	19.5	23.1	6.8	3.3665
Improved image of the area	43.0	32.6	12.7	7.2	4.5	4.0226

Table 4 Continued

SDG factors	Strongly agree	Agree	Don't know	Disagree	Strongly disagree	Mean value
Positive image of the area in the media	26.2	45.2	15.4	9.0	4.1	3.8054
Improved safety	23.5	12.7	27.6	29.9	6.3	3.1719
Development of new skills in people	25.3	16.7	27.1	25.3	5.4	3.3122
Developed a creative feeling	27.6	45.2	15.8	5.9	5.4	3.8371
Increased entrepreneurial start-up	23.5	16.7	29.9	24.9	5.0	3.2896
More positive attitude	26.7	21.3	24.0	23.1	5.0	3.4163
Greater sense of pride	25.3	43.9	17.2	8.6	5.0	3.7602
More confidence	21.7	18.1	28.1	28.1	4.1	3.2534
Greater health benefits	21.3	14.9	30.3	26.7	6.8	3.1719
Positive impact on People's health	21.7	13.6	30.3	27.6	6.8	3.1584
People are less anxious	22.2	8.6	30.3	32.1	6.8	3.0724
Improved mental health	21.7	10.4	30.8	29.4	7.7	3.0905
Improved motivation and energy	23.1	36.7	24.0	10.9	5.4	3.6109
Positive mood	20.8	42.1	22.6	9.5	5.0	3.6425
More interests in art-based education	25.3	43.9	19.9	6.8	4.1	3.7964
Added well-being	24.0	19.9	21.3	30.3	4.5	3.2851
Enhanced opportunity for enjoyment	20.8	50.2	17.2	6.8	5.0	3.7511
Contributed to community social value	23.5	41.2	25.3	5.9	4.1	3.7421
More social inclusion	20.4	23.5	27.1	24.0	5.0	3.3032
Enhanced social contact and interaction	18.1	31.2	23.1	23.5	4.1	3.3575
Social issues which improve life	20.8	20.4	30.8	24.0	4.1	3.2986
Better opportunities for young people	23.5	32.6	28.1	10.0	5.9	3.5792
Increase in property values in the area	29.4	41.2	19.9	5.0	4.5	3.8597
People feel more affluent	22.6	14.9	28.5	28.1	5.9	3.2036
Attraction and retention of skilled people	25.3	37.1	23.1	10.0	4.5	3.6878
More investment in the area	25.8	41.6	24.0	4.5	4.1	3.8054

national media attention to Margate find support in extant academic literature (see, for example, Phillips, 2011; Emms, 2011; Graham-Dixon, 2012). The findings reflect an improved image of Margate from the previous image of the fish and chip and Stag party location. Turner Contemporary with the Turner Awards located there in 2020 has contributed to the improvement of the image of Margate as a cultural town with a centre of artistic and cultural significance. Tracy Emin and other iconic established Artists have moved their studios to Margate since the Turner Contemporary has been established, further adding to the positive image of Margate, this has had a positive impact on tourism with visitors being attracted to the Turner Contemporary and then using the other developing facilities. The positive image has had a positive effect on the development of the creative feel of the town, increased the art-based education with increased interest in art courses in the college in Margate. It has also resulted in a surge of lifelong learning classes. This positive cycle has encouraged people visiting Margate to invest in the location which is less than 2 h by train from London resulting in increasing property prices and increased overall investment in Margate which is seen by many people as a growth area.

The SDG factors were cross-tabulated with the income of the participants to identify if any association exists between income and social development factors. Income is adopted as the basis for evaluating the present socio-economic situation of the population in Margate and the effect that Turner Contemporary has on the different socio-economic brackets of the population. This is because income reflects on the positive and negative experience of gentrification of the location. The higher socio-economic bracket benefits positively from gentrification due to increased property value and cultural and social opportunities. While those in lower socio-economic brackets are negatively affected by increased rental prices, costs of local facilities as well as changes in their opportunities for socializing with the closing of traditional pubs and cafes. The Chi-square non-parametric statistical test was carried out to ascertain the strength of association between the dependent variable 'income' and the independent variable 'SDG factors'. The result obtained from the analysis is presented in Table 5 below. The frequency distribution shows that majority of those in the low-income category (<£20,000) 'disagree' with the following: economic improvement (48.1%); development of local businesses (45.5%), employment opportunities (49.4%), increased prosperity (50.6%), nicer place to live (50.6%), increased facilities (44.4%), improved safety (55.6%), development of new skills (48.1%), increased entrepreneurial start-up (49.4%), more positive attitude (44.4%), more confidence (51.9%), greater health benefits (53.1%), positive impact on people's health (53.1%), people are less anxious (60.5%), improved mental health (55.6%), added well-being (58.0%), more social inclusion (48.1%), enhanced social inclusion (49.4%), social issues which improve life (48.1%), and people feel more affluent (53.1%). Some of the dissatisfaction from the participants could stem from the fact that policy interventions introduced in Margate in recent times are more focused on attracting tourism, customers, and investors, failing to support social relations and local labour thereby marginalizing the inhabitants of the local community, resulting in increased cost of living. The findings to an extent align with the research work carried out by Ward (2018). Ward (2018) notes that

Table 5 Frequency distributions and Chi-square non-parametric statistical test

						Chi-Square		
SDG factors and income	Strongly disagree	Disagree	Neutral	Agree	Strongly agree	Value	*df*	Asymptotic significance (2-sided)
Economic improvement								
<£20,000	3.7%	48.1%	32.1%	4.9%	11.1%	77.878	12	<0.001
£20,000–£40,000	5.1%	9.0%	34.6%	24.4%	26.9%			
£41,000–£100,000	2.0%	8.2%	12.2%	22.4%	55.1%			
Over £100,000	15.4%	15.4%	7.7%	23.1%	38.5%			
Increased tourism								
<£20,000	4.9%	2.5%	25.9%	46.9%	19.8%	25.777	12	0.012
£20,000–£40,000	6.4%	1.3%	17.9%	42.3%	32.1%			
£41,000–£100,000	0.0%	6.1%	8.2%	38.8%	46.9%			
Over £100,000	15.4%	7.7%	0.0%	30.8%	46.2%			
Development of local businesses								
<£20,000	6.2%	45.7%	24.7%	9.9%	13.6%	63.615	12	<0.001
£20,000–£40,000	5.1%	11.5%	20.5%	33.3%	29.5%			
£41,000–£100,000	0.0%	8.2%	8.2%	34.7%	49.0%			
Over £100,000	15.4%	15.4%	7.7%	23.1%	38.5%			
Employment opportunities								
<£20,000	6.2%	49.4%	29.6%	6.2%	8.6%	72.185	12	<0.001
£20,000–£40,000	6.4%	16.7%	26.9%	26.9%	23.1%			
£41,000–£100,000	2.0%	6.1%	10.2%	34.7%	46.9%			
Over £100,000	15.4%	7.7%	15.4%	23.1%	38.5%			
Increased prosperity								
<£20,000	6.2%	50.6%	24.7%	13.6%	4.9%	78.071	12	<0.001
£20,000–£40,000	6.4%	9.0%	26.9%	32.1%	25.6%			
£41,000–£100,000	0.0%	10.2%	12.2%	24.5%	53.1%			
Over £100,000	15.4%	30.8%	7.7%	15.4%	30.8%			

(continued)

Table 5 (continued)

SDG factors and income	Strongly disagree	Disagree	Neutral	Agree	Strongly agree	Chi-Square Value	*df*	Asymptotic significance (2-sided)
Nicer place to live								
<£20,000	4.9%	50.6%	27.2%	9.9%	7.4%	69.066	12	<0.001
£20,000–£40,000	6.4%	16.7%	17.9%	32.1%	26.9%			
£41,000–£100,000	2.0%	8.2%	8.2%	34.7%	46.9%			
Over £100,000	15.4%	15.4%	7.7%	23.1%	38.5%			
Urban renewal								
<£20,000	3.7%	7.4%	24.7%	54.3%	9.9%	47.733	12	<0.001
£20,000–£40,000	5.1%	5.1%	30.8%	33.3%	25.6%			
£41,000–£100,000	0.0%	8.2%	6.1%	32.7%	53.1%			
Over £100,000	15.4%	15.4%	0.0%	30.8%	38.5%			
Increased facilities								
<£20,000	8.6%	44.4%	27.2%	14.8%	4.9%	74.291	12	<0.001
£20,000–£40,000	6.4%	12.8%	23.1%	34.6%	23.1%			
£41,000–£100,000	2.0%	6.1%	4.1%	38.8%	49.0%			
Over £100,000	15.4%	15.4%	7.7%	30.8%	30.8%			
Improved image of the area								
<£20,000	3.7%	6.2%	19.8%	18.5%	51.9%	30.336	12	0.002
£20,000–£40,000	5.1%	7.7%	14.1%	38.5%	34.6%			
£41,000–£100,000	2.0%	6.1%	0.0%	51.0%	40.8%			
Over £100,000	15.4%	15.4%	7.7%	15.4%	46.2%			
Positive image of the area in the media								
<£20,000	4.9%	7.4%	19.8%	59.3%	8.6%	35.757	12	0.001
£20,000–£40,000	3.8%	10.3%	17.9%	37.2%	30.8%			
£41,000–£100,000	0.0%	8.2%	6.1%	38.8%	46.9%			
Over £100,000	15.4%	15.4%	7.7%	30.8%	30.8%			

Table 5 (continued)

SDG factors and income	Strongly disagree	Disagree	Neutral	Agree	Strongly agree	Chi-Square Value	*df*	Asymptotic significance (2-sided)
Improved safety								
<£20,000	6.2%	55.6%	28.4%	6.2%	3.7%	81.775	12	<0.001
£20,000–£40,000	7.7%	19.2%	37.2%	14.1%	21.8%			
£41,000–£100,000	2.0%	6.1%	16.3%	22.4%	53.1%			
Over £100,000	15.4%	23.1%	7.7%	7.7%	46.2%			
Development of new skills in people								
<£20,000	7.4%	48.1%	32.1%	8.6%	3.7%	89.431	12	<0.001
£20,000–£40,000	3.8%	12.8%	35.9%	21.8%	25.6%			
£41,000–£100,000	2.0%	6.1%	12.2%	18.4%	61.2%			
Over £100,000	15.4%	30.8%	0.0%	30.8%	23.1%			
Developed a creative feeling								
<£20,000	7.4%	4.9%	21.0%	55.6%	11.1%	42.455	12	<0.001
£20,000–£40,000	5.1%	3.8%	21.8%	41.0%	28.2%			
£41,000–£100,000	0.0%	8.2%	2.0%	36.7%	53.1%			
Over £100,000	15.4%	15.4%	0.0%	38.5%	30.8%			
Increased entrepreneurial start-up								
<£20,000	6.2%	49.4%	28.4%	12.3%	3.7%	80.545	12	<0.001
£20,000–£40,000	3.8%	12.8%	43.6%	16.7%	23.1%			
£41,000–£100,000	2.0%	6.1%	16.3%	24.5%	51.0%			
Over £100,000	15.4%	15.4%	7.7%	15.4%	46.2%			
More positive attitude								
<£20,000	6.2%	44.4%	33.3%	8.6%	7.4%	80.655	12	<0.001
£20,000–£40,000	3.8%	12.8%	26.9%	28.2%	28.2%			
£41,000–£100,000	2.0%	6.1%	10.2%	24.5%	57.1%			
Over £100,000	15.4%	15.4%	0.0%	46.2%	23.1%			

(continued)

Table 5 (continued)

SDG factors and income	Strongly disagree	Disagree	Neutral	Agree	Strongly agree	Chi-Square Value	*df*	Asymptotic significance (2-sided)
Greater sense of pride								
<£20,000	6.2%	7.4%	25.9%	53.1%	7.4%	45.118	12	<0.001
£20,000–£40,000	3.8%	10.3%	19.2%	38.5%	28.2%			
£41,000–£100,000	2.0%	6.1%	2.0%	36.7%	53.1%			
Over £100,000	15.4%	15.4%	7.7%	46.2%	15.4%			
More confidence								
<£20,000	4.9%	51.9%	33.3%	7.4%	2.5%	81.738	12	<0.001
£20,000–£40,000	2.6%	19.2%	33.3%	24.4%	20.5%			
£41,000–£100,000	2.0%	6.1%	16.3%	24.5%	51.0%			
Over £100,000	15.4%	15.4%	7.7%	23.1%	38.5%			
Greater health benefits								
<£20,000	6.2%	53.1%	33.3%	4.9%	2.5%	86.969	12	<0.001
£20,000–£40,000	9.0%	15.4%	34.6%	20.5%	20.5%			
£41,000–£100,000	2.0%	6.1%	20.4%	18.4%	53.1%			
Over £100,000	15.4%	7.7%	23.1%	30.8%	23.1%			
Positive impact on people's health								
<£20,000	7.4%	53.1%	34.6%	2.5%	2.5%	82.888	12	<0.001
£20,000–£40,000	7.7%	17.9%	33.3%	19.2%	21.8%			
£41,000–£100,000	2.0%	6.1%	20.4%	20.4%	51.0%			
Over £100,000	15.4%	7.7%	23.1%	23.1%	30.8%			
People are less anxious								
<£20,000	7.4%	60.5%	29.6%	1.2%	1.2%	94.062	12	<0.001
£20,000–£40,000	7.7%	21.8%	38.5%	11.5%	20.5%			
£41,000–£100,000	2.0%	6.1%	20.4%	18.4%	53.1%			
Over £100,000	15.4%	15.4%	23.1%	0.0%	46.2%			

Table 5 (continued)

SDG factors and income	Strongly disagree	Disagree	Neutral	Agree	Strongly agree	Chi-Square Value	*df*	Asymptotic significance (2-sided)
Improved mental health								
<£20,000	8.6%	55.6%	30.9%	2.5%	2.5%	86.355	12	<0.001
£20,000–£40,000	9.0%	19.2%	39.7%	11.5%	20.5%			
£41,000–£100,000	2.0%	6.1%	18.4%	22.4%	51.0%			
Over £100,000	15.4%	15.4%	23.1%	7.7%	38.5%			
Improved motivation and energy								
<£20,000	6.2%	6.2%	25.9%	58.0%	3.7%	60.517	12	<0.001
£20,000–£40,000	5.1%	17.9%	29.5%	21.8%	25.6%			
£41,000–£100,000	2.0%	6.1%	16.3%	24.5%	51.0%			
Over £100,000	15.4%	15.4%	7.7%	38.5%	23.1%			
Positive mood								
<£20,000	6.2%	8.6%	25.9%	55.6%	3.7%	41.621	12	<0.001
£20,000–£40,000	2.6%	12.8%	28.2%	33.3%	23.1%			
£41,000–£100,000	4.1%	6.1%	10.2%	34.7%	44.9%			
Over £100,000	15.4%	7.7%	15.4%	38.5%	23.1%			
More interests in art-based education								
<£20,000	4.9%	6.2%	19.8%	65.4%	3.7%	56.439	12	<0.001
£20,000–£40,000	3.8%	5.1%	26.9%	34.6%	29.5%			
£41,000–£100,000	0.0%	8.2%	10.2%	32.7%	49.0%			
Over £100,000	15.4%	15.4%	15.4%	7.7%	46.2%			
Added well-being								
<£20,000	6.2%	58.0%	24.7%	6.2%	4.9%	88.840	12	<0.001
£20,000–£40,000	3.8%	17.9%	25.6%	28.2%	24.4%			
£41,000–£100,000	0.0%	8.2%	12.2%	24.5%	55.1%			
Over £100,000	15.4%	15.4%	7.7%	38.5%	23.1%			

(continued)

Table 5 (continued)

SDG factors and income	Strongly disagree	Disagree	Neutral	Agree	Strongly agree	Chi-Square Value	*df*	Asymptotic significance (2-sided)
Enhanced opportunity for enjoyment								
<£20,000	7.4%	3.7%	19.8%	65.4%	3.7%	37.736	12	<0.001
£20,000–£40,000	2.6%	10.3%	19.2%	42.3%	25.6%			
£41,000–£100,000	2.0%	6.1%	12.2%	38.8%	40.8%			
Over £100,000	15.4%	7.7%	7.7%	46.2%	23.1%			
Contributed to community social value								
<£20,000	4.9%	2.5%	30.9%	54.3%	7.4%	40.743	12	<0.001
£20,000–£40,000	2.6%	9.0%	28.2%	35.9%	24.4%			
£41,000–£100,000	2.0%	6.1%	12.2%	30.6%	49.0%			
Over £100,000	15.4%	7.7%	23.1%	30.8%	23.1%			
More social inclusion								
<£20,000	6.2%	48.1%	33.3%	8.6%	3.7%	81.594	12	<0.001
£20,000–£40,000	3.8%	12.8%	30.8%	32.1%	20.5%			
£41,000–£100,000	2.0%	6.1%	12.2%	32.7%	46.9%			
Over £100,000	15.4%	7.7%	23.1%	30.8%	23.1%			
Enhanced social contact and interaction								
<£20,000	4.9%	49.4%	27.2%	16.0%	2.5%	84.284	12	<0.001
£20,000–£40,000	2.6%	10.3%	28.2%	39.7%	19.2%			
£41,000–£100,000	2.0%	6.1%	10.2%	38.8%	42.9%			
Over £100,000	15.4%	7.7%	15.4%	46.2%	15.4%			
Social issues which improve life								
<£20,000	4.9%	48.1%	35.8%	6.2%	4.9%	83.128	12	<0.001
£20,000–£40,000	2.6%	11.5%	38.5%	26.9%	20.5%			
£41,000–£100,000	2.0%	6.1%	14.3%	30.6%	46.9%			
Over £100,000	15.4%	15.4%	15.4%	30.8%	23.1%			

Table 5 (continued)

SDG factors and income	Strongly disagree	Disagree	Neutral	Agree	Strongly agree	Chi-Square Value	*df*	Asymptotic significance (2-sided)
Better opportunities for young people								
<£20,000	7.4%	7.4%	30.9%	49.4%	4.9%	53.158	12	<0.001
£20,000–£40,000	5.1%	14.1%	34.6%	23.1%	23.1%			
£41,000–£100,000	2.0%	6.1%	14.3%	24.5%	53.1%			
Over £100,000	15.4%	15.4%	23.1%	24.5%	53.1%			
Increase in property values in the area								
<£20,000	6.2%	2.5%	28.4%	53.1%	9.9%	49.633	12	<0.001
£20,000–£40,000	3.8%	3.8%	21.8%	35.9%	34.6%			
£41,000–£100,000	0.0%	6.1%	8.2%	36.7%	49.0%			
Over £100,000	15.4%	23.1%	0.0%	15.4%	46.2%			
People feel more affluent								
<£20,000	8.6%	53.1%	29.6%	4.9%	3.7%	81.321	12	<0.001
£20,000–£40,000	3.8%	16.7%	38.5%	17.9%	23.1%			
£41,000–£100,000	2.0%	8.2%	14.3%	24.5%	51.0%			
Over £100,000	15.4%	15.4%	15.4%	23.1%	30.8%			
Attraction and retention of skilled people								
<£20,000	7.4%	7.4%	27.2%	53.1%	4.9%	55.453	12	<0.001
£20,000–£40,000	2.6%	14.1%	26.9%	30.8%	25.6%			
£41,000–£100,000	0.0%	6.1%	12.2%	28.6%	53.1%			
Over £100,000	15.4%	15.4%	15.4%	7.7%	46.2%			
More investment in the area								
<£20,000	6.2%	1.2%	29.6%	55.6%	7.4%	51.688	12	<0.001
£20,000–£40,000	2.6%	6.4%	28.2%	38.5%	24.4%			
£41,000–£100,000	0.0%	6.1%	10.2%	32.7%	51.0%			
Over £100,000	15.4%	7.7%	15.4%	7.7%	53.8%			

'while cultural policy in Margate effectively leverages local characteristics to create an artful brand for the town and has stimulated a local milieu of artistic and cultural activity, it fails to effectively consider how local networks of cultural and creative practices can be maintained to provide a novel and sustainable economic base for the location'. In contrast, those in the high-income category £41,000–£100,000 and over £100,000 show remarkable differences from those in the low-income category. The findings show that those in high-income 'agree' or 'strongly agree' with all the SDG statements in the questionnaire. The findings obtained from those in the high-income category about the impact of the Turner Contemporary in not only attracting tourism but also resulting in enhanced economic development and creation of a novel entrepreneurship spirit find support in the research work carried out by Kampfner (2011). Those in the high-income category are more positive about the changes because gentrification of a location with more skilled wealthier people moving to the location benefits wealthier people who own their own homes and therefore see increased value in their homes, their wages are rising, businesses are more successful and there are more shops, cafes, restaurants and cultural opportunities for them to enjoy. The less wealthy, especially those who are renting accommodation are negatively impacted by the increase in property prices because that results in rising rents, the properties that are being rented are often sold to realize the increased values so rental accommodation is more difficult to find and more expensive. The facilities in the location will increase in price and some of the locations such as local pubs and cafes may change to suit the increasingly affluent people in the area so pushing the less wealthy locals out.

The participants were asked about their perception of how Turner Contemporary could improve. Some of the responses obtained include:

'More diverse activities and events...more exhibitions aimed at younger generation' (Female Participant).

'Be more advertised locally in neighbouring towns...more local art exhibitions for local schools to get involved...making public more aware of the facilities available in the area' (Male Participant).

'Classes for local people and advertising more about what is going on...Workshops engaging families to create things together' (Female Participant).

'Interactivity for children and more inclusive exhibitions for those not as big on art' (Female Participant).

'Better visibility of opening times. The last time we went, it was closed' (Female Participant).

'A more varied and less specialised, more local showing of art and live demonstrations could be directed to the ordinary person rather than the specialist' (Female Participant).

'More work with local schools...Outreach activities to families and children who may not otherwise get this experience' (Female Participant).

'Make it more available to local people—involve the community, local schools and colleges' (Male Participant).

'By really engaging with local artists—there are many art groups with very good talent' (Male Participant).

'By having more exhibitions of more contemporary artwork and sculptures. Have art workshops. Event weeks based on different artists or styles' (Male Participant).

There is however a consensus among all the participants regardless of income that the Turner Contemporary is of added value. In particular, there is high representation in the 'strongly agree' and 'agree' categories for the following statements: increased tourism, urban renewal, improved image of the area, positive image of the area in the media improved safety, developed a creative feeling, a greater sense of pride, improved, positive mood, more interests in art-based education, enhanced opportunity for enjoyment, contributed to community social value, better opportunities for young people, increase in property values in the area, attraction and retention of skilled people and more investment in the area. This finding helps confirm the positive aspect of gentrification—that there are more opportunities for people in the area for employment. As the investment in the area increases the population becomes more vibrant and entrepreneurial as the 'Creative Classes' drive the economy (Florida, 2003). The increase in economic development increases the wealth in the locality that contributes to social facilities such as health care, education, and security which further benefits local people and in the long term makes the location attractive and younger people are less likely to move away since they identify opportunities locally.

The results obtained from the open-ended questions in the questionnaire support the findings above. The participants were asked 'What Turner Contemporary or other cultural centres contribute'. Responses include:

'Turner Contemporary has made Margate a tourist attraction based on the old seaside town' (Female Participant).

'It brings art and creativity to the population. This coincided with media/local government increase in spending, therefore improving the area' (Male Participant).

'They bring art and culture to areas that do not necessarily have them. Everyone can benefit, young or old, they help with the learning process' (Female Participant).

'Great to have the Turner Contemporary. It was one of the things that attracted me to the area. Margate has always been a day-trip town. There has never been an arts and crafts monument' (Male Participant).

'Brings artistic people to the area. The new attraction has added to the attraction of Margate' (Female participant).

'Contributes to the community by hiring free space for events. A safe space, a nice encouraging environment, inspiring people of all ages' (Female Participant).

'A free inspirational environment to stimulate and offer experiences that some families would otherwise not become familiar with. Increased interest and appreciation of art—it creates a sense of local pride' (Male Participant).

'Draw more visitors, increasing tourism, encourages local art and educates people' (Female participant).

The study's null hypothesis is that income and perception of SDG factors are associated. The result of the Chi-Square shows significant associations with income and all the SDG factors except 'increased tourism'. Since the p-value (0.001) of the Chi-Square test is less than 0.05, the null hypothesis is accepted; therefore, there is sufficient evidence that suggests that there is an association between income and SDG factors—except for tourism. An explanation for this could be that although tourism increased, the low and seasonal wages associated with tourism do not always reflect (as in this case) an increase in socio-economic positive development for the people involved in the tourism sector. Those involved in more lucrative sectors, for example, estate agents, selling more and increased value properties will experience positive associations with gentrification.

8 Conclusion and Recommendations

This study sought to explore the perceptions of key stakeholder groups on the impact of the Turner Contemporary in Margate, the economic and social benefit of the Turner Contemporary to people who live, work, study, or visit Thanet and Kent and to identify whether any association exists between the development of an iconic cultural centre and social development factors. In addition, the study sought to identify strategies that can be adopted to improve the contributions of Turner Contemporary to the economic and social development of Margate. The findings show that while many of the participants agreed that the Turner Contemporary has changed how Margate is viewed and has been effective in creating a unique synergy between leisure, tourism, and attracting cultural customers to Margate, there is some doubt as to the overall benefits of the centre in promoting well-being for all, providing inclusive and sustainable economic growth. Some of the participants had some reservations about the effectiveness of the icon flagship centre in stimulating independent cultural and entrepreneurial activity, thereby resulting in economic regeneration and development of Margate for all. There is a need to ensure that key policy initiatives targeted at social development goals are of added value regardless of income and social status and that these initiatives help develop economic regeneration for all, create inclusive social networks, and provide physical infrastructure for the local communities to be sustainable and thrive. The public sector needs to collaborate with organizations such as housing associations to identify where social housing can be made available to those who are disadvantaged by gentrification and increased rental costs. There is also a need to ensure increased participation with local art groups to make Turner Contemporary more accessible to all in the community. Based on our findings, the following strategies can be adopted to improve the contributions of Turner Contemporary to the economic and social development of Margate:

- Empowering People—Increasing community involvement with the changes and opportunities developing in Margate as a result of the increased property values and increased availability to the public bodies of funds to support groups to make them feel included in the changes and to give them a voice to influence further developments.
- Environmental, Social, and Governance—Increase public consultation on environmental, social, and governance issues. Collaboration with the third sector and voluntary groups to ensure their needs are part of the strategy.
- Resilient Communities—Increase local community collaboration to maintain a sense of community. The local authority should look to facilitate community groups by working with present established groups to encourage greater participation and ensure communication with public sector decision-makers are clear and accessible to these community groups.
- Inclusive Societies—Increase the public facilities available for vulnerable marginalized groups. Increase support through collaboration with the public and private sectors to provide accessible accommodation, health care, and social support to ensure an inclusive society.

Further research is needed which takes into consideration a wider range of demographic factors such as gender, education, and life stages in gaining a more robust assessment of the influence of these factors on social development. In addition, our study shows no significant association between income and 'increased tourism'—further research may be needed to gain a better understanding. In addition, research based on the socio-economic impact of other iconic cultural centres developed, e.g. Tate in Albert Docks Liverpool built in 1989; Victoria and Albert, Dundee 2018 and Guggenheim Museum Bilbao opened in 1997 and The Louvre Abu Dhabi opened in 2017, would shed light on the effect in different places with different cultures and help highlight commonalities and differences these developments bring.

References

Abbing, H. (2022). Exclusion and exclusivity. In *The economies of serious and popular art: How they diverged and reunited*. Springer Nature Switzerland.

Bansal, P., & Song, H. C. (2017). Similar but not the same: Differentiating corporate sustainability from corporate responsibility. *Academy of Management Annals, 11*(1), 105–149.

Beatty, C., Fothergill, S., & Wilson, I. (2008). *England's seaside towns: A "benchmarking" study*. Department for Communities and Local Government.

Beatty, C., Fothergill, S., and Gore, T., (2014). Seaside towns in the age of austerity: Recent trends in employment in seaside tourism in England and Wales.

Bern, M., Townend, A., Khayat, Z., Balagopan, B., Reeves, M., Hopkins, M., & Kruschwitz, N. (2009). Sustainability and competitive advantage. *MIT Sloan Management Review, 51*(1).

Bowen, H. R. (1953). *Social responsibilities of the businessman*. Harper & Row.

Bryman, A., & Bell, E., (2018). Business Research Methods. Oxford, England: Oxford University Press.

Burritt, R., Christ, K., Rammal, H., & Schaltegger, S. (2020). Multinational Enterprise strategies for addressing sustainability: The need for consolidation. *Journal of Business Ethics, 164*, 389–410.

Carroll, A. B. (1979). A three-dimensional model of corporate performance. *Academy of Management Review, 4*(4), 497–505.

Carroll, A. B. (1983). Corporate social responsibility: Will industry respond to cutbacks in social program funding? *Vital Speeches of the Day, 49*, 604–608.

Carroll, A. B. (1991). The pyramid of corporate social responsibility: Toward the moral management of organizational stakeholders. *Business Horizons, 34*, 39–48.

Comunian, R., & Mould, O. (2014). The weakest link: Creative industries, flagship cultural projects and regeneration. *City, Culture and Society, 5*(2), 65–74.

Corlett, A., Odamtten, F., & Try, L. (2022). *The living standards audit 2022*. Resolution Foundation. Retrieved from https://www.resolutionfoundation.org/publications/the-living-standards-audit-2022

Davis, K. (1975, June). Five propositions of social responsibility. *Business Horizons, 18*(3), 19–24.

Elkington, J. (1994). Towards the sustainable corporation: Win-win strategies for sustainable development. *California Management Review, 36*(2), 90–100.

Elkington, J. (2018). 25 years ago I coined the phrase "triple bottom line." Here's why it's time to rethink it. *Harvard Business Review*.

Emms, S. (2011, August 6). Sea change: East Kent reinvents itself with art beside the seaside. *The Guardian*.

Florida, R. (2003). Cities and the creative class. *City & Community, 2*(1), 3–19.

Graham-Dixon, A. (2012, February 6). Turner & the elements, at Turner Contemporary, Margate. *The Daily Telegraph*. Retrieved fromhttp://www.telegraph.co.uk/culture/art/artreviews/9064397/Turner-and-the-Elements-at-Turner-Contemporary-Margate-Seven-magazinereview.html

Hah, K., & Freeman, S. (2014). Multinational enterprise subsidiaries and their CSR: A conceptual framework of the management of CSR in smaller emerging economies. *Journal of Business Ethics, 122*(1), 125–136.

Jamali, D. (2010). The CSR of MNC subsidiaries in developing countries: Global, local, substantive, or diluted? *Journal of Business Ethics, 93*(2), 181–200.

Kampfner, J. (2011, December 26). Call it the Tracey Emin effect: Art overcoming austerity outside London. *The Guardian*. Retrieved from http://www.guardian.co.uk/commentisfree/2011/dec/26/traceyemin-art-austerity-outside-London

Lee, R. J., & Newman, G. (2021). The relationship between vacant properties and neighbourhood gentrification. *Land Use Policy, 101*, 105185.

Lefebvre, H. (1991). *The production of space*. Blackwell.

Málovics, G., Nagypál Csigéné, N., & Kraus, S. (2008). The role of corporate social responsibility in strong sustainability. *The Journal of Socio-Economics, 37*(3), 907–918.

Mathews, V. (2014). Incoherence and tension in culture-led redevelopment. *International Journal of Urban and Regional Research, 38*(3), 1019–1036.

Michener, J., & Wong, D. (2018). Gentrification, demobilization, and participatory possibilities. In *Neighbourhood change and neighbourhood action: The struggle to create neighbourhoods that serve human needs* (pp. 123–146). Sage.

Moon, J. (2002). Corporate social responsibility: An overview. In C. Hartley (Ed.), *The international directory of corporate philanthropy* (1st ed., pp. 3–14). Europa Publications.

Phillips, J. (2011, April 13). Will Turner gallery help Margate to a brighter future? *The Guardian*. Retrieved from http://www.guardian.co.uk/society/2011/apr/13/turner-contemporary-gallery-margate-brighterfuture

Pratt, A. (2011). The cultural contradictions of the creative city. *City, Culture and Society, 2*(3), 123–130.

Schaltegger, S., Beckmann, M., & Hockerts, K. (2018). Collaborative entrepreneurship for sustainability. Creating solutions in light of the UN sustainable development goals. *International Journal of Entrepreneurial Venturing, 10*(2), 131–152.

Schuerman, M. L. (2019). *Newcomers: Gentrification and its discontents*. University of Chicago Press.

Turner Contemporary. (2024). *About us*. Retrieved from https://turnercontemporary.org/about/

United Nations. (2015). *Transforming our world: The 2030 agenda for sustainable development*. United Nations (UN) A/RES/70/01. Retrieved from http://www.un.org/ga/search/view_doc.asp?symbol=A/RES/70/1&Lang=E

United Nations Industrial Development Organization. (2023). *What is corporate social responsibility?* Retrieved from https://www.unido.org/our-focus/advancing-economic-competitiveness/competitive-trade-capacities-and-corporateresponsibility/corporate-social-responsibility-market-integration/what-csr

Visser, W. (2006). Revisiting Carroll's CSR pyramid: An African perspective. In E. R. Pedersen & M. Huniche (Eds.), *Corporate citizenship in developing countries* (pp. 29–56). Copenhagen Business School Press.

Ward, J. (2018). Down by the sea: Visual arts, artists and coastal regeneration. *International Journal of Cultural Policy, 24*(1), 121–138.

World Bank. (2020). *Five things you need to know about social sustainability and inclusion*. Available at Five Things You Need to Know About Social Sustainability and Inclusion (worldbank.org)

World Bank. (2023). *Social sustainability and inclusion*. Available at Social Sustainability and Inclusion Overview (worldbank.org).

World Commission on Environment and Development. (1987). *Our common future*. WCED, Oxford University Press.

Yin, J., & Jamali, D. (2016). Strategic corporate social responsibility of multinational companies' subsidiaries in emerging markets: Evidence from China. *Long Range Planning, 49*(5), 541–558.

Adebimpe Adesua Lincoln holds an LLB and an LLM in Commercial Law from Cardiff University and an MBA in International Business and PhD in Entrepreneurship and Business Development from the University of South Wales. She holds a Postgraduate Diploma in Legal Practice and also holds various teaching qualifications in Higher Education. She is a Fellow of the Higher Education academy and an Associate Member of the Chartered Institute of Personnel Development. She has 20 year's experience in Higher Education. She has held positions in Cardiff University, University of South Wales and was a Senior Lecturer in Law at Cardiff Metropolitan University before taking up a position as an Assistant Professor in Saudi Arabia. Adebimpe Lincoln is a legal practitioner. She also works with the University of Liverpool online Master's in Law Programmes, Kaplan Online Learning and is an Adjunct Professor with Robert Kennedy College, Switzerland. Adebimpe Lincoln has vast experience supervising research students. She has supervised a wide range of Master's level research including M.Sc., MBA, and LLM dissertations. She has also supervised students at Doctoral Level in the area of Entrepreneurship and Corporate Governance. Her research interests lie in the area of Female Entrepreneurship, Small and Medium Sized Enterprise (SME) Sector, Leadership Practices of Nigerian Entrepreneurs, Corporate Governance and Board Diversity, Sustainability and Corporate Social Responsibility among Nigerian SME's. She has published and presented several articles on Leadership, CSR, Governance, and Entrepreneurship in Nigeria.

Dr Jane Croad is a researcher and education consultant. Before commencing her academic career, Dr Croad enjoyed a successful corporate career and held senior management positions with Gwalia Housing, Johnson and Johnson, and the Wicker Group. Dr Croad was a Senior Lecturer in Marketing at the Cardiff School of Management, before joining Robert Kennedy College as an Adjunct Professor. Dr Croad's research focus is on urban development; sustainable city branding; corporate social responsibility and Sustainability.

Gamification Approach in CSR Communication: Case of Jotun Türkiye

Bayram Bilge Sağlam and Egemen Ertürk

1 Introduction

In recent times, communities expect businesses to develop strategies that would contribute to efforts toward social and environmental issues, instead of implementing pragmatic strategies that are profit oriented (Dawkins & Lewis, 2003; Podnar & Golob, 2007). As a prevailing concept in this context, corporate social responsibility (CSR) is an activity that needs to be managed carefully, since it does not only affect the legitimacy and the image of businesses, but also present a new ground for competition (Ryu et al., 2016). When the scientific findings regarding the benefits of CSR applications for businesses are analyzed, the rising interest of businesses on the subject can easily be understood. The meta-analysis study of Vishwanathan et al. (2020) which was done on 344 papers which researches the relation between CSR applications of businesses and their financial performance clearly states the strong relationship between these two concepts. According to the findings of the study, CSR applications provide positive effects on financial performance through increasing business reputation, increasing innovation capacity, increasing stakeholder benefits, and decreasing business-specific risks. The attraction of these benefits, combined with the communities' expectations from businesses morphed CSR related activities from being a choice, to an obligation; which led the way to budget allocations for these activities increasing considerably (Zbuchea & Pînzaru, 2017). For this reason, in order to reap the benefits in question, businesses should design CSR activities that can stand out and back those up with innovative communication strategies.

B. B. Sağlam · E. Ertürk (✉)
Maritime Faculty, Dokuz Eylul University, İzmir, Turkey
e-mail: bayram.saglam@deu.edu.tr; egemen.erturk@deu.edu.tr

S. O. Idowu, S. Vertigans (eds.), *Sustainability in Global Companies*, CSR, Sustainability, Ethics & Governance, https://doi.org/10.1007/978-3-031-77971-8_9

CSR 2.0 concept, which was introduced by Visser (2016), underlines the need to revisit and revise the CSR activities, in line with the above-mentioned changes in community-business relationship. In this new CSR approach, which was designed considering the differences between Web 1.0 and Web 2.0, prevailing concepts can be defined as increased stakeholder participation and innovative partnerships (Visser, 2016). Creativity plays a key role in businesses successfully implementing these two concepts in their CSR activities. The study of Quairel-Lanoizelée (2016), which presents the finding that CSR activities must be impossible to imitate in order to provide competitive advantage to the business, shows how crucial creativity is, in the context of CSR 2.0. The importance of creativity has also become critical in the field of environmental sustainability, which businesses are expected to attach great importance to, and as a result, the concept of green creativity has emerged. Developed by Chen and Chang (2013), this new concept of creativity is defined as the root of environmentally friendly products/services and organizational practices. In this context, it is possible to evaluate that green creativity is exactly what businesses with the desire of running efficient CSR campaigns on environmental issues need.

Green creativity could go beyond the product or service itself and find a use in CSR communication as well. In that sense, Gamification approach can present an illustration of green creativity in CSR communication. While the number of businesses adopting this approach in their CSR communications is relatively low, it is expected to increase in the following years as a new inclination (Bradley, 2014). This expectation is based on the view that the new generation that is joining the workforce is both more socially conscious and they like to play games (Baer, 2022). Similarly, when the CSR communication literature is analyzed, it can be seen there are limited number of studies on the matter.

The focus of this study is the Jotun Türkiye case, which is a localized application of the global "Jotun GreenSteps" initiative. The project, which implements the gamification approach in the CSR communication has been up and running since 2017. The study aims to analyze in-depth the benefits the project has provided to the business and the design process of the project with the help of primary and secondary data. This type of case studies, which are used commonly in CSR literature, aims to reflect transferrable experience gained from successful applications by providing extensive information on said case. The findings of the study could help businesses in achieving desired outcomes from gamification applications and help this innovative CSR communication approach proliferate.

The following sections of the study include the analysis of current literature on gamified CSR communication, general information on Jotun GreenSteps project, and findings on the interviews conducted. The study is concluded with managerial implications supported by the findings.

2 Background: CSR Communication and Gamification

Role of communication strategies cannot be underestimated when it comes to the success of CSR activities of businesses. The positive effect of communication on establishing legitimacy (Farache & Perks, 2010; Lock & Schulz-Knappe, 2019), positive attitudes of shareholders (Du et al., 2010), and creating a strong bond with community (Hall, 2006) have been empirically proven. The increasing mistrust among communities toward multi-national corporations in today's setting enhances the importance of communication. However, there are studies which show that effective CSR communication could eliminate public skepticism (i.e., Skarmeas & Leonidou, 2013; Kim & Rim, 2019). In light of this information, it can be concluded that it is a requirement for businesses to design and carry out these CSR activities with transparency, cooperation and by making a difference, all the while using the right communication strategy to reach their shareholders.

In order for CSR activities to succeed, all aspects of communication must be carefully handled. These aspects are named as "message context" and "message channel" in a CSR communication perspective in the work of Du et al. (2010). Regarding the context of the message, the importance of the issue for the community, commitment of the business to the issue, emphasizing the motivation of the business, evaluation of the desired/achieved impact, and the compatibility of the businesses core operations and its CSR activities are important aspects that should be considered. When it comes to the message channel, important issues can be listed as reporting, web site of the business, public relations, commercials, points of purchase, and effective and purposeful usage of external news.

Outside the framework put out by Du et al. (2010), the extent of the CSR activities will be held (message intensity) is also a topic up for discussion on the path to a successful outcome. In the related literature, there are findings that show while the stakeholders require the businesses to show that they care about the CSR communication, when they feel the businesses are "trying too hard," it would result in a negative perception (Coombs & Holladay, 2015). In that light, Sen et al. (2006) define CSR communication as a "double-edged sword" and point to the paradoxical nature of CSR communications.

In order to create a difference in the vast and complex field of CSR communications, fields of creativity must be explored for each of the elements stated above (Dawkins, 2004). As an instance, in the experimental research of Pérez et al. (2020), it is found that instead of thoroughly explaining the elements regarding the context of the message, adapting a storyteller approach to transfer the message to the stakeholders was more effective. Whereas in the intercultural research of Kim and Bae (2016) which was carried out on the USA and South Korea samples, it is stated that intangible CSR messages (even though it paves the way for creativity) are perceived more negatively in South Korean audiences, compared to the USA.

Due to its conceptual nature, it can not be said that there is a single method to be creative in CSR communications. However, it can be observed that due to the digitalization, Web 2.0 applications have a critical role and create new fields for creativity in CSR communications (Knaut, 2017). Web 2.0 applications, which at the core, bring about advantages regarding stakeholder participation (Troise & Camilleri, 2021), are researched in the literature on various aspects. Capriotti (2017) evaluates the CSR communication that had been updated with social media and web site usage as "conversational" in regard to forming of the message, "symmetric" in regard to sharing of the information, and "in form of a network" in regard to the direction of the sharing. Numerous academic research carried out in various geographic locations and industries state that this transformation of CSR communications has had positive effects on the success of CSR activities such as increase in message reach, stakeholder engagement, and employee attraction (i.e., Ali et al., 2015; Galati et al., 2019).

Even though social media and websites are the most common medium for CSR communications regarding Web 2.0, gamification applications are rapidly increasing their preferability as a new message channel (Harwood & Garry, 2015). It is without a doubt that video game industry being the most income generating among the entertainment industries and also continuing to grow is one of the most important reasons behind this new communication channel reaching the limelight. Data collected from Statista (2022) show that the video game industry, which have generated 93.2 billion dollars in 2021, has reached three times the size of the movie industry and seven times the size of the music industry. When the recent data regarding the video game industry are analyzed in detail, it can be seen that income generated from mobile games has the biggest share with 45% and it is followed by console games (32%) and PC games (23%) (Statista, 2022). The fact that mobile games' share is this high also points to the change in who is called as a "gamer" and the change and variety in game playing habits as well.

Another source of variety in the video game industry is among game genres and their objectives. In that sense, social responsibility fields became a new market for game developers. These games which are classified as "games for impact" or "social impact games" aim to generate or improve social awareness of the community on a social issue (LaPensée, 2014). Efforts of video game industry regarding creating social impact are not limited to these kinds of games. As an instance, the game studio ZeniMax Online donates 1 dollar to selected nongovernmental organizations for each 5 dragons players defeat in the game Elder Scrolls Online (Chalk, 2019). Similarly, the highly popular simulation game Farmville applied the "cause-related marketing" model and donated a portion of the income from the transactions of players in the game to charities (Maltseva et al., 2019).

In parallel with these developments in the video game industry, businesses in other industries realized they can tap into a new communication channel with gamification. These businesses that are not in the video game industry, with the help of game studios or advertising agencies, present games with the objective of advertisement to their stakeholders. In the related literature, this new type of game has been named "advergame" and became the subject of many studies in the field of strategic communication (e.g., Waiguny et al., 2013; Aymankuy et al., 2016; Sucu, 2020). One other field of application for the aforementioned advergames is CSR communication. In this context, the aim behind the design of advergames goes beyond the

Table 1 Advergames around the world as a communication tool

Business	Field of operation	Advergame	Purpose of the game	Impact
Kraft (USA)	Food Manufacturer	Two-Minute Drill	Raising awareness toward famine	Food donation for every point players get
Alipay (China)	Online Payment Service Provider	Ant Farm	Reducing carbondioxide emission	Tree planting for every point players get by reducing their carbon emission.
Double A (Thailand)	Paper Manufacturer	Tree Planet	Increasing green field	Tree planting for every tree planted by players in-game
WeSpire (USA)	Consulting	–	Shaping the behavior of the players in line with sustainability goals	Players to learn how to recycle and be green consumer
SAP (Germany)	Software Developer	TwoGo	Encouraging employees to carpool as an environmental action	Donation to charities in exchange for the points players get by carpooling.

Source: Compiled from Coombs and Holladay (2015), Jun et al. (2020), Sung and Lee (2020)

promotional perspective and helps to carry out CSR activities and create a social impact. Due to their design, these games have short playthrough times, are easier to play (Leclercq et al., 2018), and are easily accessible through mobile apps or web browsers (Terlutter & Capella, 2013). It is found that these games have positive impact on creating social values together with stakeholders (Jun et al., 2020), shaping consumer behavior in line with the sustainability plans of a business (Maltseva et al., 2019) and stakeholders showing a positive attitude toward the business (Sung & Lee, 2020). Table 1 shows advergame examples around the world that has been used in CSR communications.

According to the study of Wünderlich et al. (2020) which was carried out with scholars and operators in the industry, it is expected that in the next 5 years, 80% of the businesses in the Forbes 200 list will implement gamification in to their communication. It is without a doubt that these expectations are formed in the light of examples around the world showing positive impact toward the brand image and reputation as well as positive reinforcement on CSR goals of said businesses.

Specific to Türkiye, apart from the Jotun Türkiye's "GreenSteps" project, which is the subject of this study, the "I Protect My Energy" project of EnerjiSA is another prime example of gamified CSR projects. However, as it stands, gamification cannot be considered as a common application in the CSR activities of businesses in Türkiye. With the opportunities presented by gamification in CSR communication, it is important to analyze the existing examples, in order for it to become more widespread and also to contribute to the academic literature regarding this matter.

3 Jotun Türkiye and GreenSteps Campaign

Jotun Türkiye is a Norway based business that produces decorative paints, marine coatings, protective coatings, and powder coatings. The organizational structure of Jotun is divided into 7 regions and Jotun Türkiye is located in Eastern Europe and Middle Asia region. Jotun Türkiye operates in this region as both a producer and a seller.

Jotun distinguishes itself in the industry as a brand that develops their products with responsible innovation, always keeping environmental sustainability in mind. Using less volatile organic compound in these products and development of re-paintable paint show the emphasis on being environmentally friendly. Similarly, Jotun shares on their website that they go above and beyond the legal requirements when it comes to fulfilling their environmental responsibilities regarding production processes such as solid waste management, energy savings, and recycling (Jotun, 2022).

Jotun, who underlines that it sees sustainability as a long-term competitive advantage, is also contributing to their stakeholders on different sustainability fields with CSR projects both on a domestic and global scale. With the GreenSteps project, Jotun Türkiye has been trying to contribute to the environment and the society with applications they design with "environmental sustainability" in mind, since 2017. The project has been run by Jotun Eastern Europe and Middle Asia Marketing Director and Jotun Türkiye Marketing Manager. Other parties involved in the project include the advertisement agency (Kompüter Ad Agency), public relations agency (Bernaylafem Communications and Brand Management), and nongovernmental organizations that align with the aims of the project. In that line, TEMA, The Wheat Movement, and The Underwater Cleaning and Awareness Movement have been among the collaborators of this project since 2017.

The theme of the gamification approach has differed in different years of the project with several "environmental sustainability" issues in focus. The first 2 years presented a mobile application where users can plant trees by walking with their phones. In 2019, the same application was redesigned to allow the users to contribute to cleaning the sea bed from hazardous waste. Players who follow the map on the mobile application and travel 4 km on roads that were not previously traveled, or travel 6 km on roads that were previously traveled, would be eligible for the benefits. For the first 2 years, this benefit was planting a tree, and for 2019 it was cleaning an area of 10 m^2 on the sea bed. Figure 1 includes a visual for the aforementioned mobile application.

While in 2020, users were presented with a game which they can access through a web browser and which has platforming and trivia mechanics. In this game, each complete run was awarded with a seed planting. In the game, the aim is to finish multiple levels of platforms, using the gyroscope of the phone, in 5 min. If the users correctly answer the trivia questions that measures environmental consciousness that pops up every now and then, they get extra time. Figure 2 includes a visual for the aforementioned platforming game.

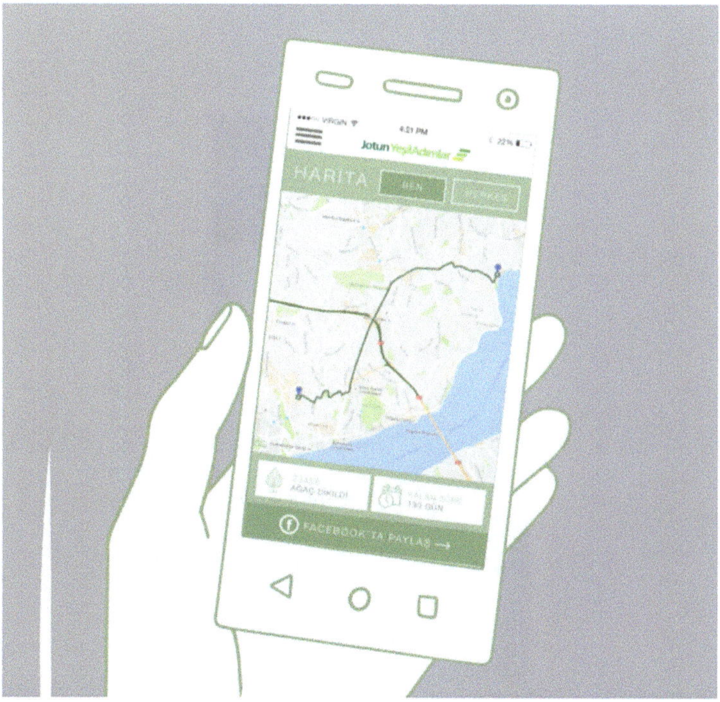

Fig. 1 Jotun GreenSteps Mobile Application Visual

There is no limit to the number of times players can complete the game, which encourages participation throughout the entirety of the campaign. In order to reach a broader audience, Jotun Türkiye social media accounts were actively used for the promotion of the campaign.

As they differ in terms of their gamification approach, in the following parts of the study the mobile application-based games that were released between 2017 and 2019 will be called "mobile games" and the browser-based game released in 2020 will be called "platformer game" as it stands out with its platforming mechanics (navigating a character through two- or three-dimensional environments).

In this case study, major actors involved in the gamification process are Jotun Türkiye as the brand and Kompüter Ad Agency as the game developer. The remaining actors (public relations agency and related NGOs) have limited contributions to the gamification aspect of this CSR initiative and thus were not included in the study. Their contributions to the initiative were asked to be reported by Jotun representatives.

In order to gain extensive information regarding the gamification approach, semi-structured interviews were carried out with Jotun Türkiye Marketing Executive on 04.02.2021 and Kompüter Ad Agency Founder on 09.02.2021 with the help of online meeting software. The selection of semi-structured interviews as data collection method allowed researchers to ask additional questions whenever there is a

Fig. 2 Jotun GreenSteps Platforming Game Visual

need for clarification of the responses and whenever the researchers had the opportunity to gain deeper insights on the phenomenon of "gamification in CSR communication." From the interviewees' perspective, this data collection method was useful as well, since interviewees had the chance to guide the interview whenever the predetermined questions seemed to fall short while explaining their experiences on the phenomenon from the angles that they believe to matter (Berg, 2000).

The predetermined questions (see Appendix) were mainly formed to gather information regarding the strategic approach of Jotun Türkiye on CSR initiatives and the benefits reaped, the conceptual inception of the project, and the decision of implementing gamification along with the benefits and hardships it brings. Apart from the online interviews, infomercials and other promotional materials (such as the official website news and updates, promotional social media posts), has been used as secondary data. These materials involve information regarding the stated aim of the CSR initiative, its application, and the timeline of its execution.

4 Findings

4.1 Strategic Approach of Jotun Türkiye Toward CSR Initiatives

The findings gathered from the interviews show that Jotun Türkiye forms their CSR initiatives to gain a competitive advantage for the brand, in accordance with Jotun's sustainability policies. In this context, Jotun assigns a special CSR budget for all 7 regions under their roof and asks them to carry out domestic project that aligns with Jotun's global sustainability goals. The process and the results of the projects are reported to the headquarters in two separate reports, former being sent at the development phase of the project and the latter being sent after the conclusion of the project. The Marketing Executive describes Jotun Türkiye putting environmental sustainability in the focal point in their CSR projects with the following words:

> Our CSR related activities are a part of Jotun's global strategy. Each region owns up a field of sustainability, while keeping the realities of their region in mind. We, as Jotun Türkiye owned up the field of environment. In other regions there are projects on animal rights, kids in need, and aids etc. We also see CSR activities as a competitive advantage. One other reason for choosing environment is that Jotun has a global environmental policy within the framework of GreenSteps project. This environmental policy is embedded in our production and waste management processes from start to finish. That is why we are here, running a leg of this project in Türkiye. We are actually harmonizing our CSR activities and sustainability approach and transforming it into a competitive advantage.

In line with this strategic approach, Jotun Türkiye underlined various communication elements that come to the forefront for the CSR project to be successful. These elements can be summed up as follows:

- Correctly identifying the importance of the environmental issues'—that are subject to the CSR project—importance in the public's eyes, with the help of public relations agency and relevant NGO's
- Reflecting the harmony between Jotun Türkiye's core functions and the CSR activities
- Solidifying the commitment to the cause in the eyes of the stakeholders by carrying out the project regularly each year
- Adapting communication methods that allows maximized stakeholder participation

The strategic approach of Jotun Türkiye toward these elements is reported by the Marketing Executive as follows:

> The most important issue is whether the CSR projects are compatible with the core values of the brand or not. Owning up a certain field and nativizing your brand image with that field would be beneficial for both the success of the project and your brand. Apart from that, we value stakeholder participation in every level. As an example, technical issues such as choosing the seedlings and when to plant those seedlings are among the decisions we make with the help of the NGO we are partners with in this project. After the launch, you are as

successful as the crowd you can reach, and people you can motivate. Attracting a wider base of people by making a game is helpful for spreading the word about the project.

When the target business outputs of Jotun Türkiye from this CSR project are analyzed, it can be seen that sales performance is not in the focal point, rather it is aimed to increase brand image/perception. GreenSteps project being open to anyone to participate without requiring any form of purchase or financial payment of any kind is an important aspect toward this goal. The following statement by the Marketing Executive explains the strategic goals of Jotun GreenSteps project:

Up until this point we did not consider an approach which required participants to make a purchase from our brand in order to participate in a social responsibility project. GreenSteps is a project we expect to contribute to the nature and raise awareness on environmentalism. We are able to observe, especially when the launch date of the campaigns near, stakeholders are intrigued and waiting for the project to start. We always launch the GreenSteps project in June. Towards March, we start to get questions about the project we will be launching later that year.

4.2 *Idea Production Process of Jotun Türkiye GreenSteps Project*

Findings of the interviews show the idea production process of Jotun Türkiye's GreenSteps project was managed together with Kompüter Ad Agency, in which ideas were bounced off between the parties. In this process, after the environmental issue to be handled for that year is decided by Jotun Türkiye, the idea production is shaped depending on the tool of communication to be used. The founder of Kompüter Ad Agency describes the idea production process in between 2017 and 2019 which saw mobile games used as a tool with the following statements:

GreenSteps project is a result of an idea which is almost like the literal translation of Jotun's global GreenSteps project from 4-5 back. We thought together with Jotun on how we can bring this to life. An idea based on mobile app gamification had come up. This idea motivated people to walk, for their own health, meanwhile it also presented the opportunity to do good for the environment as well. Users were able to plant a seedling for each 4 km of roads they have walked which was not previously walked by other users, and each 6 km walked on roads that have been previously walked. That meant they used their cars less, which meant less carbon emissions, while doing something healthy for themselves. It was a win-win project.

As it can be understood from the statement above, the most important element that was considered in the idea production process of the project is designing the game in a way that allows the users to explore an individual motivation along with the environmental motivation. In a sense, the CSR strategy of Jotun Türkiye, which was stripped from sales-oriented goals, was reinforced by avoiding a singular benefit approach. This greatly helped the project to kick skepticism out of the equation, which can be a real threat for a CSR project. Below statement of the founder of

Kompüter Ad Agency, underlines what this approach brought to the project, with its effect on minimizing skepticism toward CSR initiatives:

> There is a barrier in CSR projects. People tend to expect "the taking hand to not to see the giving hand" (a Turkish proverb) in CSR initiatives. It is in our culture. We put win-win in the center of our project. We did not go for a one-sided campaign like "share our message with your friends" or something like that. We put a win for every party involved in this design. We considered this balance, with Jotun, throughout the entirety of the project. We should not be in a position where we ask people to advertise Jotun.

It can be seen that same approach was adapted in the platforming game presented in 2020. Specific to this year, due to Covid-19 outbreak, the mobile app that encouraged people to walk the streets of their cities lost its applicability, and a need for a new idea for the project has arisen. At this point, the platforming game, which allowed users to participate from the comfort of their houses, was designed. The founder of Kompüter Ad Agency describes the win-win design for both the user and the brand that was created with the game in question as follows:

> This game was a distraction and a source of entertainment for users while we were couped up at our houses, and while it might not be as much as walking but we also see a win in this as well. Feeling good is also a win. As we aimed for—entertaining—this time around, the game itself had to be a—real game—and an entertaining one at that.

The Marketing Executive of Jotun Türkiye notes changing user preferences are another reason—apart from the pandemic—behind the decision to develop a platforming game:

> The idea of developing a platforming game was on the table before the pandemic, and it just fit so well to this period of time. Following the changing consumer preferences and adapting quickly are really important in special projects such as this one.

It is stated that another important decision point that was in consideration through the idea production process of the platforming game was the story telling and game mechanics. The game mechanics and the cinematic animation at the beginning and the end of the game are designed to raise awareness on environmental issues, which was the main aim of this project. The founder of Kompüter Ad Agency lists the decisions on this matter as follows:

> We thought—Are we really fully informing people on the downfall of the environment, or is it lacking?—. That is why we added trivia questions to the game, which will grant the user extra time if answered correctly. Actual reason behind those questions is to inform the users. I mean when you answer those questions, you think about global warming, etc., and assure that the game has an environmental-conscious side. If we designed it strictly as a Jotun advertisement, it would hurt the CSR initiative. This is the reason why we have an informative narrative on the CSR message at the start and the end of the game. It is essential that whoever plays this game, gain a strong sense of awareness of the CSR message, whether it be through the narrative, or the game mechanics.

Sharing the results of this gamified CSR activities with the stakeholders has been an important part of the idea production process since the project's inception in 2017. Animated shorts and other kind of informative posts are prepared and served through the official website and social media accounts of the company for the consideration of the stakeholders, especially users that participated in the games. Following part of the transcription of the animated shorts prepared after the conclusion of 2020s platforming game and Jotun Türkiye's tweet (as seen in Fig. 3) regarding the outcomes of 2017s mobile application game present examples for this approach.

> …15.000 seedlings have been delivered in a short time. 51.850 individual users have spent an average of 3 minutes and 31 seconds on the website. So we kept our promise, and planted 15.000 seedlings on a barren land close to Ayvalık. Thus, people who have played the game contributed to a greener land, without even leaving their houses. The campaign is over but the game is still up. Go to jotunyesiladimlar.com to try it.

4.3 *Benefits and Challenges of Gamification Approach (Fig. 3)*

Findings from the interviews show that gamification approach, due to its dual communication channel which allows for easier stakeholder participation, is a better alternative to conventional media tools. It is also found that due to advergames being "fun," it has a potential for word-of-mouth advertising, which allows the method to have a bigger impact with lower budgets. The founder of Kompüter Ad Agency states the effects of the campaigns as follows:

> Even though it was promoted with a limited budget, due to word-of-mouth, the game was downloaded 35.000 times in its first year. We had a target of 3 months to hit 10.000 seedlings, but we reached this target in half the time we expected. Literally there were no streets

Fig. 3 Jotun Türkiye Social Media Post on 2017 CSR Project*. *Translation of the tweet: "We have reached 10,000 trees with Jotun GreenSteps campaign. Now we are at Gönen with TEMA Foundation to plant them!"

> left which had not turned green, not only in Istanbul, but in every city center around Türkiye. Even some Turkish people who heard about the app somehow in Europe and USA have downloaded the app and walked in their respective cities. We actually painted a lot of the cities in the world.

The great interest shown in Jotun GreenSteps projects also benefits Jotun Türkiye on brand awareness and brand image. The founder of Kompüter Ad Agency summarizes the role of gamification approach in this matter with the following statement:

> You get to spend some time with the brand and be an active part of this experience through the game. This unwillingly forms a lasting bond between you and the brand. You become the—choice—when that bond can be formed. If you can not manage that, it is hard to stay in the competition. Gamification is much more impactful in forming that bond, when compared to a regular ad spot. Users remember a brand that provided them an interactive experience where they can be a part of it all.

The Marketing Executive of Jotun Türkiye's comments on the bond between Jotun Türkiye and people who participate in their projects, back up the role of gamification in forming that bond:

> In the middle of the year, we start getting the—what would GreenSteps do this year?—questions. We are in close contact with our customers through our retailers, call center, and social media platforms. They can reach us easily and that allows us to directly collect feedback.

Apart from all these benefits, the challenges faced by Jotun Türkiye and Kompüter Ad Agency with their gamification experience should also be evaluated. It is stated that most of the challenges related with the gamification process are related with the technical aspects of the game (i.e., Game mechanics, difficulty level, operating system compatibility). In the process of making these decisions, considering the benefits of the CSR project, projecting of the brand image, and choosing the target audience correctly are named as important elements that need to be considered. The decision-making process regarding the difficulty level of the platforming game between Jotun Türkiye and Kompüter Ad Agency can be viewed as an example from the following quotes:

> We designed the game as if it was a real game. We did not want to make a game just for the sake of it. We wanted to present a gaming experience that is actually fun to play. Of course it serves a specific purpose, but in the meantime it was important for us that the game felt like a—real-game, with a certain level of difficulty.—The Marketing Executive of Jotun Türkiye
>
> In order to decide on the difficulty of the game, you have to contemplate on issues such as the benefits of the game, the brand, and the target audience. When you first start the game, you roll the seedling on land, then comes a level with bridge mechanics. We added that level so the game would not be monotonous, and it would have various mechanics. Some users had really hard time on that level with bridges, that's why we made it easier half-way through the project. It was still relatively difficult, but considerably easier than its original state. Because, yes, people want to contribute to planting a tree as fast as they can but if the game was really easy, then the users would feel deceived.—The founder of Kompüter Ad Agency

5 Conclusion

In today's business scene, organizations have to be in touch with communities in order to attract attention in their fields, and they can do it so with the help of CSR initiatives on top of their core business functions.

In this light, CSR activities became a competitive field much like the core production/service activities of businesses and these CSR activities can only be competitive as much as they are innovative. Innovative CSR activities should include stakeholders in value creating process (Leclercq et al., 2018) and differ from cut-and-dry CSR activities with their creative aspects (Visser, 2016). Gamification in CSR communication checks both of these boxes, and presents an opportunity for businesses in this field. Gamification can be a tool to create a platform where green creativity can bloom. Best practices in this field that has green creativity at the heart of it, can be a great guide for businesses that try to make a difference with their environmentally conscious CSR initiatives.

In this research, different examples around the world on gamification in CSR communication was presented, the newly forming literature was evaluated, and an example from Türkiye, which is Jotun GreenSteps, was presented in light of interviews carried out with involved parties. While it is a rather recent trend to employ gamification in businesses' CSR communications, it is not hard to predict it will become a more common practice in the near future. Apart from the benefits listed above, the fact that "gaming" as a culture rapidly increasing worldwide would provide a leveraging effect. However, a business adapting gamification approach in their CSR communications does not mean it would automatically make these efforts successful on its own. As it is the case in Jotun GreenSteps project, the game to be developed in accordance with the CSR initiative should be in line with the sustainability strategies of the organization on both idea production level and technical aspects. This requires the consideration of CSR goals, target audience, and compatibility with core functions of the business. With the proliferation of gamification approach, carrying out comparative case studies would benefit both the literature and the industrial applications.

Appendix

Predetermined Semi-Structured Interview Questions

1. Corporate social responsibility (CSR) practices are an issue that needs to be carefully managed in terms of the legitimacy and image of businesses, but also emerges as a new area for competition among businesses. In this sense, how would you evaluate Jotun Türkiye's strategic approach to CSR activities?
2. How Jotun Türkiye's CSR activities are organized within the business (e.g., involvement of different business functions; involvement of Jotun headquarters)?

3. Gamification approach in CSR communication emerges as a very up-to-date method. In this context, could you briefly describe the idea production stage of the Jotun GreenSteps project?
4. What benefits do you think the Jotun GreenSteps project brings to your business' brand image and performance outcomes?
5. In the relevant literature, it has been revealed that CSR activities should be developed in a structure that supports stakeholder participation and creates space for creative elements in order to produce successful outputs. In this context, what advantages does the gamification approach provide compared to conventional CSR communication models in achieving the expected goals?
6. Can you evaluate the feedback from your stakeholders (customers, employees, media, suppliers) regarding the Jotun GreenSteps project?
7. What are the elements that you think come to the forefront, in line with your experience in order for the gamification approach to yield successful results?
8. What technical problems did you encounter during the process of launching the Jotun GreenSteps project?
9. Could you briefly describe the process carried out between Jotun Türkiye and Kompüter Advertising Agency in terms of the story, gameplay and CSR message of the game in the Jotun GreenSteps project?
10. What are your future predictions regarding the use of gamification approach in CSR communication?

References

Ali, I., Jiménez-Zarco, A. I., & Bicho, M. (2015). Using social media for CSR communication and engaging stakeholders. In A. Adi, G. Grigore, & D. Crowther (Eds.), *Corporate social responsibility in the digital age (Developments in corporate governance and responsibility)* (Vol. 7, pp. 165–185). Emerald.

Aymankuy, Y., Demirbulat, Ö. G., & Saatçi, G. (2016). An assessment of game based advertisements (Advergame) applicability for tourism sector. *Journal of Tourism Theory and Research, 2*(2), 77–88.

Baer, S. (2022). Gamification turning corporate responsibility into an authentic and fun experience. *Forbes*. Retrieved from https://www.forbes.com/sites/forbeshumanresourcescouncil/2022/04/05/gamification-turning-corporate-responsibility-into-an-authentic-and-fun-experience/?sh=1301bca8354e

Berg, B. L. (2000). *Qualitative research methods for the social sciences*. Allyn & Bacon.

Bradley, D. (2014). The value of gamified CSR. *PRWeek*. Retrieved from https://www.prweek.com/article/1314544/value-gamified-csr

Capriotti, P. (2017). The world wide web and the social media as tools of CSR communication. In S. Diehl, M. Karmasin, B. Mueller, R. Terlutter, & F. Weder (Eds.), *Handbook of integrated CSR communication* (pp. 193–210). Springer.

Chalk, A. (2019). *Slaughter dragons to save cats in this elder scrolls online charity campaign*. PCGAMER. Retrieved from https://www.pcgamer.com/slaughter-dragons-to-save-cats-in-this-elder-scrolls-online-charity-campaign/

Chen, Y.-S., & Chang, C.-H. (2013). The determinants of green product development performance: Green dynamic capabilities, green transformational leadership, and green creativity. *Journal of Business Ethics, 116*(1), 107–119.

Coombs, W. T., & Holladay, S. J. (2015). Two-minute drill: Video games and social media to advance CSR. In A. Adi, G. Grigore, & D. Crowther (Eds.), *Corporate social responsibility in the digital age (Developments in corporate governance and responsibility)* (Vol. 7, pp. 127–142). Emerald.

Dawkins, J. (2004). Corporate responsibility: The communication challenge. *Journal of Communication Management, 9*(2), 108–119.

Dawkins, J., & Lewis, S. (2003). CSR in stakeholder expectations: And their implication for company strategy. *Journal of Business Ethics, 44*(2–3), 185–193.

Du, S., Bhattacharya, C. B., & Sen, S. (2010). Maximizing business returns to corporate social responsibility (CSR): The role of CSR communication. *International Journal of Management Reviews, 12*(1), 8–19.

Farache, F., & Perks, K. J. (2010). CSR advertisements: A legitimacy tool? *Corporate Communications: An International Journal, 15*(3), 235–248.

Galati, A., Sakka, G., Crescimanno, M., Tulone, A., & Fiore, M. (2019). What is the role of social media in several overtones of CSR communication? The case of the wine industry in the southern Italian regions. *British Food Journal, 121*(4), 856–873.

Hall, M. R. (2006). Corporate philanthropy and corporate community relations: Measuring relationship-building results. *Journal of Public Relations Research, 18*(1), 1–21.

Harwood, T., & Garry, T. (2015). An investigation into gamification as a customer engagement experience environment. *Journal of Services Marketing, 29*(6/7), 533–546.

Jotun. (2022). *About Jotun*. Retrieved from https://www.jotun.com/ww-en/about-jotun

Jun, F., Jiao, J., & Lin, P. (2020). Influence of virtual CSR gamification design elements on customers' continuance intention of participating in social value co-creation. *Asia Pacific Journal of Marketing and Logistics, 32*(6), 1305–1326.

Kim, S., & Bae, J. (2016). Cross-cultural differences in concrete and abstract corporate social responsibility (CSR) campaigns: Perceived message clarity and perceived CSR as mediators. *International Journal of Corporate Social Responsibility, 1*(1), 6.

Kim, S., & Rim, H. (2019). The role of public skepticism and distrust in the process of CSR communication. *International Journal of Business Communication, 2019*, 1–21.

Knaut, A. (2017). How CSR should understand digitalization. In T. Osburg & C. Lohrmann (Eds.), *Sustainability in a digital world. CSR, sustainability, ethics & governance* (pp. 249–256). Springer.

LaPensée, E. (2014). Survivance among social impact games. *Loading... The Journal of the Canadian Game Studies Association, 8*(13), 43–60.

Leclercq, T., Hammedi, W., & Poncin, I. (2018). The boundaries of gamification for engaging customers: Effects of losing a contest in online co-creation communities. *Journal of Interactive Marketing, 44*(2018), 82–101.

Lock, I., & Schulz-Knappe, C. (2019). Credible corporate social responsibility (CSR) communication predicts legitimacy: Evidence from an experimental study. *Corporate Communications: An International Journal, 24*(1), 2–20.

Maltseva, K., Fieseler, C., & Trittin-Ulbrich, H. (2019). The challenges of gamifying CSR communication. *Corporate Communications: An International Journal, 24*(1), 44–62.

Pérez, A., Baraibar-Diez, E., & García de los Salmones, M. D. M. (2020). Stories or expositive messages? Comparing their effectiveness in corporate social responsibility communication. *International Journal of Business Communication, 2020*, 1–25.

Podnar, K., & Golob, U. (2007). CSR expectations: The focus of corporate marketing. *Corporate Communications: An International Journal, 12*(4), 326–340.

Quairel-Lanoizelée, F. (2016). Are competition and corporate social responsibility compatible? *Society and Business Review, 6*(1), 77–98.

Ryu, D., Ryu, D., & Hwang, J. H. (2016). Corporate social responsibility, market competition, and shareholder wealth. *Investment Analysts Journal, 45*(1), 16–30.

Sen, S., Bhattacharya, C. B., & Korschun, D. (2006). The role of corporate social responsibility in strengthening multiple stakeholder relationships: A field experiment. *Journal of the Academy of Marketing Science, 34*(2), 158–166.

Skarmeas, D., & Leonidou, C. N. (2013). When consumers doubt, watch out! The role of CSR skepticism. *Journal of Business Research, 66*(10), 1831–1838.

Statista. (2022). *Games market revenue worldwide in 2021, by device*. Statista. Retrieved from https://www.statista.com/statistics/278181/global-gaming-market-revenue-device/

Sucu, I. (2020). Advergame uygulamalarinda popüler kültür ve tüketim kültürü etkisi [Popular culture and consumption culture effect in advergame]. *Dijital Çağda İşletmecilik Dergisi, 3*(1), 1–10.

Sung, Y. H., & Lee, W. N. (2020). Doing good while playing: The impact of prosocial advergames on consumer response. *Computers in Human Behavior, 106*(2020), 1–11.

Terlutter, R., & Capella, M. L. (2013). The gamification of advertising: Analysis and research directions of in-game advertising, advergames, and advertising in social network games. *Journal of Advertising, 42*(2–3), 95–112.

Troise, C., & Camilleri, M. A. (2021). The use of the digital media for marketing, CSR communication and stakeholder engagement. In M. A. Camilleri (Ed.), *Strategic corporate communication in the digital age* (pp. 161–174). Emerald.

Vishwanathan, P., van Oosterhout, H., Heugens, P. P., Duran, P., & Van Essen, M. (2020). Strategic CSR: A concept building meta-analysis. *Journal of Management Studies, 57*(2), 314–350.

Visser, W. (2016). The future of CSR: Towards transformative CSR, or CSR 2.0. In A. Örtenblad (Ed.), *Research handbook on corporate social responsibility in context* (pp. 339–367). Edward Elgar.

Waiguny, M. K., Nelson, M. R., & Marko, B. (2013). How advergame content influences explicit and implicit brand attitudes: When violence spills over. *Journal of Advertising, 42*(2–3), 155–169.

Wünderlich, N. V., Gustafsson, A., Hamari, J., Parvinen, P., & Haff, A. (2020). The great game of business: Advancing knowledge on gamification in business contexts. *Journal of Business Research, 106*(2020), 273–276.

Zbuchea, A., & Pînzaru, F. (2017). Tailoring CSR strategy to company size? *Management Dynamics in the Knowledge Economy, 5*(3), 415–437.

Bayram Bilge Sağlam is an Assistant Prof. Dr. in Dokuz Eylul University, Maritime Faculty in Izmir, Turkey. He has completed his PhD in Dokuz Eylul University Graduate School of Social Sciences, Maritime Business Administration in 2019. He has presented several papers on port management and sustainable supply chains in international conferences, has articles in journals and in edited books. He has participated in transport and port related projects in Turkey.

Egemen Ertürk is an Assistant Prof. Dr. in Dokuz Eylul University, Maritime Faculty in Izmir, Turkey. He has completed his PhD in Dokuz Eylul University Graduate School of Social Sciences, Maritime Business Administration program in 2020. He has several papers published on conflict management, port management, and corporate social responsibility both in international conferences and peer reviewed journals.

Part II
Sustainability in African Companies

Ghana
Nigeria 1
Nigeria 2
Nigeria 3

Sustainability Issues in the Cocoa and Chocolate Industry: Building a Promising Future for a Beleaguered Industry

Sam Sarpong

1 Introduction

Chocolate is one of the world's favourite treats, worth over **$100 billion in retail sales globally** (Maximise Market Research, 2024). Western Europe, North America and Asia Pacific (including Australia) are the biggest chocolate consumer markets accounting for 70 per cent of sales worldwide (Barry Callebaut, 2024). The key ingredient in chocolate, cocoa, however, typically grows far from where most chocolate lovers enjoy their treat. The vast majority of cocoa still originates from family-run small farms and in countries within the Global south (Odijie, 2023). For years, the cost and supply of cocoa had remained stable, making chocolate an accessible luxury for consumers around the world (Gitau et al., 2024). However, a drastic shift has occurred, leaving the industry grappling with skyrocketing prices and severe shortages of cocoa beans (Moreno, 2024). The rippling effects of the cocoa crisis have been quite profound, impacting not just cocoa farmers but the entire global chocolate production line (Gitau et al., 2024).

Fluctuating rainfall patterns, longer dry spells and higher temperatures are impacting the growth and yield of cocoa crops leading to the current shortfall (Gitau et al., 2024; Odijie, 2023). The development has led to a reduction in cocoa production, thereby, pushing the market into a volatile state where prices are at the mercy of speculative trading and reactive supply chain moves (Gitau et al., 2024). Many farmers have struggled to produce the amount of cocoa beans required as aging cocoa trees in West Africa, where the bulk of cocoa production takes place, have for decades been bearing fewer and less healthy pods (Moreno, 2024). Some of the trees have become susceptible to black pod and swollen shoot diseases, hence,

S. Sarpong (✉)
University of Central Lancashire, Preston, UK
e-mail: SSarpong@uclan.ac.uk

S. O. Idowu, S. Vertigans (eds.), *Sustainability in Global Companies*, CSR, Sustainability, Ethics & Governance, https://doi.org/10.1007/978-3-031-77971-8_10

requiring high maintenance costs. This has been exacerbated by seasons of poor weather that further diminished the harvests from farmers. Following from this, many farmers are also stepping away from the cocoa trade and focusing, instead, on crops that yield higher payouts, meaning demand simply is not aligning with supply (Gitau et al., 2024). Subsequently, prices have soared as a result of the supply-demand imbalance.

The historic cocoa shortage also reveals other problems that have plagued the West African region. For decades, farmers in Cote d'Ivoire and Ghana were chronically underpaid. The majority of them do not earn a living income, which has made it difficult for a number of them to afford decent housing, food, education for their children and other basic expenses (Walker, 2020). Without this being fulfilled, the farmers who have since become poorer, are now exploring alternative sources of livelihood (Odijie, 2023). When it comes to the companies that make and sell chocolate, some reforms have been proposed by them and even put in place; however, the problem is they are voluntary and based on an individual company's willingness to assume additional costs to, for instance, pay farmers a higher price for their cocoa (Walker, 2020).

Currently, the global market for chocolate and chocolate products is on the rise with the market projected to grow from $48.29 billion in 2022 to $67.88 billion by 2029 (Fortune Business Insight, 2024). The increasing popularity of chocolate confectioneries is primarily propelling the global market for cocoa and chocolate. New product developments in different food sectors are facilitating the market growth (Fortune Business Insight, 2024). With this also comes an inclination towards the darker varieties of chocolates as consumers prefer them over other types due to better health benefits and lower calorie intake. Additionally, the market dynamics have been influenced heavily by speculators (Moreno, 2024). This influx of speculative investment has led to a dramatic increase in cocoa prices, doubling within a single year and peaking at levels unseen before.

The high prices for cocoa have made it difficult for chocolate manufacturers to sustain production without passing costs onto consumers (Moreno, 2024). The question that arises is whether the supply side can match up with the demand being made. The growing demand for cocoa, thus, underscores the urgency in addressing the issues that relate to the industry's sustainability. Questions have been raised too about the future of the cocoa and chocolate industry in the light of all these happenings. One key issue, though, will be the sort of transformative approach that can be provided to aid a top-quality execution of a cocoa sustainability programme. Incidentally, understanding the dynamics between where cocoa is grown, and where chocolate is consumed, is also key to understanding cocoa sustainability and what can be described as 'sustainable chocolate' (Barry Callebaut, 2024).

The chapter, thus, provides an outlook of the cocoa industry, its challenges and measures needed to ensure its sustainability. It focuses on Ghana, which is one of the topmost cocoa production areas. The focus on Ghana is because it is a leading producer of cocoa and also a veritable intrant to explore conditions prevailing in the industry. The chapter draws on different data sources, including policy and document analysis of the cocoa sector, alongside focus group interviews and interviews with officials of the cocoa industry in unveiling the issues in the industry. It begins by looking at the general nature of the cocoa industry, its challenges and prospects. It then explores the issues regarding the sustainability of the industry. The next stage is the focus that is placed on Ghana as the focal point for our discussion. Here, we attempt to look into the key challenges and conditions existing in the cocoa industry in the country. This is followed by the methodology adopted for the study. The next section delves into the sustainability issues existing in cocoa systems. This will be followed by an evaluation of the measures to improve the environmental sustainability of cocoa farms. The conclusion follows next.

2 Cocoa Production and Growth

Cocoa farmlands can be located mostly in tropical geographies of West Africa, Southeast Asia and Latin America (Malan, 2013). In West Africa, cocoa is produced on small family farms, where farmers generally live on less than $2/day and rely on cocoa mainly for their income (Wessel & Foluke Quist-Wessel, 2015). Whilst cocoa has provided a critical source of revenue for these farmers and governments, cocoa-growing communities have also suffered from years of predatory market practices, declining commodity prices and unfair terms of trade (Walker, 2020). The perverse market dynamics, coupled with other debilitating factors that impact negatively on farmers, have contributed to severe problems in the industry (Odijie, 2023). For instance, West Africa has lost 90% of its original moist forest to cocoa growing and what remains is heavily fragmented and degraded (Leach & Fairhead, 2000).

Of late, concerns have been expressed regarding whether cocoa production would be able to satisfy rising demand in the long term (Casey, 2023). This follows recent studies indicating that yields are relatively lower than their potential across major cocoa-producing countries (Kozicka et al., 2018). The problem of low production is compounded by unfavourable weather conditions being experienced in the producing areas (Mera et al., 2021). High temperatures are increasing seedling mortality, decreasing leaf lifespan, and speeding up pod ripening, all of which can affect cocoa bean yields and quality (Climate-Smart Cocoa, 2022). Changes in weather conditions are causing some production areas to become less productive and unsuitable for growing cocoa (Bunn et al., 2017).

The complex problems facing the cocoa sector have become too fragmented and large for any one entity to address on its own (Walker, 2020). Given this and other concerns, an increasing number of stakeholders, from both the private and public sectors and civil society alike, have begun a number of initiatives to ensure the industry stays afloat (Barry Callebaut, 2024). The recent developments in both consuming and producing countries point to a strong desire for more systemic changes within the industry. Industry players now understand that they ignore sustainability at their own peril and at the risk of higher prices, lower production and adverse supply shocks in the years to come. It is expected that the urge to make the cocoa industry more sustainable and resilient can lead to the optimisation and maximisation of benefits to farmers and their families and the industry at large (Walker, 2020).

3 The Quest for Sustainability

In the broadest sense, sustainability refers to the ability to maintain or support a process continuously over time (Mollenkamp, 2023). It is based on what people value or find desirable. Sustainability can be seen as the capacity to maintain or improve the state and availability of desirable materials or conditions over the long term (Harrington, 2016). *It serves as a basis to ensure that* **the needs of the present are met without compromising the ability of future generations to meet their own needs** (WCED, 1987). In everyday use, sustainability often focuses on major environmental problems, including climate change, loss of biodiversity, loss of ecosystem services, land degradation, air and water pollution. The idea of sustainability can guide decisions at the global, national and individual levels (Berg, 2020). In business and policy contexts, sustainability seeks to prevent the depletion of natural or physical resources (Mollenkamp, 2023).

3.1 Sustainability in the Cocoa and Chocolate Supply Chain

Sustainability in the cocoa and chocolate supply chain refers to a situation whereby cocoa and its derivatives such as chocolate can be available for and enjoyed by future generations. According to Barrientos (2014), 'cocoa sustainability' emanates from a perceived business imperative to safeguard the industry's long-term viability. Ensuring sustainability also means that cocoa ought to be produced in a manner that ensures a regard for a **diverse environment** and the **farming communities** (Barry Callebaut, 2024). Sustainable cocoa supply chain, therefore, calls for high standards of quality and productivity. It makes provision for a situation whereby cocoa is safe for consumption, complies with manufacturers' quality requirements and meets the growing global demand. The concerns being expressed within the industry are strengthening a huge resolve to ensure cocoa production is safeguarded in the long term, a situation which is prompting shifts towards sustainability in the sector (Walker, 2020).

In the contemporary era, the importance of the ideation of business models and products that generate positive social impact and present alternatives for sustainable environmental solutions becomes increasingly noticeable (Sarpong & Alaurussi, 2022). This is also indeed the case of cocoa. Consumers of cocoa are now particular about sustainable stewardship within the industry following recent reports of certain unsustainable practices. The desire to ensure the industry lives up to expectation has mainly been driven by conscious consumers and environmental advocacy organisations who are putting growing pressure on commodity traders, manufacturers and other stakeholders to find solutions to the myriad of problems confronting the industry (Bartley et al., 2015). Younger generations too are also deeply involved as they openly align with brands that reflect their values (Sarpong, 2023). They investigate where their food comes from and how it was made and are, therefore, particular about sustainable products. Cocoa and its allied products like chocolate are experiencing this scrutiny from various groups, hence the current apprehensions.

3.2 *Challenges in the Industry*

In terms of what the industry offers, so much too can be said too about its adverse effects. Cocoa sourcing has ramifications for the planet, people and businesses. The industry has been associated with the improper use of agrochemicals, water waste and loss of biodiversity, albeit to a varying degree. Furthermore, many social issues such as child labour, unsafe working conditions, human rights violations, lack of schools and healthcare facilities tend to accompany cocoa sourcing (Fountain & Huetz-Adams, 2020; Lambin et al., 2018). These social conditions underscore the economic reality of cocoa production, including unproductive farming practices, the lack of management skills, volatile prices and the resulting low yields and inadequate and unstable income.

Factors ranging from the natural cyclical production of trees to weather conditions such as heavy rains, higher temperatures or droughts also affect the supply of cocoa (Wong, 2021). There are also issues with low yields which can be attributed to pests, aging trees and diseases that attack the trees (Wessel & Foluke Quist-Wessel, 2015). Estimates hold that as much as 30–40 per cent of the crop is lost annually to pests and diseases (Duffey, 2009). Incidentally, many farmers cannot afford to buy fertilisers, pesticides and herbicides. In Ghana, although the government provides subsidised fertilisers, they are woefully inadequate and farmers have to contend with smaller quantities for their crops. High volatility and continuous price drops also create risks for all cocoa value chain participants, especially the farmers. They bear the greatest risk as they cannot influence market prices. Increased volatility causes uncertainty for farmers and affects their incomes (Hütz-Adams & Schneeweiß, 2018). Aside from that, political and civil unrest in producing countries tend to affect production and supply (Anderson, 2011). It is pertinent to note that cocoa production dropped considerably in the Cote d'Ivoire from 2002–2007 when the country experienced political unrests.

Other challenges in the industry pertain to the limited knowledge of farmers in modern farming techniques and difficulties in getting access to credit, a vital part in cocoa farming. The low levels of adult literacy among cocoa farmers make it impossible for some farmers to adopt new technologies that they are taught. Health risks and lack of access to education and quality healthcare also remain a constraint too. These and other factors mean that the cocoa industry's long-term viability and production is seriously threatened. Indeed, the nature and scope of the cocoa industry's issues requires that it addresses the diverse socio-economic, environmental and commercial issues.

The next section focuses on Ghana's cocoa production and its related issues. This is to enable us to gain an insight into what has besieged the industry and how this can be remedied.

4 Ghana's Cocoa Production

Cocoa is iconic for Ghana not just in terms of the money it brings to the nation's coffers. It is indeed one crop that Ghana greatly identifies with and places much premium on. Ghana prides itself as the producer of the best cocoa in the world, a condition which has been accomplished by family-run small farms dotted round the forest area of the country. Cocoa employs approximately 800,000 farm families and generates about $2 billion in foreign exchange annually and is a major contributor to the Ghana government revenue (GCB Bank, 2022). Production of cocoa has helped to open up Ghana's rural economy for integration into the national economy and has also enhanced trade. It has also been a major contributor to Ghana's educational system, providing scholarships to children of cocoa farmers and other brilliant students alike. Through this effort, many needy students have been able to complete secondary schools and have become skilled personnel in various disciplines.

In 2022, the export of cocoa beans and cocoa products from Ghana amounted to about $2.3 billion, which was a decrease of 19.0% from the preceding year (Sasu, 2024). The highest value of the exports, in recent years, was achieved in 2021, with approximately $2.84 billion. Ghana produced a record 1.4 m tonnes of cocoa beans in the 2020/21 season, but it has been tough for the country in recent years as illegal mining, known in the local parlance as 'galamsey', has taken portions of cocoa lands. Some cocoa farmers have also sold their farms to the illegal miners to mine on them. Some, too, have had their lands encroached upon by the miners, jeopardising efforts at ensuring that Ghana's place among the top producers remains intact. Between 2019 and 2020, around 19,000–20,000 ha of cocoa farms across the country were destroyed by illegal miners in the search for gold, according to a Ghana Cocoa Board (COCOBOD) official. This longstanding friction over land is currently exacerbating the problems of an already volatile cocoa market.

Cocoa-growing communities in Ghana have also suffered from years of predatory market practices, declining commodity prices and unfair terms of trade (Barry Callebaut, 2024). These perverse market dynamics, coupled with low productivity by poor farm management practices, have contributed to severe problems in the cocoa sector. Poverty is rampant among the farmers; there is also a high level of deforestation, whilst accusations of persistent child labour have been made against the farmers (Barry Callebaut, 2024; Ollendorf, 2021). Alongside these, frequent pesticide poisonings and declining food and livelihood security remain major challenges facing the industry (Sasu, 2024). The use of fertilisers and pesticides is quite common on cocoa farms but this is expensive for farmers even with government subsidies.

Currently, there is a fear that the frequent droughts and rising temperatures affecting Ghana and other West African countries may turn large swathes of the region to savanna by 2050 (Hudson, 2022). In order to halt this, the United Nations Development Programme (UNDP) is working with stakeholders to train many farmers to grow 'shade trees' over their farms to protect cocoa trees from heat and water stress. According to Hudson (2022), this practice has the added benefit of increasing cocoa yields in the short term and overall sustainability in the long term in the country. For a country that heavily depends on cocoa, such assistance is needful to revamp the cocoa industry.

The importance of the cocoa industry is not lost on Ghanaians. It is highly regulated due to its economic importance as an export revenue generator. COCOBOD, a state-controlled institution, is responsible for regulating the prices of cocoa and coordinating the marketing activities.

4.1 The Role of COCOBOD

COCOBOD is the main player in Ghana's cocoa industry, performing various critical functions. Back in the 1930s, when Ghana started selling and exporting cocoa in commercial quantities, it had many European merchants who were buying directly from cocoa farmers for export. By 1937, the farmers reported that the price being paid to them was quite low, so they decided not to sell their cocoa. A committee was set up by then government which recommended the establishment of a marketing board in 1940. That was reconstituted in 1984 as COCOBOD with the mission to promote, facilitate and encourage the production, marketing and processing of cocoa, and more recently that of coffee and shea nut. COCOBOD is the only institution permitted to sell Ghanaian cocoa to the world market. COCOBOD's functions include quality control, provision of seeds and the marketing of cocoa beans and related products. It also creates employment in its own right via its 'rehabilitation programmes'—i.e., cutting down and replacing diseased and overaged cocoa trees—programmes which now account for 40,000 jobs. Besides, it provides farmers with subsidised fertilisers and extension services.

Through COCOBOD, the government remains a key player in the cocoa value chain, selling beans in the central market and regulating the price for producers and buyers. Cocoa is grown and harvested by farmers who ferment and dry the cocoa beans before sale to the licensed buying companies (LBCs). Regulated by COCOBOD, these privately owned and operated businesses are responsible for purchasing the cocoa at farm gate at a guaranteed floor price (i.e., the 'producer price') and for transporting the cocoa to one of three takeover points to sell at a fixed price to COCOBOD for export. They also share the responsibility of delivering cocoa that only meets COCOBOD's stringent quality standards. Strict rules govern their buying activities. LBCs are required to grade the beans for size and quality (Ollendorf,

2021). Once the bag is sealed, the cocoa remains in the custody of the LBCs until it is taken over by COCOBOD.

Over 80% of beans purchased by the LBCs are shipped abroad in raw form, meaning value is added in other markets other than Ghana. Local processing is largely limited to the production of semi-finished goods or chocolates which are then exported or consumed locally.

5 Methodology

In order to know the sustainability issues in the cocoa industry, particularly in Ghana, the chapter delved into a number of literatures on the issue. The perspectives of some industry players, such as farmers, extension officers, researchers, licensed cocoa buying companies and policy makers, were also elicited. The chapter built on this to assess the sustainability issues in the industry. Focus group interviews were used to gather the views of the farmers and clarifications on issues raised by them were sought with industry players. Forty-five farmers were interviewed, as well as three extension officers of COCOBOD, two COCOBOD managers and two policy analysts.

The study organised focus group interviews for two groups of 12 and 15 farmers, respectively, at Konongo, a cocoa-growing area in the Ashanti region of Ghana, and 18 people in Tarkwa-Breman, in the Western region of Ghana, to get a better understanding of issues pertaining to the industry. The number of respondents chosen was arbitrary. The respondents for the focus group interviews were those who came and expressed interest in speaking to the issue. The study felt that the number of respondents was sufficient enough to have the information needed to have an informed knowledge of the happenings within the industry. The participants provided a good understanding of the history of cocoa production in their area and Ghana as a whole, which was important in getting to terms with the issues. Of the 45 farmers who took part in the study, 17 were females.

Interviews were also held with three extension officers and two managers of COCOBOD and two cocoa policy analysts as well. Most of these discussions were to clarify issues the farmers had raised and also to have an insight into COCOBOD's operations and Ghana's strides in the cocoa industry. Some information was gathered during visits to cocoa farms through participant observation. These were supplemented with documentary analysis. Data collection was implemented between October 2023 and January 2024, which mostly coincided with the main harvesting season. The experts, i.e. policy analysts and extension officers, were identified by contacting the main government ministries and administrations of relevance to the cocoa sector. The field work included travel to key cocoa-producing area, where focus group interviews were carried out with cocoa farmers as well as extension officers in the cocoa-producing regions. Policy analysts and COCOBOD officials were also roped in to offer their opinions. In all, there were 52 participants. This was scaled up with secondary research alongside observations.

These interviews with the respondents focused on the general outlook of the cocoa industry and were meant to understand the challenges, opportunities and other issues in the cocoa industry. The data were analysed through inductive coding and memoing, which Miles et al. (2013) have noted provides an iterative process of reviewing the extensive field and interview notes to extract key emerging themes. Data analysis was a continuous process during and after the field work, as the key themes emerging from the data were presented, discussed and further refined during the focus group sessions. External validation was sought during subsequent expert interviews after the field work.

5.1 Measuring Sustainability in Cocoa Production

Measuring sustainability is difficult as the metrics are evolving. Currently, they include certification systems, types of corporate accounting and some indices. The chapter took into account the multiple dimensions contributing to farm level sustainability by looking at particular themes that can be captured within the industry and within the environmental, social and economic domains.

In considering pathways that could lead to the adoption of more sustainable farming practices, the chapter explored, in part, how these can be obtained in cocoa production by examining the role of certification schemes and the significance of cooperatives as part of this. Cooperatives provide farmers with a shared space for capacity development, acquiring inputs jointly, bargaining for better market conditions and prices are, therefore, helping out in this direction (Bymolt et al., 2018). On the other hand, certification regulations and sustainability standards promote the adoption and application of practices such as intercropping, integrated pest management or circular farm management.

The continuous rise of certification as a form of sustainability governance is grounded in various factors. These include the inability or unwillingness to pass and enforce robust legislation on sustainable production at the national level, high levels of poverty among small-scale producers and poor working conditions, pervasive challenges of environmental degradation and biodiversity loss, and increasing public pressure from NGOs and consumers to combat social injustices and protect the environment (Common Fund for Commodities Annual Report, 2018). Standards and certification are continuously evolving amidst persistent struggles for legitimacy and demonstrable impact on the ground. What these developments perhaps best express is a recognition of the complexity of sustainability challenges at production level, highlighting the limits of current approaches and driving a continued search for new, improved responses.

6 Assessing the Inherent Problems in Cocoa Production

6.1 Focus on Farmers

Ghanaian cocoa farmers are quite expectant as they feel they deserve more than they are being given. They remain passionate too about the importance of their role in society. '*We want more to be done for us. Inflationary levels are high currently and therefore we feel we need to be paid quite well so that we stay committed to cocoa farming. Without us, Ghana's economy will be in shambles*'. This message found expression throughout all the focus group interviews. With poverty endemic in cocoa-producing areas, a veritable way to minimise this trend, perhaps, will be to raise the farmers and their communities from the high level of poverty which has engulfed them. The government in Ghana has, over the years, raised the producer price annually in a bid to attract more cocoa farmers. It has also been able to ensure farmers feel quite interested in cocoa production, yet, differences continue to emerge regarding the most reasonable amount that ought to be paid to the farmers and what the government can offer. Whilst the farmers think they deserve more, the government, on the other hand, feels the periodic increments are in line with its avowed policy to give farmers what they are due.

The quest to ensure farmers are paid fairer price was hugely debated. The farmers noted that the low producer price of cocoa leads to low investment in crop management, labour shortage and high cost of labour, and poor infrastructure in farming communities. '*We always want the government to come to our aid by providing us with a fairer rice for our produce'*, many of the farmers indicated. It is, however, pertinent to note that the government considers many factors when setting the producer price of cocoa. Among these factors are: world market price trends, the objective to establish a price stabilisation fund, the general expectation of farmers that the producer price should only be increased or at least maintained irrespective of the trend of world market prices, and the anticipated effect of producer price on the farmers' morale (Amoah, 1998).

With the endemic poverty running through the farming communities, the farmers also stated that it affects their ability to hire more workers to work on their farms, a tendency that results in these farmers drafting in their family members, including children, into working on farmlands. Whilst some scholars have denounced this by saying this development in itself leads to a situation where the future generation of cocoa farmers are deprived of their childhoods and education (Walker, 2020), many of the farmers disagreed with that. To them, using their children on their farms does not amount to child labour, rather, it is a way to prepare their kids to become future cocoa farmers. A visibly upset farmer asked: '*What is wrong with having my children working with me during the weekends? They don't do much here except to help in cutting the pods or spreading out the beans on the mat. They go to school during the week days. I live on my farmland so you can see my children running around the entire place any time they are not in school. So, how can this be tagged as child labour? They are here always, not because we force them to do so'*, the farmer explained.

Many farmers believe people in the Global north especially and international child rights advocacy groups in particular, are not familiar with the family system in developing countries, especially Africa. '*When they see our own kids helping out on farms, they brand that as child-labour. Our cultural system is different from theirs, so they cannot understand how we do things here*', a farmer also indicated. Another noted, '*we are a close-knit society and all of us, from kids to the elderly, contribute to the family upkeep in different ways. Our children don't do tedious work as we do, but they help out because they want to. Is there anything wrong with that*?' he queried.

In the course of the interaction with the farmers, it emerged that many were those who felt that the definition of child labour was inappropriate in the Ghanaian context. '*They are killing our industry with the child labour tag*', one farmer retorted, noting that, '*it could be a conspiracy to give a dog a bad name so that they can avoid paying the fairer prices we are demanding for our cocoa*'. This issue was bandied about for a long time in view of its importance to the farmers. However, defined, many farmers felt this is not an apt description of what goes on their farms. Though some did not rule out pockets of the so-called child labour phenomenon, they believe it cannot be a major practice. They felt, on the whole, the concept needs some clarification. '*We sometimes outsource out farming activities to some labourers, which in that case could lead to these labourers bringing their children in to work with them, but we don't think this is so rampant here*', one farmer explained. She explained further: '*The government's new feeding programme in which children of school-going age are fed in primary schools has served as an incentive for kids to stay in school, so hardly would you find children of school-going age idling about these days. Even if there have been some incidences of child labour, I don't think it is a major phenomenon here in this country like the way child rights advocates are portraying it'*.

Another issue that engaged us in the interviews was the use of agrochemicals in farming. The farmers called for more training on how to handle the chemicals as poor handling of these chemicals has proved fatal for some of their colleagues. Many deaths and cases of poisoning have been caused by the mishandling of pesticide wastes and containers (Boateng et al., 2023). The level of awareness and knowledge of pesticide risks among certified cocoa farmers is essential for improving safety in all aspects of pesticide handling. Lately, more agricultural extension officers have been employed to provide farmers with the requisite knowledge to tackle such issues. These officers have since intensified the organisation of periodic workshops for the farmers on the use of agrochemicals and related issues. The extension officers also provide the farmers with new information on cocoa production which is helping the farmers to put up with best practices.

Other issues that came up were the lack of community resources, such as financial support, funding for fertilisers, quality education and healthcare, which the respondents said demand some intervention from the government. They seemed to have trouble accessing the infrastructure, training and financing that they need to invest in their farms. Most of the farmers complained that it is not easy to be a cocoa farmer, especially with dying crops and shrinking harvests. As a means to overcome that, a farmer suggested that: *'farmer cooperatives could serve as key partners to help ensure farmers get the best from the authorities'*. Cocoa cooperatives provide a means by which members of farmer groups are able to have access to agricultural services and for institutional and value chain actors to reach and engage with farmers. Through this means, the farmers are able to collectively seek certain products and services like finance, which otherwise would have been difficult to have individually. Issues regarding the aging, low-producing trees and their changeable incomes have led to a number of these cocoa farms being abandoned, sold or turned into something else.

6.2 The State of Cocoa Financing in Ghana

The issue regarding cocoa financing was also debated on quite extensively. Ghana requires external support for cocoa purchases or financing facilities, so, each year COCOBOD pursues international syndicated loans to finance licensed buyers who purchase cocoa from smallholder farmers for export. A consortium of banks is involved in providing funds for cocoa purchases. When the syndication process unfortunately drags on longer, it affects cocoa purchases which, in effect, leads to farmer frustrations. According to a policy analyst, *'through their acts, the financial institutions are playing a yeoman's role in safeguarding jobs and also supporting an iconic industry'*. But farmers do not get paid until this facility is obtained. Sometimes it takes a long time for the government to access such a facility and that means farmers would have to wait endlessly before they are paid for their produce. What makes this disconcerting is that COCOBOD has, in recent times, been saddled with a myriad of challenges including its overwhelming debt levels which from all accounts is making farmers quite uneasy.

A policy analyst explained, *'COCOBOD has to source for funds, oftentimes from foreign sources, before purchasing cocoa. This leaves farmers in a state of impoverishment. After going through the entire processes to harvest their cocoa, they then have to wait until money comes through to government coffers before they would be paid for their produce'*. The extension officers also noted that, *'in view of that, Ghanaian farmers who are close to the border areas are sometimes compelled to smuggle their cocoa to either Cote d'Ivoire or Togo to sell them'*. The famers also attributed the smuggling of cocoa to neighbouring countries by their colleague farmers to the low producer price.

Many of the farmers the study interacted with were of the opinion that the producer prices are too low and felt the government underpays them by offering them low prices. This, they believe, is leading to low investment in crop management, labour shortages and poor infrastructure in farming communities. They questioned why they are not allowed to sell their cocoa freely on the international market, as is the case with crops like yam and pineapples. What emerged from the interactions was that the farmers do not have adequate knowledge of the complex nature of international trade in primary commodities like cocoa and, therefore, do not realise that they cannot easily sell their produce directly on the international market.

Efforts to increase cocoa prices have, meanwhile, been underway since 2018, with Côte d'Ivoire and Ghana, taking concrete steps towards a more collaborative system to exert more control on global prices. To increase returns to local producers, the two countries set a floor price for beans at $2600 per tonne in 2019 and implemented a fixed premium on cocoa sales. A living income differential (LID) added $400 on every tonne of cocoa, effective from the 2020/21 season to protect farmers' income, with the aim of giving 70% of the free-on-board threshold price to farmers. This led to a 28% increase in the price of cocoa in the 2020/21 season, with the farm gate price set at GHS10,560 ($1837 at the time). This price was maintained in the following season, despite a dip in global cocoa prices. However, the LID has, in some cases, been cancelled out by the introduction of negative origin differentials, a premium on quality which sometimes falls below zero in response to commodity price pressures. That aside, enforcement of the LID has been lax, with some cocoa buyers refusing to pay despite rising prices of chocolate to consumers.

In spite of that, efforts are being made to redress the poor pricing for farmers. In early October 2023, Ghana raised its farm gate price for the 2022/23 season to GHS12,800 ($1249), an increase of 21%. As well as the higher price, on-going collaboration by Ghana and Côte d'Ivoire to coordinate origin differentials could support the meaningfulness of the LID, while Nigeria' decision to apply the LID to its price setting further strengthens the programme.

The farmers also complained about the lack of social amenities like electricity and portable water in their community. According one extension officer, '*the issues raised by farmers as constraints are quite peculiar to the rest of the cocoa growing communities in Ghana, hence, a coherent package of social and technical solutions for cocoa production in Ghana will have to include arrangements and strategies for tackling these problems*'. COCOBOD officials provided assurances of providing water to such communities and explained that this is on-going.

6.3 *Illegal Mining on Cocoa Farms*

A major problem in cocoa-growing areas of late is the influx of illegal small-scale miners who are encroaching on cocoa farms. The encroachment stems from the realisation that a number of these farms have gold deposits underneath them. Some farmers have received large amounts to part with their farmlands, whilst others have had their lands encroached upon, a situation which the authorities are finding quite difficult to stop. Some farmers who bemoaned the threat posed by the activities of illegal miners explained that their farmlands are being destroyed by the day because of the greed of illegal miners. The lure of gold is strong among many Ghanaian youths and this is pushing some small-scale miners to resort to mining on cocoa farms, Artisanal mining accounts for about a third of the country's gold output and farmers talked about individuals especially those fronting for Chinese businessmen persistently offering them large sums of money to give up their lands.

Some farmers, meanwhile, cannot resist the allure of quick money that is being offered them. They explained that the proceeds from cocoa farming take much longer to come through, so a number of them accept the monies and then hand their farms over. Ghana has since lost some arable lands because of this tendency. *Between 2019 and 2020, around 19,000–20,000 ha of cocoa farms across the country were destroyed by illegal miners*, according to one extension official. This long-standing friction over land is exacerbating the problems of an already volatile cocoa market.

6.4 Giving Cocoa Farming a Boost

It is apparent that many Ghanaian youths have a negative perception of agriculture. A general perception in Ghana is that those who go into agriculture are rural folks as well as young persons who are school dropouts. Younger generations, therefore, increasingly seek alternative careers in cities rather than following their families into farming and specifically cocoa farming which takes a longer time to reap. A huge debate on why farming is not so attractive to the youth also took centre stage in our discussions. The extension officers and farmers who were interviewed felt the need for farming to be made more attractive in order for cocoa production to gain a strong foothold. An extension officer stated that, '*Ghana stands to benefit immensely from agriculture if the right policies are adopted in making farming attractive*'. Another extension officer noted that, '*in order to make farming attractive, the success stories of those who have really made it ought to be told. Besides recognition should be given to those who have made it through farming*'.

Ghana has realised the need to acknowledge its farmers. It has since set the first Friday of December each year as a day to celebrate the hard-working farmers in the country. It has also given recognition to cocoa by observing a chocolate day on every 14 February (Valentine's Day). COCOBOD has also implemented a number of programmes and policies to attract the youth to cocoa farming, according to a COCOBOD manager. Among some of the schemes that COCOBOD recently launched are 'Cocoa Pension Scheme' as a means to attract the youth to the industry. COCOBOD is also encouraging young people to form 'young farmers associations' to give them help to realise their vision. It has also set up a gender desk and research department, along with implementing measures to ensure gender equality and youth inclusion.

Touching further on what can be done to attract the youth especially to cocoa farming, an extension officer said farming can be attractive if and when government lays a strong foundation and ensures that agriculture becomes the backbone of Ghana's economy. It was acknowledged by the extension officer that, '*the foundation for this will be for training programmes to be stepped up as well as providing access to inputs*'. She went on to declare that, '*helping farmers to implement efficient, environmentally friendly and safe practices in their daily operations can boost their productivity'.* Meanwhile, as a sign of government's determination to revamp the cocoa industry, a number of extension officers have been trained and deployed in cocoa-growing areas of late to take farmers through the rudiments of cocoa production, a step which is seemingly bearing fruit.

Despite such laudable measures, the extent to which the government's cocoa sector development strategy adequately meets the needs and aspirations of farmers remains to be seen. Although the strategy attempts to tackle both economic (liberalised market and pricing policy) and technical issues, the overall strategy remains essentially a top-down linear approach. Much needs to be done to engage farmers before certain policies are rolled out. The farmers complained that sometimes they are made to adopt certain practices or technologies without being consulted first. It was acknowledged during the interviews that such engagements need a genuinely participatory approach in which farmers are not passive recipients of plans or technology, but key players in identifying, analysing, designing and implementing these activities. This, it was felt, could indeed pave the way for a high rate of adoption by the farmers.

6.5 Controversies in Supply Chain

The growth and production of cocoa has been marred by a lot of controversies. These include issues bordering on the inappropriate use of agrochemicals, the alleged use of child labour, the inadequacy in benefits to cocoa farmers, amongst others. The management of COCOBOD insists they are doing much to improve the image of the industry and the lot of cocoa farmers, yet a lot still needs to be done in that direction. The lack of funds to buy cocoa from the farmers, in recent times, for instance, is acting as a serious disincentive to cocoa farming. A cocoa analyst stresses that, '*all the industry stakeholders should be able to create lasting benefits for cocoa farmers, their families and their communities—empowering them to own their futures and achieving their own success*'. The analyst indicated further that, '*every responsible company with a footprint in a cocoa-growing region needs to have a plan that can bolster the long-term viability of the commodity they trade in*'.

The most fundamental recommendation for the industry, meanwhile, would be to help farmers earn more money and reduce their risks, as many of the respondents alluded to in the focus group interviews. Incidentally, this has not been the case. According to Oxfam (2023) the major chocolate manufacturers are rather making huge levels of profits whilst failing to pay prices that support a living income for cocoa farmers in Ghana. Oxfam has revealed that the world's four largest chocolate manufacturers, namely, Hershey, Lindt, Mondelēz and Nestlé, made nearly $15 billion in profits from their confectionery divisions alone since the onset of the pandemic. In a survey it conducted of 400 Ghanaian farmers supplying cocoa corporations, Oxfam found that the farmers' net incomes had fallen on average by 16% since 2020. Oxfam also analysed the wealth of the two biggest private chocolate corporations, Mars and Ferrero, which, it claimed, has risen by $39 billion since 2020, giving them a combined net worth of around $157 billion. As such, Oxfam concluded in its report that there is big money for chocolate manufacturers but definitely not for farmers.

6.6 Bolstering the Cocoa Industry

Lately, a number of companies have been helping to bolster the long-term viability of the commodity they trade in. One of them, Olam, for instance, has trained farmers in Cote d'Ivoire and Ghana in good agricultural practices, hosting coaching sessions to encourage the adoption of farming techniques that sustainably increase yields. The company also provides interest-free financing to farmers too. Other companies such as Forbes Chocolates have worked through local foundations. Forbes Chocolates, for instance, works with people in Ghana to create a better life by building schools, focusing on health issues and providing A few have.

Ghana is now quite keen to add value to its cocoa. It is, therefore, encouraging investments from both local and international players in the building processing facilities in the country. Some have since answered the call. Ghana's Afrotropic Cocoa Processing commissioned a new cocoa processing factory in 2019 in an investment valued at $30 million. In November 2021, Cargill completed the $13 million expansion of its cocoa processing plant in Tema, bolstering its annual production capacity to 90,000 tonnes. Most recently, in May 2022, Koa, a Swiss-Ghanaian start-up, received at $3.5 million investment from the Landscape Resilience Fund, which mobilises finance for communities, conservation and commerce, and the IDH Farmfit Fund, an Utrecht-based Sustainable Trade Initiative that promotes sustainable trade to build a new cocoa processing facility. These investments support Ghana's ambitions to reshape its role in the global cocoa market and retain more value from its leading commodity crop.

Small and middle-scale entrepreneurs like TearAfrica have also heeded the call to add value to cocoa and are producing chocolates in the country. Many, though, have complained about the constraints they face competing internationally. *TearAfrica*, which has a facility in rural Ghana and produces quality chocolates comparable to those produced in Europe is, however, constrained by the set of regulations it has to navigate before it exports its chocolates. Despite its rich products, competing with chocolates produced in say Belgium and Switzerland, which are renowned chocolatiers, seems quite a huge task for some of these local firms. According to a representative of the company, this is because 'Made in Ghana' chocolates do not resonate with many consumers in Europe.

6.7 *Ghana's Attempt at Cocoa Farming Sustainable*

Ghana is trying desperately to make cocoa farming quite sustainable. It has initiated some measures like the mass-spraying of all cocoa farms under the Cocoa Diseases and Pests Control programme since 2001 at no direct cost to the farmer. The government has also initiated an interest-free credit scheme called the Cocoa 'Hi-Tech' programme since 2003, which aims at increasing productivity by providing fertilisers and pesticides to farmers. That aside, it is also looking at improving the productivity of cocoa farmers. COCOBOD officials interviewed claimed the government is implementing productivity enhancement programmes with the aim of increasing productivity from 600 kg per hectare to 1000 kg per hectare by 2026/27.

More importantly, as one of the officials put it, *'with consumers' preference for sustainable chocolate and cocoa product concepts, it seems sustainability stewardship has become quite an important element in the industry*'. In line with that, the respondent said, the government's approach is to ensure transparency in cocoa production. He noted that sourcing sustainable cocoa does not only pay dividends in customer loyalty and risk reduction, but it is also critically important to other stakeholders and helps to future-proof one's business. The official added that the government is also working extensively with stakeholders to ensure cocoa farming does not pose a risk to the environment.

In spite of that, the officials were of the view that the price of cocoa constitutes the biggest threat to the long-term sustainability of cocoa farming. This is because the farmers turn their attention elsewhere when the producer price is low. An indication of this is that, in the early 2000s many farmers began to focus on oil palm trees instead of cocoa when prices of palm nuts went up than cocoa. The seemingly low producer price offered to farmers tends to demoralise them, leading them to shift to better paying crops or jobs and that is exactly the feeling among the farmers interviewed.

7 Addressing the Challenges

Currently, many initiatives are underway by NGOs, chocolate companies, significant global actors like the European Union and other stakeholders to find solutions to the challenges in the industry. Stakeholders in the cocoa industry are increasingly partnering now to dedicate funding and expertise to improve sustainable cocoa. The world's two largest producers of cocoa, Côte d'Ivoire and Ghana, have also formed a partnership to drive up the price for cocoa farmers to ensure farmers live off their produce quite well (Fountain & Huetz-Adams, 2020). Such approaches have sought to address particularly salient issues like child labour or deforestation in West Africa (Carodenuto, 2019; NORC, 2020). But these initiatives and processes in the past did not always include all stakeholders. This, therefore, led to a situation where some failed or were not successfully implemented (Nelson & Phillips, 2018). Initiatives meant to curb the so-called child labour, for instance, need to be done thoughtfully by looking into the context in which families use their children on the farms. Many farmers see their denigration on this issue as an affront to their sensitivities and as a sense of betrayal and lack of knowledge about their cultural orientation.

The Cote d'Ivoire-Ghana Cocoa Initiative (CIGCI), which was established in March 2018 to improve farmer pay and financially improve the lives of cocoa farmers is a step in the right direction. The CIGCI was set up to bring pressure to bear on international cocoa buyers and trade houses for a minimum floor price of $2600 per tonne for cocoa beans produced in the two respective countries. Meetings with stakeholders from trade houses, cocoa purchasing companies, chocolatiers, the World Cocoa Foundation, and the International Cocoa Organisation culminated in a decision to introduce a new trading mechanism with the Living Income Differential (LID), set at US$400.00 per tonne, for cocoa sold by both countries starting from the 2020–2021 season.

8 Debating Sustainability in the Cocoa Industry

The cultivation, processing and use of cocoa, oftentimes and to some degree, fly in the face of sustainability issues. The reality is that cocoa farming as it exists today needs to address certain public concerns whilst providing a clear business incentive for continued production.

The fluctuating nature of cocoa production has always been a subject of concern. There are situations where cocoa farmers are driven to produce cocoa beans on a large scale to ensure they keep pace with high demands for the produce. Subsequently, when production goes up, the producer price goes down and this affects the farmers badly whilst the economies of exporting countries also receive a jolt, throwing out of gear their economic projections.

The cocoa price hike is just one example of how climate change has far-reaching impacts on society and the economy. Extreme weather and changing climate patterns have upended crop harvests. Crops are increasingly under threat from heatwaves, intense rains and other climate-related risks whilst pests continue to ravage whole cocoa farms. These have negatively impacted thousands of smallholder producers who have seen their harvests diminish. Excessive rainfall in Ghana and Côte d'Ivoire during the fourth quarter of 2023, for instance, led to a flare-up of swollen shoot virus and black pod disease, a condition that causes cocoa pods to rot and harden. Unfortunately, fighting this plague has been an uphill battle. The mealy bugs are highly resistant to pesticides, relegating farmers to try to curb the spread by destroying infected plants, breeding disease-resistant trees and even inoculating crops with cocoa swollen shoot virus disease vaccines.

8.1 Issues Regarding Sustainable Initiatives

Many organisations and governments have expressed the desire to work for the fulfilment of the sustainability issues within the cocoa industry. For instance, the Côte d'Ivoire-Ghana Cocoa Initiative (CIGHCI), a joint economic pact, is working to ensure the development of a sustainable pricing mechanism for cocoa producers in Cote d'Ivoire and Ghana. The current European Union (EU) supply chain requirements are also aimed at improving human rights and environmental protection, particularly deforestation. Collective efforts taken by industry and governments, such as the Cocoa and Forests Initiative (CFI), have promised to end deforestation and to set up joint monitoring mechanisms to respond to incidents of forest clearance. Some organisations in the chocolate business have also begun to implement the cocoa life sustainability process in their chocolate manufacturing (Barry Callebaut, 2024). But some of their sustainability mechanisms, nonetheless, have also received criticisms for being out of touch with reality. For example, certification schemes initiated by some of the companies lack farmer inclusion in decision-making and the ability to trigger large-scale transformation (Glasbergen, 2018).

Certification schemes in global supply chains are usually a combination of set requirements (standards) on three main themes: environmental sustainability, social sustainability, and safety and quality. Smallholders who are inadequately supported in terms of access to finance, land, labour and other resources have trouble getting certified without external assistance and support (Common Fund for Commodities Annual Report, 2018). Even Fairtrade, with its focus on smallholders, does not appear to attract the poorest or most marginalised producers (Elliott, 2018). Aside that, much of the sustainability trend has been seen as being driven by actors from 'Northern' consumer markets, raising questions of inclusiveness and resulting in challenges to legitimacy and effectiveness of the sustainability standards (Common Fund for Commodities Annual Report, 2018). Whilst 'sustainability' is often expected to be a force for good, rectifying socio-environmental issues and promoting genuine partnerships, sustainability initiatives investigated in cocoa often neglect to redress underlying power asymmetries particularly between Northern corporate actors and Southern stakeholders (Krauss, 2017).

Indeed, much has been done in recent years to put the industry on the path of sustainability (Barry Callebaut, 2024). This, in itself, is quite a good thing. Despite these initiatives, the collective efforts have been slow to yield results on the ground. The reasons for these include structural problems in cocoa commodity markets that fail to adequately incentivise sustainable cocoa, and a failure to listen to the needs of farmers and bring them into the development of industry-wide solutions as equal partners. It can be argued too that even within one initiative, the diverse actors involved and their differing understandings of sustainability in socio-economic, commercial and environmental terms offer ample opportunity for tensions (Krauss, 2017). A systematic, equitable exchange on and analysis of the commensurability of socio-economic, environmental and commercial priorities across different actors and contexts could be an initial move towards negotiating between different stakeholders.

It needs to be reiterated that much remains to be understood too about how to create more synergistic effects between sustainability dimensions as well as the impacts of sustainability initiatives, which are essential to guide the future development of supply chain mechanisms and policies to improve sustainability in cocoa production. Currently, there are a wide range of interventions at the industry's disposal. The question that arises, though, is to what extent do these schemes or initiatives promote equity and change deep-seated snags that have contributed to the cocoa industry's current predicament? This is where much efforts ought to be made collectively to ensure that stakeholders are genuinely committed to making the cocoa industry what it should be like in contemporary times.

9 Conclusion

The chapter explored issues pertaining to the cocoa and chocolate industry. It delved into the various aspects of the cocoa industry in order to determine its sustainability. The effects of the current cocoa crisis have been quite profound, impacting not just cocoa farmers but the entire global chocolate production line. The operational viability of many chocolate producers has been threatened as a result of this. In the face of this on-going crisis, the industry has, in recent years, recognised the need to pivot towards long-term sustainability and resilience. The chapter found that although there are some initiatives underway that are helping to deal with sustainability issues, there are, indeed, some complications to navigate. This, to some extent, provides an illustration of the gravity of problems inherent in the industry today. A number of issues within the industry remain contested or unaccomplished, making further attempts to find solutions to the myriad of the problems, a near impossibility.

One notable point is that the cocoa and chocolate industry lacks fairness in the eyes of many people. The fundamental characteristic of fair trade is that of equal partnership and respect—partnership between the Southern producers and Northern importers. However, to many observers, this has not been the case. Hence, it is essential that chocolate companies forge very fruitful collaborations with farmer groups and governments in cocoa-producing countries in advancing the cause of the industry. The parties also need to develop purchasing systems for cocoa that can ensure a decent return to farmers. Rapid fluctuations in the international prices affect the farmers in so many ways. When cocoa prices change, farmers have only a very limited ability to adjust their production in line with demand. This has added to their insecurity. The current crisis in the cocoa industry has brought to the fore the need to safeguard the interest of cocoa farmers and also to support them more strongly.

It is also important to reiterate that there should be a strong commitment by industry players to act together to reverse the human and environmental failings in cocoa-growing areas and the entire supply chain. Coming out with innovative approaches that can spearhead this drive will be very important in the years to come. The industry has signalled its determination to pursue this. However, if companies and governments of cocoa-producing countries want to achieve real progress, they will need to be more specific in their commitments and do more to alleviate the sufferings of the cocoa farmers whose blood and toil has kept this industry running.

References

Amoah, J. E. K. (1998). *Marketing of Ghana Cocoa, 1885-1992 (Cocoa outline no 2)*. Jemre Enterprise.

Anderson, R. (2011). Ivory Coast crisis: Impact on the international cocoa trade. *BBC News*. Accessed January 23, 2024, from https://www.bbc.com/news/business-12677418

Barrientos, S. (2014). Gendered global production networks: Analysis of cocoa–chocolate sourcing. *Regional Studies, 48*(5), 791–803.

Barry Callebaut. (2024). *Cocoa sustainability guide: Understanding sustainable chocolate*. Accessed March 1, 2024, from https://www.barry-callebaut.com/en/manufacturers/cocoa-sustainability-guide-understanding-sustainable-chocolate#:~:text=Western%20Europe%2C%20North%20America%20and,chocolate%20lovers%20enjoy%20their%20treat

Bartley, T., Koos, S., Samel, H., Setrini, G., & Summers, N. (2015). *Looking behind the label: Global industries and the conscientious consumer*. Indiana University Press.

Berg, C. (2020). *Sustainable action: Overcoming the barriers*. Abingdon.

Boateng, K. O., Dankyi, E., Amponsah, I. K., Awudzi, G. K., Amponsah, E., & Darko, G. (2023). Knowledge, perception, and pesticide application practices among smallholder cocoa farmers in four Ghanaian cocoa-growing regions. *Toxicology Reports, 10*, 46–55.

Bunn, C., Lundy, M., Läderach, P., & Castro, F. (2017, November 13–17). Global climate change impacts on cocoa. In *Paper presented at International Symposium on Cocoa Research, Lima, Peru*. Accessed January 21, 2024, from https://www.icco.org/wp-content/uploads/T4.152.-GLOBAL-CLIMATECHANGE-IMPACTS-ON-COCOA.pdf

Bymolt, R., Laven, A., & Tyszler, M. (2018). *Demystifying the cocoa sector in Ghana and Cote d'Ivoire*. The Royal Tropical Institute. Accessed March 3, 2024, from https://www.kit.nl/wp-content/uploads/2020/05/Demystifying-complete-file.pdf

Carodenuto, S. (2019). Governance of zero deforestation cocoa in West Africa: New forms of public–private interaction. *Environmental Policy and Governance., 29*(1), 55–66.

Casey. (2023, 5 January). *Cocoa can't be sustainable without systemic changes for farmers, report says*. Food Dive. Accessed January 21, 2024, from https://www.fooddive.com/news/cocoa-sustainability-ghana-ivory-coast-africa-chocolate-farmers-living-income-report-research/639741/

Climate-Smart Cocoa. (2022). *Climate change and potential hazards*. Climatesmartcocoa. Accessed January 12, 2024, from https://climatesmartcocoa.guide/climate-change-and-potential-hazards/

Common Fund for Commodities Annual Report. (2018). *The impact of voluntary sustainability standards on small-scale farmers in global commodity chains*. Accessed March 9, 2024, from https://www.common-fund.org/sites/default/files/Publications/CFC_AR_18_paper_Certification.pdf

Duffey, T. (2009, May). *Managing pest and disease pressures—cocoa farmers perspectives*. The Manufacturing Confectioner. Accessed May 26, 2024, from https://www.gomc.com/first-page/200905055.pdf

Elliott, K. A. (2018). *What are we getting from voluntary sustainability standards for coffee? CGD Policy Paper*. Centre for Global Development.
FortuneBusinessInsight.(2024).*Cocoaandchocolatemarketsize*.AccessedJune1,2024,fromhttps://www.fortunebusinessinsights.com/industry-reports/cocoa-and-chocolate-market-100075
Fountain, A., & Huetz-Adams, F. (2020). *Cocoa Barometer 2020.* Accessed February 26, 2024, from file:///C:/Users/user/Desktop/2020-Cocoa-Barometer-EN.pdf
GCB Bank. (2022). *2022 cocoa sector report.* GCB Bank. Accessed February 29, 2024, from https://www.gcbbank.com.gh/research-reports/sector-industry-reports/120-cocoa-industry-in-ghana-2022/file
Gitau, M., Sousa, D., Peng, I, & Dontoh, E. (2024). *Why the world of chocolate is in crisis.* Accessed May 22, 2024, from https://economictimes.indiatimes.com/small-biz/trade/exports/insights/why-the-world-of-chocolate-is-in-crisis/articleshow/109488696.cms?from=mdr
Glasbergen, P. (2018). Smallholders do not eat certificates. *Ecological Economics, 147*, 243–252.
Harrington, L. M. B. (2016). Sustainability theory and conceptual considerations: A review of key ideas for sustainability, and the rural context. *Papers in Applied Geography, 2*(4), 365–382.
Hudson, H. (2022, 2 June). *Ghana is cocoa, cocoa is Ghana*. OPEC Fund. Accessed March 1, 2024, from https://opecfund.org/news/ghana-is-cocoa-cocoa-is-ghana
Hütz-Adams, F., & Schneeweiß, A. (2018). *Pricing in the cocoa value chain – Causes and effects.* Deutsche Gesellschaft für Internationale Zusammenarbeit. Accessed November 12, 2023, from https://suedwind-institut.de/files/Suedwind/Publikationen/2018/2018-13%20Pricing%20in%20the%20cocoa%20value%20chain%20%E2%80%93%20causes%20and%20effects.pdf
Kozicka, M., Tacconi, F., Horna, D., & Gotor, E. (2018). *Forecasting cocoa yields for 2050.* Biodiversity International. Accessed December 12, 2023, from https://cgspace.cgiar.org/handle/10568/93236
Krauss, J. (2017). What is cocoa sustainability? Mapping stakeholders' socio-economic, environmental, and commercial constellations of priorities. *Enterprise Development and Microfinance, 28*(3), 229–249.
Lambin, E. F., Gibbs, H. K., Heilmayr, R., Carlson, K. M., Fleck, L. C., Garrett, R. D., & Walker, N. F. (2018). The role of supply-chain initiatives in reducing deforestation. *Nature Climate Change, 8*(2), 109–116.
Leach, M., & Fairhead, J. (2000). Challenging neo-Malthusian deforestation analyses in West Africa's dynamic forest landscapes. *Population and Development Review, 26*(1), 17–43.
Malan, B. B. (2013). Volatility and stabilization of the price of coffee and cocoa in Côte d'Ivoire. *Agricultural Economics, 59*, 333–340.
Maximise Market Research. (2024). *Chocolate market: Global industry analysis and forecast (2023-2029).* Accessed March 2, 2024, from https://www.maximizemarketresearch.com/market-report/global-chocolate-market/13157/
Mera, C., Magdovitz, M., & Rawlings, A. (2021). *Outlook 2022: Hell in the handbasket (Agri Commodity Markets Research).* RaboResearch. Accessed March 30, 2024, from https://research.rabobank.com/far/en/documents/842941_Rabobank_ACMR-Outlook-2022.pdf
Miles, M. B., Huberman, A. M., & Saldana, J. (2013). *Qualitative data analysis*. Sage.
Mollenkamp, D. T. (2023). *What is Sustainability? How sustainabilities work, benefits, and example*. Investopedia. Accessed February 29, 2024, from https://www.investopedia.com/terms/s/sustainability.asp
Moreno, J. E. (2024, May 10). A failed crop rattled the chocolate industry, then speculators came. *New York Times*. Accessed from https://www.nytimes.com/2024/05/10/business/cocoa-prices-chocolate.html#:~:text=Then%20Speculators%20Came,Then%20they%20went%20nuts.&text=A%20failed%20crop%2C%20followed%20by,on%20inexpensive%20crops%20and%20labor
Nelson, V., & Phillips, D. (2018). Sector, landscape or rural transformations? Exploring the limits and potential of agricultural sustainability initiatives through a cocoa case study. *Business Strategy and the Environment, 27*(2), 252–262.

NORC. (2020). *Assessing progress in reducing child labour in cocoa production in cocoa growing areas of Cote d'Ivoire and Ghana*. Accessed January 14, 2024, from https://www.norc.org/content/dam/norc-org/documents/standard-projects-pdf/NORC%202020%20Cocoa%20Report_English.pdf

Odijie, M. (2023, October 4). Cocoa prices are surging: west African countries should seize the moment to negotiate a better deal for farmers. *The Conversation*. Accessed May 27, 2024, from https://theconversation.com/cocoa-prices-are-surging-west-african-countries-should-seize-the-moment-to-negotiate-a-better-deal-for-farmers-214305

Ollendorf, F. (2021). Corporate social responsibility in the global cocoa chocolate chain: Insights from sustainability certification in Ghana's Cocoa Communities. In *Global commodity chains and labor relations* (pp. 316–337). Leiden. Accessed February 25, 2024, from https://brill.com/view/book/9789004448049/BP000018.xml

Oxfam. (2023, 14 February). *Towards a living income for cocoa farmers in Ghana*. Oxfam. Accessed March 8, 2024, from https://oxfamilibrary.openrepository.com/bitstream/handle/10546/621485/rr-ghana-cocoa-farmers-living-income-140223-en.pdf

Sarpong, S. (2023). Work-life balance: The challenges and the search for equilibrium in Malaysia. In D. Crowther & S. Seifert (Eds.), *The Routledge companion to the future of management research* (pp. 172–187). Routledge.

Sarpong, S., & Alaurussi, A. (2022). Waste to wealth: Enhancing circularities in the Malaysian economy. *Technological Sustainability, 1*(2), 145–159.

Sasu, D. D. (2024). *Export value of cocoa beans from Ghana 2015-2022*. Statista. Accessed March 1, 2024, from https://www.statista.com/statistics/1172186/export-value-of-cocoa-beans-and-products-from-ghana/#:~:text=In%202022%2C%20the%20export%20value,approximately%202.84%20billion%20U.S.%20dollars

Walker, J. (2020, 18 November). *Cocoa farmers don't earn a living income. Where can government play a role?* Fairtrade International. Accessed May 27, 2024, from https://www.fairtrade.net/news/cocoa-farmers-dont-earn-a-living-income-government-role

WCED. (1987). *Our common future*. Oxford University Press.

Wessel, M., & Foluke Quist-Wessel, P. M. (2015). Cocoa production in West Africa, a review and analysis of recent developments. *NJAS-Wageningen Journal of Life Sciences, 74–75*, 1–7.

Wong, Y. (2021). *Cocoa price forecast: What's next after multi-year highs?* Capital.com. Accessed October 12, 2024, from https://capital.com/cocoa-price-forecast-what-is-next-after-multi-year-highs

Professor Sam Sarpong's research interests lie in the relationship between society, economy, institutions and markets. He tends to explore the nature and ethical implications of socioeconomic problems. His publications span a variety of fields including sustainability issues; management practices; CSR/Business Ethics; entrepreneurship and global marketing strategy. He is currently affiliated with the School of Business, University of Central Lancashire (UCLan), UK. Prior to joining UCLan, Sam held academic positions at Xiamen University, Malaysia; Narxoz University, Kazakhstan; University of Mines and Technology, Ghana; Swansea Met University, University of London (Birkbeck College) and Cardiff University, respectively. He obtained his Ph.D. from Cardiff University and his MBA from the University of South Wales. He is a reviewer for many leading journals and also serves as Editor (NEP/AFR—RePEc) and as an Associate Editor for the *International Journal of Corporate Social Responsibility* (Springer).

A Multi-Sector Assessment of Sustainability and Socially Responsible Practices of International Businesses in Nigeria

Adebimpe Adesua Lincoln and Brendhain Diamond

1 Introduction

Sustainability and social responsibility agendas have been driven to create a more peaceful society through mainly voluntary efforts driven by peoples' evolving views on company responsibilities toward non-owners. There is now growing consensus on the importance of adopting business practices and initiatives that adhere to international guidelines and principles on sustainability (Burritt et al., 2020; Adesua Lincoln & Diamond, 2023). The catalysts for this include pressure from academia, government, institutional investors, and shareholder activism (Burritt et al., 2020). Multinational Enterprises (MNEs), being prominent and wealthy, are seen by some as being responsible for not only working to improve the well-being of the societies in which they operate but would also be expected to behave conducive to the values and goals of each of these societies. This view would require MNEs to proactively work to better societies while also adapting internal company operations to the local sensibilities of each country in which they operate. Without a global sensibilities' standard, these intrusive obligations would be counter to company efficiency aims, as companies would be required to conduct themselves differently in each jurisdiction and have different societal goals depending upon the differing needs of the countries they operate. Those who oppose broader responsibilities argue that in doing so, companies would lose their efficiencies, which is a vital benefit to society. The efficiency criticism aside, the proposed obligations upon MNEs are based upon a perceived moral obligation conditional upon relatively high power and influence (Wettstein, 2012). The normal behavior of international businesses or MNEs might be better understood by reviewing corporate responses to sustainability and social

A. A. Lincoln (✉) · B. Diamond
University of Liverpool, Liverpool, UK
e-mail: b.diamond@liverpool.ac.uk

S. O. Idowu, S. Vertigans (eds.), *Sustainability in Global Companies*, CSR, Sustainability, Ethics & Governance, https://doi.org/10.1007/978-3-031-77971-8_11

responsibility initiatives within their shareholder ranks and government pressures from the countries in which they operate. Consequently, it seems reasonable that if the shareholders of the international business community were inclined to be positive toward sustainability and social responsibility, then the MNE leadership would also tend to embrace these initiatives. A converse argument can also be raised to the effect that if MNE leadership reacts with reluctance, then this may indicate a disinterest (Burritt et al., 2020). While not universally true, institutional investors and shareholders tend to focus on the profits of the companies in which they invest (Samat & Ali, 2015; Balp, 2018). On the other hand, an increasing number of private shareholders are interested beyond the single goal of profits and are concerned with how the companies operate. This broader interest includes organized efforts to bring social responsibility activism through stock ownership (Balp, 2018). By owning shares, shareholder activists have the right to bring resolutions forward and may be granted access to company directors for direct dialogue. The academic community is also a significant player in sustainability and social responsibility discussions. It could be a natural starting point in understanding the nature of the concept and the diversity in approaches. The academic community has a long history of addressing companies' responsibilities. Due to the expansive variety of different approaches, the results can be described as fragmented or even to the point of chaos. At the least, it can be said that the academic community has not reached a consensus as to whether companies have extended responsibilities, what form these responsibilities would take, or even the motivation behind the responsibilities. In addition, international businesses face pressures from governments dictating appropriate behavior and legislating to ensure compliance, for example the approach adopted in Denmark. The progressive models adopted in Denmark place heavy social responsibility burdens on companies even where there exists a robust social safety net and services (Vallentin, 2015). Vallentin (2015) identifies the anomaly of Danish society as having elevated consensus and high expectations of conforming to societal norms. This conclusion would indicate that when a consensus is met, laws and company behavior will reflect this consensus. This move could be seen as a signal of society's changing expectations moving in favor of socially responsible actions. While a review of laws may indicate the possible existence of a societal consensus for corporate social responsibility (CSR), a lack of CSR laws is not proof of a lack of consensus. In some countries, of which Nigeria is a prime example the government's inability to manage the rising social problems necessitates companies taking on additional responsibilities to fill the gap.

Speaking as the UN Secretary-General and about the needs of developing nations, Kofi Annan asserted that "…more and more we are realizing that it is only by mobilizing the corporate sector that we can make significant progress. The corporate sector has the finances, the technology, and the management to make this happen (Wade, 2005)." Annan's assertion recognizes both that governments of developing nations are incapable of satisfying their populations' well-being needs and that corporations could meet these needs. It follows from Annan's view that the route for progress in developing nations depends on corporations not only "doing no harm" but also "doing good for society" beyond the company's primary profit-generating aim. The different types of actors, their motivations, and the effectiveness of initiatives adopted whether it be environmental or social in nature result in either opportunities for the future or negative consequences for local communities in which the MNE operates. Burritt et al. (2020) argue that MNEs arguably are institutions whose primary economic activities have contributed significantly toward "unsustainability" and have resulted in a profound negative impact on the local communities in which they operate and the environment (Schaltegger & Burritt, 2018). In addition, MNEs have been accused of complicity and engagement in practices that exacerbate local problems such as corruption and lack of governance (OECD, 2023). According to the OECD (2023), *"the ability of multinational enterprises to promote sustainable development is greatly enhanced when trade and investment are conducted in a context of open, competitive and appropriately regulated markets with rule of law and protection of civic space."* MNEs operating in Nigeria have capitalized on Nigeria's weak legal-regulatory landscape and the inability of successive Nigerian governments to equitably protect the interests of its citizens, the local communities, and the environment. While not discounting examples of MNE social contributions in Nigeria, such projects must always be viewed through the lens of tokenism when one considers the vast resources pillaged on an annual basis and the obscene profits made by these international businesses, while the Nigerian society is left crippled at the knees. According to the Environmental Performance Index (EPI) published by researchers from Columbia University's Center for International Earth Science Information Network and Yale Center for Environmental Law & Policy, as of 2024 Nigeria currently has an environmental responsibility score of 37.5% and is ranked 141 in the world out of 180 countries. The 2024 EPI combines 58 indicators across 11 issue categories, ranging from climate change mitigation and air pollution to waste management, sustainability of fisheries and agriculture, deforestation, and biodiversity protection. Nigeria received poor scores and ranking by the center—the country had a score of 20.0 for "environmental health" and is ranked 164; 14.4 for "sanitation and drinking water" ranked 175; and 0.0 for "bioclimatic ecosystem resilience" and ranked 170 out of 180 countries. These appalling results help substantiate the argument that CSR and sustainability measures currently adopted by MNEs in Nigeria are merely an attempt to enhance MNEs' legitimacy—there is a lack of connectedness between these corporations and the local communities in which they operate other than its sole objective of profiteering and wealth maximization. It is farcical to think that MNEs have made any meaningful contributions to the life of the average Nigerian in comparison with what they could potentially achieve if these businesses were serious about making meaningful long-term impacts. It would be disingenuous not to challenge the efficacy of these international businesses as change agents in Nigeria or question their development agenda in the country. The evidence is clear, MNEs in Nigeria are more of a negative influence on the local communities and the Nigerian society, consequently, these MNEs' impact on Nigeria's national development must now be perceived in this light. Their presence has brought about systemic poverty and economic and environmental devastation. These international businesses are more interested in enriching the pockets of corrupt politicians and bureaucrats than enhancing the status of the average Nigerian. It is to their benefit for Nigeria to remain in a state of stagnation so that they can continue to drain the natural resources and continue the transfer of wealth to their "home countries," enriching themselves while leaving Nigerians impoverished. According to a recent 2022 World Bank report titled "A Better Future for All Nigerians: Nigeria Poverty Assessment 2022" 4 in 10 Nigerians live below the national poverty line. Many Nigerians lack education and access to basic infrastructure, such as electricity, safe drinking water, and improved sanitation. According to the World Bank (2024), Nigeria is considered the world's second-largest poor population after India. Nigeria offers limited opportunities to most of its citizens. As of 2023, the poverty rate in the country is estimated to have reached 38.9%, with an estimated 87 million Nigerians living below the poverty line. These MNEs have moral and ethical responsibilities to solve the problems caused by their exploitative activities and to ensure funds earmarked for the social development of the Nigerian nation are not handed down to corrupt officials but are disbursed to specific projects that add value to the economic advancement of the country (OECD, 2023).

These MNEs are complicit in the current economic and social degradation of the country—reference to token measures will no longer suffice! It is as though Nigerians are only worth the proverbial "*two goats and a toilet*" or the "*substandard school and pipe borne water*." The notion that the issues which persist in the Niger Delta region of the country where many of the MNEs in the oil sector operate are "dated" is disingenuous. The intelligent question that comes to mind is this—how can these MNEs boast of meeting UN SDG goals without restoring the region to its natural state? What becomes obvious is the fact that these UN SDG goals are now being used as false propaganda by many MNEs thereby bringing the UN initiative into disrepute. The "ongoing discussion around postcolonialism efforts of MNEs" is rooted in fallacy. The reality is that MNEs in Nigeria still operate from the same rule book that they have operated since the 1960s nothing much has changed in this regard except that now we have real evidence of their detrimental presence across various parts of Nigeria. There is a need to ensure enhanced regulatory oversight on human rights as well as the environment and social governance practices of these MNEs to make them accountable where there have been significant failings—a recent example is the February 2023 court case (brought by more than 11,000 residents in Nigeria) in the UK against an MNE over the alleged impact of its activities in the local community (Lexis Nexis, 2023). Recent court decisions which imposed liability on MNE for their reprehensible behavior in parts of the Niger Delta are a step in the right direction. These businesses in addition to paying any monetary penalties imposed by the courts must engage in voluntary acts of compliance and fix what they have broken—they should put measures in place to prevent future occurrences and actively helping to rebuild local communicates to replicate the type of local communities that these MNEs have in their home country. This perspective invites a deeper look into the dynamics within developing nations. This study therefore seeks to

(A) Identify some of the leading MNEs making the most significant contributions to sustainability and social development in Nigeria.
(B) Identify MNEs' sustainability focus areas and social development goals and agenda.
(C) Assess the potential benefits accruing to these MNEs from adopting these initiatives.

2 Sustainability and Corporate Social Responsibility

In his 1953 book, Bowen (1953) can be seen as having initiated modern CSR, wherein he recognized large companies as having relatively great power and influence in society. Significantly, Bowen (1953) assigned an obligation upon businesses to act according to the values and goals of society. Despite its nearly 70-year history—CSR remains a broad and vague concept. The Sustainability concept is analogous to CSR—it involves the impact a business operation has on the environment, local communities, and society within which it operates. Sustainability as a term is attributed to the 1987 Brundtland Report. The Report states that sustainable development can be achieved if the level of "development" meets the needs of the present without compromising the ability of future generations to meet their own needs (World Commission on Environment Development, 1987). Cruz and Marques (2014) define sustainability as a societal objective involving the need to achieve the triple bottom line performance of profit, planet, and people. Sustainability is now considered of utmost importance in terms of long-term value delivery from a financial, social, environmental, and ethical perspective (UN Global Compact, 2015). Accordingly, the UN Global Compact (2015) states that it is of utmost importance for companies to ensure that they can operate responsibly and endeavor to align their activities with universal principles and take actions that support local communities and the wider society. In so doing, companies must commit at the highest level, engage in annual reporting on their efforts, and engage locally where they have a presence. Thus, the goal of sustainable and socially responsible practices is to ensure business operations positively impact society and the environment. A detailed examination of the nuances and workings of CSR and sustainability theoretical perspectives is essential in understanding the practical applications, functionality, and limitations of sustainability and CSR models in operation. Elkington's Triple Bottom Line (TBL), Carroll's pyramid model, and the United Nations Social Development Goals are considered in this research due to their relatively long and successful history in getting corporations to adopt the approaches. These theoretical perspectives see companies as having broader responsibilities beyond what is recognized within shareholder theory and wealth maximization. Carroll and Elkington see companies as having broader responsibilities beyond what is recognized within shareholder theory. Elkington's attempt to quantify positive and negative impacts from companies has also been successful regarding companies implementing TBL, but implementation has not resulted in changing the paradigm of managers or corporate culture. A possible contributor to this is the use of a plus/minus calculation where the positive effects counterbalance the adverse effects. Due to the subjective nature of setting quantities to the impacts of each corporate action, TBL may well add to moral ambiguity. Managers set the quantitative value of each impact. Managers' subjectivity may place higher values on positive impacts and lower values on negative impacts. An additional issue with TBL is that it opens the way for moral rationalization because companies can continue actions with negative impacts and offset the negative impacts with positive actions. Carroll offers a model with the benefit of flexibility in practical situations, whereby managers are permitted to prioritize different levels of the pyramid depending upon the economic condition. This approach allows companies to prioritize the first two levels, generating profits and obeying the law, when economic survival takes primacy. Carroll's pyramid relies heavily upon managers making moral decisions when balancing the interests of different stakeholders. With company cultures being heavily focused upon the primacy of profits, under the pyramid, managers' subjective judgment may always see the economic environment as calling for profit focus. Carroll and Elkington's approaches suffer from the element of subjectivity. At a minimum, it is questionable whether management has the expertise to make the moral decisions needed to balance the needs of various stakeholders. While it might be argued that society could benefit from moral managers, there is little evidence to support that managers can fulfill this task. Elkington presents a system for making such calculations, but, in practice, the system is based upon subjective and qualitative values disguised as quantitative values.

2.1 United Nations Social Development Goals and Elkington: Triple Bottom Line

With many jurisdictions requiring companies to report their impacts on the environment and society, the Triple Bottom Line (TBL) has had a great deal of success in getting companies to adopt accounting practices (Hartmann, 2020). Via a new approach to accounting, Elkington sought the primary aim of fundamentally changing capitalism by inducing a paradigm shift. Elkington saw capitalism's single focus on profits as needing correction, and TBL accounting would divert this focus to societal and environmental concerns. TBL was to replace traditional profit-driven capitalism with a new process of keeping costs down while maximizing both environmental conservation and social equity for both narrow and broad stakeholders, also known as the 3Ps of Profit, People, and Planet. There is a direct correlation between how the sustainability concept was coined originally to include three main pillars, namely, economic, environmental, and social, and the 3Ps of Profit, People, and Planet.

1. *Environmental Pillar—Planet:* The environmental pillar focuses on the impact of business operations on the planet. This pillar is concerned with the business

making positive impacts vis-a-vis reduction in carbon footprints, recycling, air, water, and land use.

2. *Social Pillar—People:* The social pillar focuses on equality and social justice for all stakeholder groups. This pillar focuses on health and well-being, diversity and equality, local community support initiatives, human rights, education, equity, and access to social resources.
3. *Economic Pillar—Profit:* The economic pillar involves wealth maximization, profitability, financial statements, obligations, and governance.

Furthermore, a link can be seen between the TBL 3Ps and the United Nations Sustainable Development Goals (UN SDGs). The UN SDGs framework proposes THREE main sustainability pillars, namely environmental, social, and economic—these pillars incorporate 17 sustainable development goals. Figure 1 below shows the interrelationship between the three sustainability pillars and the UN SDGs.

TBL stipulates that practitioners should maintain accounting information beyond the standard financial accounting by also valuing the positive and negative impacts upon broad stakeholders. In effect, by documenting the positive and negative impacts to ascertain the net effect of a company's activities, corporate officers would seek to maximize the bottom line in these areas and not only focus on company profits. Elkington sees the use of TBL as a transformative element in that the implementation of TBL transforms the very nature and approaches of companies. In effect, when companies measure and report social impacts, company officers will be more aware and, therefore, act more responsibly toward society. In simple terms, by monitoring themselves and doing good, then company officers will become good. The process is to be based upon supposed objective measurements of social impacts called "Measurement Claims." These would include internal and external impacts of company activities that create or reduce a company's so-called "social capital." Social capital is akin to goodwill in that it reflects the faith society has in a company. Each year, a company reports its net effects via an "Aggregate Claim," which is calculated as the net effect of the various measurements. The argument is that the documentation, calculation, and reporting of the effects of company action, when net positive, will have a corresponding positive reaction within the greater society.

Environmental - Planet	SDG 12: Responsible consumption and production; SDG 13: Climate action; SDG 14: Life below water; SDG 15: Life on land
Social - People	SGD 1: No poverty; SDG 2: Zero hunger; SDG 3: Good health and well-being; SDG 4: Quality education; SDG 5: Gender equality; SDG 6: Clean water and sanitation; SDG 10: Reduced inequalities; SDG 11: Sustainable cities and communities; SDG 16: Peace, justice, and strong institutions
Economic - Profit	SDG 7: Affordable and clean energy; SDG 8: Decent work and economic growth; SDG 9: Industry innovation and infrastructure; SDG 17: Partnerships for the goals

Fig. 1 United Nations Social Development Goals and the three sustainability pillars

This positive reaction will decrease negative views amongst the public. In terms of the Legitimacy Theory, society would accept the company and see it playing a legitimate role in society.

While TBL prescribes objective measures for social effects, these measurements have not proven to be attainable as a meaningful result. Additionally, TBL reporting has been shown to have limited appeal. Bohlmann et al. (2018) found that while positive TBL reporting attracted employee candidates with high interest in environmental issues, this appeal did not extend to candidates with moderate to low interest in this issue. They concluded that the best approach would be for companies to extend employee recruitment qualifications to include like-minded candidates at the expense of only using skills and experience. The result would be companies recruiting from a smaller talent pool as it de-emphasizes the most talent qualifications in general, and they prescribe the exclusion of candidates who do not prioritize the environmental and social impacts of companies. Companies strongly committed to sustainability and social responsibility may benefit from TBL as it documents the results of company efforts and identifies its net results. TBL can be used internally by a company to improve its net effect and identify critical areas needing improvement. At the same time, TBL's flaws may make the results non-comprehendible for corporate officers; thus, there is no transformative effect. This analysis is evidenced by Elkington's 2018 article, in which he frustratingly acknowledges that despite thousands of corporations adopting TBL, it has not resulted in diverting company focus from the singular bottom line of profit as such failed to achieve its primary aim. He attributes this to the "hard-wired" business culture inspiring him to see the need for a recall and revamp of TBL (Elkington, 2018). Elkington's placement of blame can be seen as a confirmation of the earlier conclusion that industry officers and management do not recognize there is a societal norm that places CSR responsibilities upon companies. After 25 years of implementation, Elkington was forced to recognize the fundamental nature of companies as capitalistic and that TBL fails to facilitate the transformation that TBL seeks. While Elkington does not advocate abandoning TBL, he does acknowledge that TBL's lack of efficacy warrants a revamping of TBL, as TBL has failed to elicit a paradigm shift. Despite its implementation by thousands of companies, Triple Bottom Line has failed to transform companies that adopt the system. Elkington places the blame for this failure upon the nature and purpose of companies, namely, profit being the preeminent motivation and drive within company culture. This is not to say that TBL is without benefits. When applied, TBL produces quantitative information (i.e., a specific number), whereby a company can compare its performance to its own previous years' performances to evaluate potential progress or shortcomings. Unfortunately, this number is highly subjective and based on calculations of subjectively generated numbers. Therefore, comparing one company's TBL numbers to another is not likely to provide meaningful results, as other companies suffer from the subjective nature of their calculations. In the end, the subjective nature of TBL results in meaningless information to individuals outside of the company and cannot be comparable to the numbers generated by other companies.

2.2 *Carroll's Pyramid*

Carroll's pyramid proposes FOUR aspects of social responsibility for businesses, namely economic, legal, ethical, and philanthropic.

- *Economic Responsibility*: company responsibilities and actions are made within the framework of achieving maximized profits—"be profitable—profitability is the foundation upon which all others rest."
- *Legal Responsibility*: laws and legal constrictions on businesses "obey the law—law is society's codification of right and wrong; play by the rules of the game."
- *Ethical Responsibility*: the social contract of companies being allowed to operate and pursue profits, companies are also expected to adhere to the "codified ethics"—"be ethical—the obligation to do what is right, just, and fair; avoid harm."
- *Philanthropic Responsibility*: extends company responsibility beyond merely not being bad to ensuring the company contributes to improving the quality of life within the local communities in which they operate—"be a good corporate citizen—contribute resources to the community; improve quality of life" (Carroll, 2004).

It follows from Carroll's views that, in general, managers would need to act not only as moral decision-makers based upon economic criteria but also become moral creatures. Carroll points out that even Friedman recognized the need for businesses to operate within the social norms "embodied in ethical customs." Based upon this quote from Friedman (2002), Carroll asserts that Friedman meant that the pursuit of profits should be performed within both the codified ethics in the laws and the standard ethics in society. Therefore, Carroll sees Friedman as recognizing the need for companies to satisfy the first three levels of the pyramid. This interpretation of Friedman omits Friedman's specification as to what ethical norms should be followed. In the concluding sentence of the quoted article, Friedman specifies that the ethical norms to be followed are avoiding "deception and fraud." While these two moral conditions can be included within Carroll's third level, they are negative rules to "not be bad" and not the positive rules Carroll would like to place on corporations "to do good." For example, Friedman might anticipate managers not being completely forthright during a negation, whereas Carroll might expect the corporation to pay the subjective value of "just wages." Friedman's views can be seen as expecting companies to pay the legal minimum wage, the wages as stipulated in a union agreement, or whatever wages the market sets as the standard "going rate" (Friedman, 2002). Additionally, Milton Friedman (2002) differentiates between placing moral responsibility upon a corporation as an entity and upon managers as individuals. In effect, people are moral creatures, and corporations are not. Corporations are created to serve the specific interests of the owners. Because managers are to act as agents for the owners, Friedman is open to managers spending the owners' money on philanthropic endeavors on the condition that the owners wish their money to be spent as such. If the owners do not wish their money to be spent in such a manner, then managers doing so, with no associated increase in profits, would be violating their fiduciary duties, as well as performing an immoral act of "harming" the owners. Friedman argues that, while companies have many positive social impacts via the efficient use of resources and providing employment opportunities, looking after social concerns is the government's responsibility. It follows from Friedman's view that if society sees the social norm as being that individuals should contribute to the general well-being of society, with wealthy people paying more, then it is the government's responsibility to tax and spend accordingly, or individuals should spend their own money in such efforts. Managers should not go against the owners' will and tax the owners to spend on social considerations, as he sees it as a form of theft, and managers are ill-suited to decide how the money would be spent. For Friedman, if a manager's morality "to be good" would have them spend philanthropically, then they should use their own money to do so. It follows from this that Friedman does not confirm Carroll's third level and is strongly opposed to the fourth level of responsibility unless the owners themselves "will for their money to spent so."

Carroll argues for a broad view of stakeholder theory, including those with legally enforceable claims and moral claims. While Friedman's position agrees with the first group as companies must follow the law, Carroll sees moral claim stakeholders as individuals with no legal rights, despite their claims or feelings that something is due to them. With this second group comes the element of subjectivity. Carroll sees different groups' claims as having different degrees of legitimacy, thus leaving managers with the difficult task of finding the best balance, which is "a legitimate and desirable goal for management to pursue to protect its long-term interests." At this critical juncture, Carroll fails to motivate both why this is legitimate, beyond the subjective feelings of the holders of moral claims, and why this is in management's long-term interest. More critically, Carroll conflates the agent role of management with the corporation itself. This conflation results in Carroll seeing managers acting in their long-term interests and not the owners' interests. In its place, Carroll asserts that managers are responsible for going beyond amorality and normal morality behaviors to be exemplary in their morality. He motivates this upon the subjective purpose of pursuing a subjective "good society" which remains undefined. One can question whether a good role model for society promotes "Robinhoodism" by taking money from company owners, without authorization and not in the owners' interests, and then spending it as one sees fit. Beyond that criticism, Carroll fails to motivate how managers doing this action would transform society into the ideal "good society" beyond corporations being an example of this Robinhoodism behavior. At the same time, it is not typical within society to task individuals with acting as good citizens "to do good," nor to promote "Robinhoodism." Instead, the law can be seen as limiting bad actions to not conflict with societal norms and not requiring role-model behavior. Ultimately, Carroll's position is burdened by the subjective nature inherent in moral claims and the proper balance between the various broadly defined stakeholders. Carroll places the responsibility of making these decisions upon company managers, who are ill-prepared to make such subjective judgments. Furthermore, should these subjective claims and aims be tied to Carroll's sense of the ideal society or the ever-changing social norms held by society?

3 Sustainability and Corporate Social Responsibility in Nigeria

Despite coordinated efforts from the international community on sustainability and social responsibility, there is a clear lack of clarity and consistency in the approaches adopted in Nigeria vis-a-vis the interpretation and implementation of CSR and sustainability concepts. Nigeria lacks a codified law that regulates social responsibility. There is a lack of clarity on what social responsibility and sustainability mean from a Nigerian perspective—thus corporate social responsibility and sustainability as concepts remain undefined. Due to the importance placed on religion and cultural beliefs in the country, one would expect that the concepts in the Nigerian context would largely be dependent on societal and cultural practices, and moral and ethical norms. Nigeria is crippled by incoherent legal-regulatory framework vis-a-vis CSR and sustainability. There is a host of legislative provisions in Nigeria, which regulate business activities and socially responsible practices. CSR and sustainability rules are espoused in often overlapping guidelines and legislative provisions which are frequently inconsistent. Arguably, societal unpredictability makes it more challenging to enact robust and coherent regulations or ensure transparency and accountability. Table 1 below outlines the main substantive Environmental, Social, and Governance ("ESG")-related regulations in Nigeria.

Corporate law in Nigeria is regulated by the 2020 Companies and Allied Matters Act (CAMA). CAMA restates the provisions of some of the other legislative provisions with a focus on the impact of the company's operations on the community and the environment. CAMA recognizes the increasing importance of the need to ensure stakeholder engagement and the adoption of a stakeholder perspective on governance. Consequently, CAMA proposes reforms that widen the responsibility of the company to include duties to various stakeholders. CAMA states: "*Directors shall act at all times in what he believes to be the best interest of the company as a whole ...and in doing so, shall have regard to the impact of the company's operations on the environment in the community where it carries on business operations.*" This duty is codified under Section 305 (3). According to Section 305 (4) "*the matters to which a director of a company is to have regard in the performance of his functions include the interest of the company's employees in general, as well as the interest of its members.*" While the reforms brought in by CAMA 2020 enshrine the enlightened shareholder perspective into corporate decision-making, there is a move toward a "shareholder-centric" notion of "wealth maximization" as such limited in its application. Section 305 (9) of CAMA states that "*any duty imposed on a director under this section is enforceable against a director by the company,*" those with business relationships such as suppliers, customers, and others are precluded from enforcing the protections contained under Section 305. The express exclusion of other primary stakeholder groups in exercising the rights conferred under Section 305 is a significant oversight in the 2020 Act.

Table 1 Substantive environmental, social, and governance-related regulations in Nigeria

Constitution of the Federal Republic of Nigeria, 1999
Climate Change Act, 2021
Environmental Impact Assessment Act, 2004
Harmful Waste (Special Criminal Provisions) Act, 2004
National Environmental Standards and Regulation Enforcement Agency Act, 2007
Environmental Impact Assessment Act, 2004
Companies and Allied Matters Act, 2020
Petroleum Industry Act, 2021
Nigerian Sustainable Banking Principles, 2012
Nigerian Stock Exchange's Sustainability Disclosure Guidelines, 2018
Nigerian Code of Corporate Governance, 2019
Nigerian Sustainable Finance Principles, 2021
Nigerian Securities Exchange Sustainability Disclosure Guidelines
Environmental Guidelines and Standards for the Petroleum Industry in Nigeria, 2018
SEC Guidelines on Sustainable Financial Principles for the Nigerian Capital Market, 2021
Nationally Determined Contributions—UN Framework Convention on Climate Change, 2018
National Climate Change Policy for 2021–2030
National Action Plan on Gender and Climate
Environmental Sanitation Law of Lagos State
Environmental Pollution Control Law of Lagos State
Federal National Parks Act, 2004
Water Resources Act, 2004
Endangered Species Act, 2004
Sea Fisheries Act, 2004
Inland Fisheries Act, 2004

4 Research Methodology

Data for this study were obtained through an in-depth review of the annual sustainability reporting of MNEs in various sectors in Nigeria. The findings presented in this research are based on an analysis of publicly available information only. The study adopts an exploratory approach to gain new insights based on the data generated. A document analysis of the MNEs' annual CSR and sustainability reports was carried out. Document analysis is a form of qualitative research that uses a systematic procedure to analyze documentary evidence and answer specific research questions. The document analysis adopted in this research involved repeated examination and interpretation of the data to gain meaning from the constructs under investigation. Inclusionary criteria were adopted which determined the document sources yielding the greatest quantity and quality relevant to the research questions. This process helped identify the most appropriate sources to use for the sampling and ensured the authenticity and representativeness of sources identified for the sample. Inclusionary criteria enabled systematic document selection—three inclusionary criteria were used:

Table 2 MNE sector categorization

Food and beverage	Telecommunications	Pharmaceutical	Construction	Oil and gas
Cadbury	MTN	GlaxoSmithKline	Julius Berger	Chevron
Coca-Cola				Exxon Mobil
Nestle				Shell
Unilever				Total energies

- *Time Limit*: a 5-year limit was placed on the period for the CSR and sustainability reports.
- *Geographic representation*: Data collection was limited to those MNEs evidencing CSR and sustainability initiatives locally in Nigeria.
- *Documents*: only official company documents and reports were included—it was therefore imperative that the MNE activities be formally reported and visible on their website.

Deductive categorization and coding were adopted to confirm, expand, or refine the existing understanding of sustainability and social responsibility practices of the MNEs. This commenced with an initial categorization of keywords derived from the United Nations' 17 Social Development Goals (SDGs). These predetermined categories then formed the basis for the codes used to analyze the MNEs' CSR and sustainability reports. A thematic analysis involved the use of pattern recognition; and categorization of the data was then carried out. The result is presented through descriptive writing complemented with figures and tables where appropriate. An initial scan of a list of MNEs operating in Nigeria was conducted. These firms were stratified by sector and industry. The initial result revealed 41 MNEs. The second stage involved a review of each MNE website to ascertain whether they have a clear CSR, sustainability, or Environmental Social Goal (ESG) statement or report. This resulted in a total of 11 MNEs with visible internet profiles on these initiatives. The 11 MNEs were in 5 main sectors/industries, namely food and beverage, telecommunications, pharmaceutical, construction, oil and gas, and logistics (See Table 2 below).

5 Results and Discussion of Findings

The OECD 2023 guidelines respond to urgent social, environmental, and technological priorities facing societies and businesses. The OECD (2023) makes recommendations for enterprises to align with internationally agreed goals on the environment, climate change and biodiversity, employment and industrial relations, and human rights issues. The OECD (2023) calls on MNEs to conduct their business operations responsibly and ensure that they address adverse environmental impacts of their business activities and contribute to the wider goal of sustainable development by taking into consideration the need to protect the environment, workers, the local communities, and society in which they operate. In addition,

MNEs should act within the framework of laws, regulations, and administrative practices in the home countries in which they operate and ensure compliance with relevant international agreements, principles, and standards (OECD, 2023). According to the OECD (2023), MNEs should have respect for human rights and seek ways to "prevent or mitigate adverse human rights impacts that are directly linked to their business operations, products, or services by a business relationship, even if they do not contribute to those impacts." Avoid unlawful employment and industrial relations practices and ensure that their business operations are within the framework of applicable local laws, regulations, and prevailing labor relations and employment practices as well as applicable international labor standards. When dealing with consumers, MNEs are encouraged to take reasonable steps to ensure goods and services that they provide are of good quality and reliable and that they act according to fair business, marketing, and advertising practices (OECD, 2023). The OECD (2023) recognizes that MNEs play a key role in advancing sustainable economies and environmental challenges (these include climate change; biodiversity loss; degradation of land, marine, and freshwater ecosystems; deforestation; air, water, and soil pollution; mismanagement of waste and hazardous substances). The OECD echoes the spirit of the UN SDG goals, many of its recommendations are closely linked with those proposed by the United Nations Sustainable Development Goals (SDGs) which focus on a sustainable agenda and action plan for people, planet, and prosperity (UN, 2015). According to the UN (2015), "the Sustainable Development Goals will stimulate action over the next 15 years in areas of critical importance for humanity and the planet." The UN recognizes the important role stakeholders play in implementing sustainable development and in shifting the world onto a sustainable and resilient path (UN, 2015). The UN proposes 17 Sustainable Development Goals and 169 targets. The universal Agenda proposed balances the three dimensions of sustainable development in extant academic literature, namely economic, social, and environmental dimensions. The 17 Sustainable Development Goals include elimination of poverty; zero hunger, sustainable agriculture; good health/well-being; quality education, life-long learning; gender equality, female empowerment; clean water/sanitation; affordable clean energy; decent work/economic growth; industry, innovation/infrastructure; reduced inequalities; safe sustainable cities/communities; responsible consumption/production; climate action; life below water; life on land, ecosystem; peace, justice/strong institutions; partnership for the SDGs. The SDGs encompass CSR, sustainability, and ESG concepts, and as such, offer a comprehensive categorization of pertinent initiatives against which the MNEs' activities can be benchmarked. A review of the MNE's CSR and sustainability report showed commitment to various SDGs globally as shown in Table 3 below. MNEs engage in multiple initiatives globally, all MNEs stated that they engage with climate action and responsible consumption and production—the level of engagement with some of the SDGs in Nigeria is trivial.

Cadbury/Mondelez: Cadbury's sustainability and social development initiatives focus on FIVE main aspects, i.e. environment, social, governance, sustainable agriculture, people, and communities (see Table 4 below).

Table 3 MNE global social development goal commitments

	Partnership for the SDG	Peace, Justice/ Strong Institutions	Life on Land, Ecosystem	Life Below Water	Climate Action	Responsible Consumption/ Production	Safe Sustainable Cities/ Communities	Reduced Inequalities	Industry, Innovation/ Infrastructure	Decent Work/ Economic Growth	Affordable Clean Energy	Clean Water/ Sanitation	Gender Equality, Female Empowerment	Quality Education, Life-Long Learning	Good Health/ Wellbeing	Zero Hunger, Sustainable Agriculture	Elimination of Poverty
Cadbury/ Mondelez	X	X	X	X	X	X	X	X	X	X	X	X	X	X	X	X	X
Coca-Cola	X		X	X	X	X		X		X	X	X	X		X	X	X
Chevron		X	X	X	X	X			X	X	X	X			X		
Exxon Mobil					X	X					X						
Glaxo (GSK)	X	X	X		X	X		X	X	X		X	X		X		X
Julius Berger	X	X	X		X	X	X	X	X	X	X	X	X	X	X	X	
Nestle	X	X	X	X	X	X	X	X	X	X	X	X	X	X	X	X	X
Shell	X	X	X	X	X	X		X	X	X	X	X	X		X		
Total energies	X	X	X	X	X	X	X	X	X	X	X	X	X	X	X	X	X
Unilever	X		X	X	X	X					X	X		X	X	X	X

Table 4 Cadbury's CSR and sustainability goals

Environment	Greenhouse gas, energy, waste, water, packaging
Social	Human capital, human rights, well-being, occupational health and safety, product quality and safety, community engagement
Governance	Diversity (gender, ethnic, age, tenure)
Sustainable agriculture	Sustainably source ingredients: Increases resilience of the supply chain to conserve nature and empower producers and farm workers; encourages and supports ingredient suppliers to drive continuous improvement in sustainable farming.
People and communities	Human rights; diversity, equity & inclusion; community development.

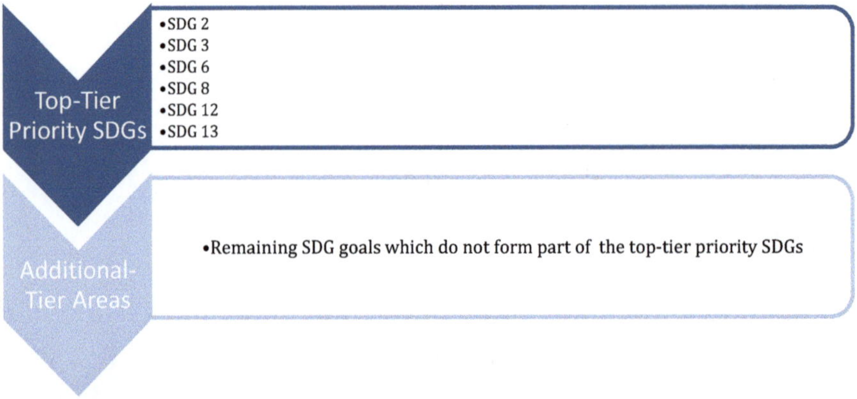

Fig. 2 Cadbury's priority areas aligned to the UN SDGs

As shown in Fig. 2 below, Cadbury also distinguishes these initiatives into TWO main categories, namely (a) top tier—priority SDGs where their impact is most closely linked, i.e. SDG 2, 3, 6, 8, 12, and 13; (b) additional tier—areas where it seeks to positively impact the remaining SDG goals (Cadbury MDLZ, 2022).

A notable example of Cadbury's CSR initiative in the country is the "Bourn Vita Teachers Award"—an annual award that recognizes the best teachers in Nigeria. Other initiatives include a four-year project which cost $ 1 million aimed at tackling malnutrition and obesity in selected primary schools within its host communities in Lagos State. Cadbury Nigeria Plc's Annual report and financial statements for the year ending 31 December 2023 highlight its community involvement. The annual report shows that Cadbury members of staff:

(a) Carried out volunteering activities as part of World Environment Day in June 2023 to clean up the environment around the piped water donated to the Agidingbi host community.
(b) Held career and mentoring sessions and donated branded company items, safety, and hygiene materials to the pupils at Agidingbi Primary School.

Donated toiletries and food items to a care home in Yaba, Lagos (Cadbury MDLZ, 2023).

Chevron: Chevron's SDG agenda focuses on THREE main areas, namely, climate, nature, and people (see Fig. 3 below).

Chevron contributes to sustainable development in health, education, and economic prosperity within the communities. Table 5 below illustrates Chevron's main CSR and sustainability initiatives and how these align with the three areas of focus (Chevron, 2022).

Chevron's social initiatives have ranged from initiatives devised to respond to needs in the Niger Delta, fighting HIV/AIDS in Nigeria, caring for the environment, supporting public health promoting education, and helping students. An example of Chevron's sustainability agenda in Nigeria is the diversion of "incinerator ash destined for landfill disposal to a third-party waste management facility" since 2021—the facility processes the ash into base material used for cement production (Chevron, 2023). In Lagos, Chevron funded the "Lekki Conservation Centre, a 190-acre (0.8-sq-km) sanctuary that protects the Lekki Peninsula and promotes environmental practices through research and education." Chevron's website does not provide any recent sustainability projects. Health, education, and economic development projects include:

- 2005: $118 million on 600 programs that have provided scholarships; built new schools, medical facilities, and housing; and supported agriculture development and infrastructure improvements as part of the community engagement in the Niger Delta.
- 2006: "Roll Back Malaria initiative" provides support to pregnant women and children. As of 2017, the program is estimated to have served more than 60,000 people.

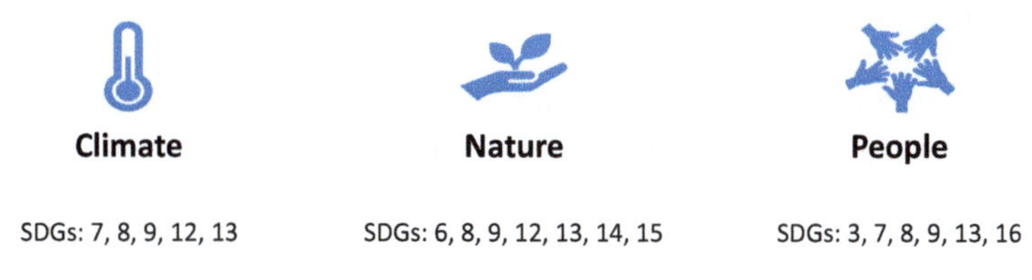

Fig. 3 Chevron's CSR and Sustainability Goals aligned to the UN SDGs

Table 5 Chevron's CSR and sustainability goals

Climate	Nature	People
Innovative approach to operations and climate change	Resource management Biodiversity, land and water stewardship Environmental risks	Governance and transparency Communities, people, and culture Thriving workforce, creating prosperity Health and safety management Human rights

- 2008: $5 to Fight AIDS, Tuberculosis, and Malaria. Donation of X-ray units, consulting rooms, laboratories, and wards to support early diagnosis and treatment of tuberculosis and other chest and lung diseases.
- 2009: Scholarships that aid visually challenged students and scholarship programs for medical and engineering professionals.
- 2011: $50 million alliance with the U.S. Agency for International Development and the Niger Delta Partnership Initiative (NDPI) Foundation to address socio-economic challenges in the Niger Delta region implemented through the Foundation for Partnership Initiatives in the Niger Delta (PIND).
- 2012: Support provided to 670 health facilities in Nasarawa, Bayelsa, and Rivers states for those with HIV.
- 2014: $1.7 million donation to the "PROMOT" project initiative in Bayelsa State to provide education and HIV testing services for pregnant women. Donation of equipment to the University of Lagos Teaching Hospital Molecular Biology Research Laboratory helps diagnose and treat people with genetic abnormalities.
- 2015: $5 million contribution to "Global Fund" to support the prevention of mother-to-child transmission of HIV.
- 2016: $1.43 million donated to the "PROMOT" project in Bayelsa state.

Coca-Cola: Coca-Cola's strategy is "centered around people, consumers, employees and driving sustainable solutions that build resilience into the MNE's business to respond to current and future challenges while creating positive change for the planet" (Coca-Cola, 2022). In particular, Coca-Cola's CSR and sustainability goals are achieved under SIX main areas of focus, namely, water leadership, portfolio, packaging, climate, sustainable agriculture, and people and communities. These initiatives are closely aligned with the UN SDGs and include, for example, initiatives tailored toward water conservation, waste recycling, and climate agenda through active engagement with greenhouse gas emissions and human rights issues (Fig. 4).

Coca-Cola's 2022 CSR and sustainability report shows a corporate mission that seeks to target the following social development issues, namely: Anti-competitive Behavior—Anti-corruption—Biodiversity—Child Labor—Customer Health and Safety—Customer Privacy—Diversity and Equal Opportunity—Emissions—Employment—Forced or Compulsory Labor—Labor/Management Relations—Local Communities—Marketing and Labeling—Materials—Occupational Health and Safety—Supplier Environmental Assessment—Public Policy—Supplier Social

Fig. 4 Examples of some of the initiatives adopted by Coca-Cola

Assessment—Water and Effluents—Waste. Table 6 below illustrates Coca-Cola's SIX main CSR and sustainability goals and specific areas of focus.

Recent sustainability initiatives include the 2024 collaboration with the United States Agency for International Development (USAID) and "Techno Serve" Nigeria, to launch the Nigeria Plastic Solutions Activity (NPSA). NPSA is tasked with recovering approximately 49,000 metric tons of plastic waste in Nigeria. The initiative valued at $4 million seeks to tackle plastic waste challenges in Nigeria through innovative recycling solutions. The program aims to create employment and drive circularity in Nigeria.

ExxonMobil: ExxonMobil seeks to create "sustainable solutions that improve quality of life and meet society's evolving needs." It strives to do this in ways that help protect people, the environment, and the communities where it operates. ExxonMobil's main agenda includes climate action, clean energy, and responsible consumption and production—activities that are closely aligned to the SDGs as depicted in Fig. 5 below (ExxonMobil, 2023a). ExxonMobil's 2023a annual report contains a summary of its corporate social responsibility initiatives. The MNE's primary areas of target were security, education, and health. The MNE contributed 25,285,608 million Naira in 2022 and 3,000,750 million Naira toward various

Table 6 Coca-Cola's CSR and sustainability goals and focus areas

CSR/ sustainability agenda	Area of focus
Water leadership	Regenerative water use; improve the health of watersheds; return 2 trillion liters of water to nature and communities globally.
Portfolio	Offering drinks with reduced added sugar; introducing more brands with nutrition and wellness benefits; providing small package options and clear nutrition information on packaging, communications; and marketing.
Packaging	Collect and recycle a bottle or can for each one sold in 2030
Climate	Reduce absolute greenhouse gas emissions by 25% by 2030.
Sustainable agriculture	Sustainably source 100% of priority agricultural ingredients. Engagement with suppliers to implement principles for sustainable agriculture framework and drive progress on water sustainability.
People and communities	Working toward achieving 35% of senior leadership positions held by women. Mentoring programs designed to promote inclusion for people with disability.

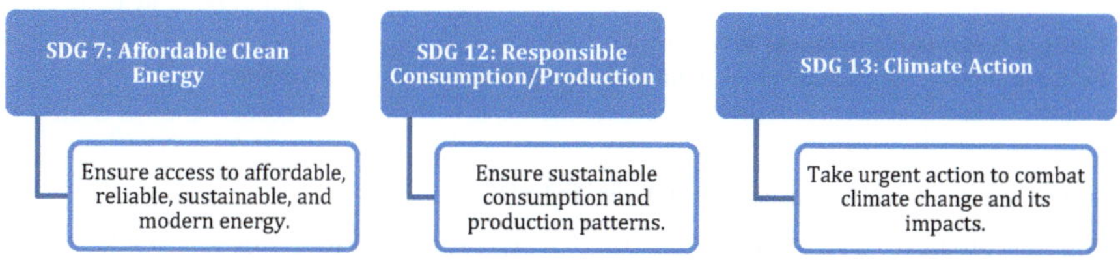

Fig. 5 ExxonMobil's SDG Targets

projects, gifts, and sponsorship programs. Initiatives include generator purchase, purchase of food items for widows and the less privileged, purchase of clinical materials, purchase of electrical and food items, a Down Syndrome, and renovation of a primary school (ExxonMobil, 2023b).

GLAXO (GSK): Glaxo recognizes the importance of health as a central aspect of the 17 SDGs. Accordingly, Glaxo's CSR and sustainability 2019 report states "Although each of the 17 global goals is focused on a different issue, we believe that health underpins almost every development theme, each of which enables, or is enabled by, advances in population health" (Glaxo, 2019). Glaxo's approach to responsible business is based on THREE main pillars, namely, innovation, performance, and trust. Glaxo's (2021) annual report highlights its commitments to underprivileged communities across Nigeria through health and education initiatives and improving quality of life by providing support to local cultural institutions. Glaxo made donations to the Nigerian Association of Resident Doctors and Medical Guild Lagos to protect health workers in 2021. The MNE also partnered with Save the Children International through the "INSPIRING" project to donate medical equipment and products worth over 100 million Naira to thirty health facilities in Jigawa and Lagos State. In 2023 Glaxo announced plans to cease operations in Nigeria due to declining revenue and worsening economic conditions. Glaxo fulfills THIRTEEN sets of social development commitments across these THREE pillars as depicted in Table 7 below.

Shell: The United Nations Goals are part of Shell's Powering Progress strategy which directly contributes to THIRTEEN SDGs (Shell, 2022). Shell groups these SDGs into FOUR main areas of focus, namely, achieving net-zero emissions; powering lives, respecting nature, and core values—these projects are often implemented in partnership with local authorities (see Table 8 below).

In 2023, Shell invested $3 million in education programs. In its 2023 sustainability report, the MNE states that it contributed $42.2 million in direct social

Table 7 Glaxo's CSR and sustainability focus

Innovation	Performance	Trust
Using science and technology to address health needs through: New medical innovation Global Health Health security	Making products affordable and available through: Pricing Product reach Healthcare access	Being a modern employer through: Employee engagement Inclusion and diversity Health and well-being and development
Being a responsible business through: Reliable supply; ethics and values; data and engagement; environment impact		

Table 8 Shell's CSR and sustainability focus and SDGs

Powering progress goal	Achieving net-zero emissions	Powering lives	Respecting nature	Core values
UN SDG	7, 9, 13	5, 7, 8, 10	6, 12, 14, 15	3, 16, 17

investment to various projects in Nigeria which focus primarily on enterprise programs, health, education, and road safety (Shell, 2023). In January 2024, Shell announced the sale of Shell Petroleum Development Company of Nigeria Limited (SPDC), its Nigerian onshore subsidiary.

Total Energies: Total Energies builds its sustainability approach on FOUR core areas of action—climate and sustainable energy; people's well-being; care for the environment; and creating value for society. In keeping with the United Nations Sustainable Development Goals (SDGs) 11—"Sustainable Cities and Communities," SDG 13—"Climate Action," and SDG 17—"Partnerships for the Goals," Total Energies engaged in various initiatives. For example, in March 2020, Total Energies entered a partnership with Rite Foods, a food and beverage company in Ogun State, within the Southwestern region to create the first Back-to-Back (B2B) energy operational unit in Nigeria. The initiative is "endowed with a hybrid off-grid installation with the photovoltaic solution integrated into the diesel and gas energy mix" (Total, 2020). The development helps the MNE avoid 14,000 tons of CO2 emissions for the next 25 years and results in cost savings of around $250,000 Return on Investment in five and a half years to customers. Another climate action initiative is the solarization of the MNE blending units to reduce their CO2 emissions. Total Energies equips its blending plant unit in Lagos with solar panels (360 Sun Power Modules) for an installed capacity of 118 kW. It is anticipated that by 2023, Total Energies will solarize eight other operating sites for an estimated capital expenditure of $1.8 million, thereby reducing its Scope 2 Co2 emissions by 18%—around 1.2KT per year. In keeping with its objective of "abating emissions at each source" Total Energies leak reduction initiatives involve an annual campaign to identify and repair leaks at all operated sites—resulting in a 4KT emissions decline in 2021 (Total, 2022b). As part of its initiative to champion a legacy of sustainable education, Total Energies pays 64 beneficiaries' full tuition fees for 6 years; provides 21 internet-ready laptops for secondary and university children with internet subscriptions; 44 desktops for 4 sponsored houses for children in primary schools. Total Energies also provides charitable donations to sponsored charity organizations in Lagos, Oyo, Port Harcourt, and Kaduna States. Total Energies budding entrepreneurs' initiative seeks to provide five vulnerable young residents of the Delta State region with skills acquisition and the opportunity to learn a profession and create their businesses. The initial step involves "training in the profession of the participant's choice, followed by support for business creation and two years of pre-paid rent on business premises." Since its introduction in 2016, 64 people young people have been trained (Total, 2022b). As shown in Table 9 below, Total Energies distinguishes its level of social development contributions. Total Energies categorizes its contributions according to those it considers—core contributions, direct contributions, and indirect contributions.

Furthermore, as depicted in Fig. 6 below, Total Energies assesses the FOUR core areas of action and the firms' level of contribution against relevant SDGs (Total, 2022a).

Julius Berger: Julius Berger carries out its social development activities under FOUR main pillars, namely, economic, environmental, social, and governance.

Table 9 Total CSR and sustainability contributions aligned to SDGs

Core contributions through its mission	Direct contributions through its responsible business approach	Indirect contributions through its responsible business approach
SDGs 7–8–9–13	SDGs 3–4–5–10–12–14–15–16	SDGs 1–2–6–11

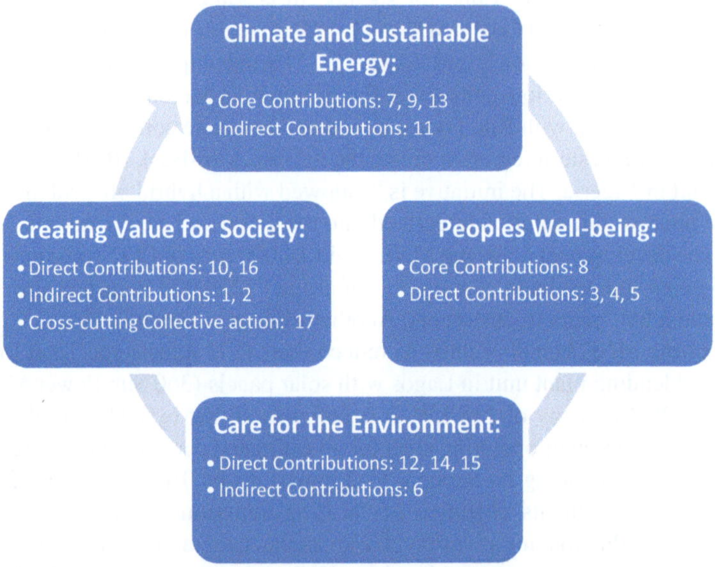

Fig. 6 Total's four core areas of action and level of contribution aligned to the SDGs

Table 10 below shows the MNE's CSR and sustainability activities grouped under the relevant pillars (including material topics of importance vis-à-vis each pillar) and how these align with the SDGs. The MNE groups its social development activities into THREE main parts:

- Local Communities—community development projects; employment and skill development; stakeholder engagement.
- Safety—safety training; occupational health; collaborative safety initiatives.
- Social investment activities—education and skills development; health and well-being; employment creation (Julius Berger, 2022a, 2022b).

During the year 2022, Julius Berger undertook CSR initiatives valued at 507.20 million Naira and 560.38 million Naira in 2021. The MNE also made 24.98 million Naira in charitable donations in 2022 and 28.85 million Naira in 2021. CSR initiatives focused on education, human capital development, community development and inclusivity, philanthropic and social welfare, and emergency response. Charitable donations were made to schools, sporting associations, orphanages, and religious institutions (Julius Berger, 2022a, 2022b). Of importance are some of the projects identified in its CSR and sustainability report vis-à-vis its social

Table 10 Julius Berger's CSR and sustainability pillars and how these align with the SDGs

Environment pillar	Social pillar	Governance pillar	Economic pillar
Materials (SDG 12) Energy usage and efficiency (SDG 7) Environmental compliance (SDG 15) Supplier environmental assessment (SDG 12, 15) Water and effluents (SDG 6, 13) Environmental impacts of products and services (SDG 12, 13, 15)	Diversity and inclusion (SDG 5,10 Employment (SDG 8) Labor relations (SDG 8,10) Occupational health and safety (SDG 8) Training and education (SDG 4) Security practices (SDG 16) Human rights assessment (SDG 10) Local communities (SDG 11,17) Customer health and safety (SDG 3) Socioeconomic compliance (SDG 8) Supplier social assessment (SDG 9,12)	Anti-corruption (SDG 16) Anti-competitive behavior (SDG 16) Public policy (SDG 16) Whistleblower policies and protection (SDG 16)	Economic performance (SDG 8) Market presence (SDG 9) Indirect economic impacts (SDG 11, 17) Procurement practices (SDG 9, 12) Tax (SDG 8, 16) Prefabricated and modular construction (SDG 9, 11) Digitization (SDG 9)

development agenda. Four main projects are identified which are also clearly aligned to specific SDGs:

(a) *Julius Berger Nigeria Plc's Science, Technology, Engineering, and Mathematics (STEM) Scholarship Awards:* promoting inclusion and diversity in the engineering industry and contributing to reaching both Sustainable Development Goal (SDG) 4 and 5, which focus on inclusive and equitable quality education, achieving gender equality and empowering women in Nigeria.
(b) *The Bonny Consulate Building:* increased tourism activities in the community and provided productive employment and decent work for residents of the community contributing to reaching both Sustainable Development Goals (SDG) 8 and 11, which focus on inclusive and sustainable economic growth and providing sustainable cities and communities.
(c) *Toilets and borehole construction*: seeks to create a hygienic and conducive learning environment for the students, teachers, and staff of the schools contributing to reaching Sustainable Development Goals (SDG) 3, 4, and 6, which focus on providing clean water and sanitation, quality education, and good health and well-being.
(d) *Farmers training project*: these projects seek to contribute to Sustainable Development Goals (SDG) 2, 6, 8, and 12 which seek to promote sustainable

agriculture; gender equality and empowerment for women and girls; promote inclusive and sustainable economic growth; and ensure sustainable consumption and production patterns (Julius Berger, 2022a, 2022b).

Nestle: Nestle's approach to social development focuses on creating shared value. The main areas of focus are: Nutrition—Health—Wellness; Rural Development; Water; Environmental Sustainability People—Human Rights and Compliance, all of which are firmly embedded in a policy drive that puts individuals, families, communities, and the planet at its core. Table 11 below illustrates a link between the core areas of focus and Nestle's social development contributions.

Through 42 specific commitments, Nestle seeks to meet its ambitions for 2030 in line with the timescale of the Sustainable Development Goals (Nestle, 2023). Social initiatives and charitable gifts and donations made by the MNE during the year 2022 amounted to 235,789,213 million Naira in the year 2022 and 552,100,014 million Naira in 2021 (Nestle, 2022). The initiatives include water projects, educational projects and technical training centers, scholarship schemes, and women empowerment projects. In 2021, Nestlé launched the "Developing Inclusive Grain Value Chains project to help 5,000 smallholder farmers earn better livelihoods by supplying high-quality maize, soybeans, millet, and sorghum to Nestlé." In 2024, to enhance climate resilience and mitigate the impact of adverse weather conditions on farms, Nestlé Nigeria collaborated with volunteers from Convention for Business Integrity Innovations Nigeria Limited to plant 1000 trees at the Nestlé Dairy Demonstration Farm in the Paikon-Kore grazing reserve Abuja.

Unilever: Unilever adopts a THREE-pronged approach to social development. The THREE approaches align to relevant SDGs as shown in Fig. 7 below.

Unilever focuses on climate action, protection, and regeneration of nature, a waste-free world, positive nutrition, health and well-being, equality, diversity and inclusion, raising living standards, and finally future work which seek to help equip young people with essential skills (Unilever, 2021). The Unilever (2022) sustainability report highlights specific socially responsible and sustainability initiatives,

Table 11 Nestle's CSR and sustainability focus

Nutrition—health—wellness	Rural development	Water	Environmental sustainability	People—human rights and compliance
Food and nutrition security Undernutrition Food and product safety Responsible marketing and influence	Rural development and poverty alleviation Responsible sourcing and traceability Animal welfare Women's empowerment	Water stewardship Water, sanitation, and hygiene	Natural resources stewardship Climate change Resource efficiency (food) waste and the circular economy	Fair employment and youth employability Employee safety, health, and wellness Human rights Business ethics

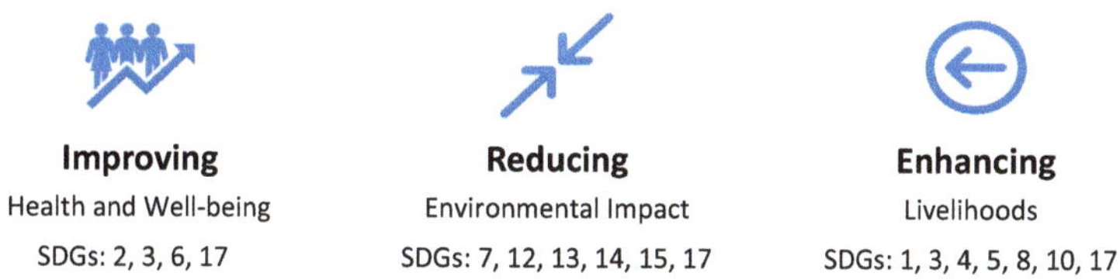

Fig. 7 Unilever's social development agenda aligned to SDGs

Table 12 Unilever sustainability focus and SDGs

Improving health and well-being	Reducing environmental impact	Enhancing livelihoods
Health and hygiene; safe drinking water; sanitation; nutrition	Greenhouse gases; water usage; waste reduction; sustainable sourcing	Sustainable sourcing; fairness in the workplace; fair compensation; health, safety, and well-being; gender-balanced organizations; women's safety; skills and training; empowering women; inclusive business for smallholder farmers and small-scale retailers, etc.
SDGs: 2, 3, 6, 17	SDGs: 7, 12, 13, 14, 15, 17	SDGs: 1, 3, 4, 5, 8, 10, 17

for example a three-week program across 30 schools in Lagos to educate students on plastic recycling and littering. In addition, Unilever engaged in a campaign aimed at empowering vulnerable groups within the community through the OMO's "Lend a Hand of Care" campaign, Unilever collated over 10,000 pieces of clothing and other household items across the nation from donations. These items were donated to NGOs in Abuja, Kwara, and Lagos. The MNE's initiative to contribute to a "Fairer, More Socially Inclusive World" shows a female employee representation ratio of 41.1%. The Unilever Secondary Internship (USSIP) summer internship program for secondary school students provides pupils with the opportunity to learn basic workplace-ready skills such as Presentation; Teamwork; Marketing; Communication skills, Sales, and Project planning proficiency. Table 12 below illustrates Unilever's THREE main approaches, specific initiatives, and how these align with the SDGs.

6 Conclusion

With developing nations' governments unable to meet the needs of their populations, multinational enterprises have, both voluntarily and by mandate, taken on many governmental tasks in many developing countries around the world. A result is that companies become political actors within developing nations via their CSR and sustainability activities. The preceding analysis shows that many of the initiatives that the MNEs report that they engage in are in the form of philanthropic endeavors, e.g. building schools, paying teachers' salaries, buying school supplies, providing utilities such as water and electricity supplies and systems, health care, road construction, small business development, and helping the agricultural community. Despite this list of efforts, there is a divergence between the perceptions of success between the communities and the MNEs, wherein MNE executives have been satisfied with the results, while the community members may not be (Adesua Lincoln and Diamond, 2023). Societies hold expectations that companies do more for the greater good. Consequently, the local communities' perception and consumer behavior and reactions to sustainability and social responsibility activities may provide insight into what societies hold as expectations. It is unclear from the MNE sustainability report how the initiatives they support were identified and whether the projects carried out were undertaken following consultations with the local communities. To be effective, initiatives must be influenced by the specific needs voiced by each local community in which the MNEs operate, rather than a one-size-fits-all approach—this is because local sensibilities and needs differ from the Western paradigm. These differences might cause conflicts of prioritization between the international businesses and the local community in which they operate—this will enable the international businesses to adapt their Western perspectives to the local communities to provide a framework under which a meeting of minds can be facilitated. Sustainability and social responsibility differ between nations and their internal cultures. Even similar cultures can differ significantly. Sustainability and social responsibility expectations can also differ between regions inside a nation, with each region having different types of sustainability and social responsibility expectation activities depending upon the local culture and leading economy. Variation in expectations can also be found to extend to different industries, wherein societal views seem to reflect higher expectations upon some industries than others. Consequently, MNEs may find themselves with many differing sustainability and social responsibility policies specific to each country of operation, as rules and expectations can differ between jurisdictions—this has resulted in MNEs adopting a lax approach to sustainability and social responsibility in some developing countries compared to the approaches adopted in their "Home" countries.

While MNEs are colloquially understood as a single entity, the conventional arrangement comprises a hierarchy with the mother company at the top owning and controlling dozens of subsidiaries for which the mother company has little legal liability. In addition to these subsidiaries, there can be thousands of suppliers within the hierarchy spread across dozens of countries. At the same time, individual operations are governed by the different domestic governments within each jurisdiction, which allows companies different degrees of influence and autonomy depending upon the power balances within those jurisdictions. While a multinational organization can extend a great deal of control worldwide as a single organization, national and international legal systems only govern the individual parts within specific jurisdictions and, generally, do not recognize MNEs as being a single unit. This overall dynamic of the power imbalance between MNEs and the limited capacity of developing nations to regulate and enforce rules results in the general lack of sustainable development and well-being within developing nations. Ruggie (2018) recognizes the distinction between CSR and sustainability on a national level versus the global level based to a high degree upon the differing relative power between companies and their domestic Western governments, in contrast with these same companies and the developing nations' governments. Ruggie (2018) sees the Western governments as more effectively satisfying the well-being needs of their population and as being more capable of meeting companies on an equal power balance. Therefore, Western governments have less need for companies to step up and take on governmental roles, but they are also less vulnerable to the power and influence wielded by corporations. As such, Western MNEs have their base in countries where there is a distinction between government-type functions and company responsibilities. The result of the power imbalance is that many international businesses have caused severe environmental damage, which has negative consequences on the population's health and dissonance within society. Meyer (2015) recognizes that MNE's behavior in developing nations can have spillover effects. He sees spillovers occurring across a broad array of interactions with other businesses, governmental institutions, and private citizens. These spillovers can be beneficial, such as the transfer of advanced technology and management practices. The presence of MNEs can, however, have destabilizing effects on a nation, local community, and government institutions when the scale of the MNE operation competes with the government itself and can cause the redistribution of the population, which could contribute to urbanization and the breakdown of traditional family and community dynamics. Being a positive role model can be in the best interest of the MNE when there is a positive effect on government institutions and business practices as it may contribute to a stable environment to operate. A partnership relationship with government institutions can enhance the results of education programs by the parties coordinating with each other, as the MNEs can fill in the gaps left by domestic institutions. This consideration is particularly critical as the educational needs of an MNE's home country are likely to differ significantly from the developing nation's needs. The differing ethical values between the two jurisdictions are a significant obstacle anti-corruption education programs may face. These differences would mean that the education programs would need to be customized to each community's sensibilities, and MNEs need to be mindful of not abusing their relative power in the dynamic by expecting the developing nation's norms to adhere to the MNE's home country's norms. MNEs need to work closely with the local community to better understand their values (Idemudia, 2011), or else the CSR actions risk marginalization and ineffectiveness. Differing underlying ethical assumptions results in socially responsible activities that may be incompatible with local values and could not only lead to failure of the activities but also cause harm to the local communities. Thus, the need for companies to break from an approach with interactions limited to the government and the MNE (Meyer, 2015) and be engaged with the local communities is further emphasized. It follows from the above discussion that MNEs can enrich themselves and operate with a high degree of autonomy and influence, with little need to consider the long-term well-being of the population. The poor within society are especially vulnerable and hard-hit within developing nations as they often have little power or avenues to voice themselves. For example, when MNEs come into Nigeria to extract natural resources, the lax nature of CSR and sustainability rules and regulatory frameworks resulted in a situation where the activities of these international businesses are left unchecked. Even when these regulations are in place, the Nigerian government is unlikely to have the will or the ability to enforce them. While MNEs may provide employment opportunities during the extraction period, once the resources are exhausted, the country no longer has long-term natural resources in the ground. They also no longer have employment benefits and may be left with long-term environmental damage, degradation of the ecosystem, or even disasters, which may offset the short-term wage benefits the nation has gained. For example, as stated earlier, with regard to the Niger Delta region, concerns of the local communities dependent upon farming and fishing for their livelihood were pushed aside by petroleum extractors (Idemudia, 2009). When the environmental damage from the extraction process ruined their fields, fishing waters, and the roofs of their homes, then the multinational enterprises hired researchers who reported finding no evidence of these claimed impacts. In effect, the country has traded away a long-term asset for short-term benefits. In the end, it is a very legitimate question for Nigerians to ask, "What do we get out of this dynamic? When these companies have exited, then what are we left with?"

7 Recommendations

MNES must begin to consider expanding their CSR and sustainability initiatives to actively tackle real issues in host countries by providing long-term practical skills training and industrial training facilities and centers to graduates to make them more competitive in the job market. In addition, these initiatives should look to fostering creative and entrepreneurial skills in women and children. It is also imperative that funds set aside for such endeavors are not given directly to the Nigerian government or its parastatals—instead, the MNEs need to take ownership of these initiatives themselves. There seems to be more emphasis by these MNEs on how they perceive their activities fit in with the UN SDGs rather than actual empirical evidence of "impact." There is a mismatch between the financial figures touted by MNEs as their actual contributions to CSR and sustainability initiatives and the results they present in evidencing the money spent—it is ludicrous for these MNEs to think they can continue with this illusion unchallenged! The SDGs have now become a "tick box" exercise and a badge of honor even where no meaningful environmental and social development agenda is achieved. There is an urgent need to address this issue pragmatically. MNEs need to be willing to be agents of positive change in Nigeria by effectively engaging in long-term sustainable initiatives. A robust mechanism is needed to evidence the application of financial resources to specific societal needs in Nigeria—benchmarked against the MNC's approach in its home country and other Western countries in which they operate to ensure equitable practices across countries. Participatory research of MNE CSR and sustainability activities is required in the local communities in which these MNEs operate to effectively evaluate the efficacy of MNE CSR and sustainability contributions.

References

Adesua Lincoln, A., & Diamond, B. (2023). Contribution of sustainable development goals and corporate social responsibility initiatives of multinational enterprises (MNEs) to social development in Nigeria: A critical assessment of the different parties and the dynamic involved in mandating CSR to identify best practices for developing nations. In S. O. Idowu & L. Zu (Eds.), *Elgar companion to corporate social responsibility and the sustainable development goals* (pp. 190–220). Edward Elgar.

Balp, G. (2018). The corporate governance role of retail investors. *Loyola Consumer Law Review, 31*, 47.

Bohlmann, C., Krumbholz, L., & Zacher, H. (2018). The triple bottom line and organizational attractiveness ratings: The role of pro-environmental attitude. *Corporate Social Responsibility and Environmental Management, 25*(5), 912–919.

Bowen, H. R. (1953). *Social responsibilities of the businessman*. Harper & Row.

Burritt, R., Christ, K., Rammal, H., & Schaltegger, S. (2020). Multinational enterprise strategies for addressing sustainability: The need for consolidation. *Journal of Business Ethics, 164*, 389–410.

Cadbury MDLZ. (2022). *Environmental social and governance (ESG) datasheet*. Available at Cadbury MDLZ_ESG_Datasheet_2022.pdf

Cadbury MDLZ. (2023). *Cadbury Nigeria Plc Annual report and financial statements for the year ended 31 December 2023*. Retrieved from https://africanfinancials.com/document/ng-cadbur-2023-ar-00/

Carroll, A. (2004). Managing ethically with global stakeholders: A present and future challenge. *Academy of Management Executive, 18*(2), 114–120.

Chevron. (2022). *Enabling human progress, 2022 corporate sustainability report*. Retrieved from chevron-sustainability-report-2022.pdf

Chevron. (2023). *Energy for a growing World, 2023 corporate sustainability report*. Retrieved from https://www.chevron.com/newsroom/media/publications/corporate-sustainability-report

Coca-Cola. (2022). *Business and sustainability report*. Retrieved from coca-cola-business-sustainability-report-2022.pdf

Companies and Allied Matters Act (CAMA) 2020.

Cruz, N. F., & Marques, R. C. (2014). A multi-criteria model to determine the sustainability level of water services. *Water Asset Management International, 9*(3), 16–20.

Elkington, J. (2018). 25 years ago, I coined the phrase 'triple bottom line.' Here's why it's time to rethink it. *Harvard Business Review*.

ExxonMobil. (2023a). *Sustainability report*. Retrieved from exxonmobil-sustainability-report.pdf

ExxonMobil. (2023b). *Annual report and account*. Retrieved from https://11plc.com/wp-content/uploads/2024/04/2023-Annual-Report.pdf

Friedman, M. (2002). The social responsibility of business is to increase its profits. In L. P. Hartman (Ed.), *Perspectives in business ethics* (2nd ed., pp. 260–264). The McGraw-Hill Companies.

Glaxo Smith Kline. (2019). *Our contribution to the social development goals*. Retrieved from Glaxo our-contribution-to-the-sdgs.pdf

Glaxo Smith Kline. (2021). *Annual report financial statement*. Retrieved from https://ng.gsk.com/media/6364/gsk-annual-report-2021.pdf

Hartmann, B. (2020). Triple bottom line. In G. Rimmel (Ed.), *Accounting for sustainability* (pp. 101–110). Routledge.

Idemudia, U. (2009). Oil extraction and poverty reduction in the Niger Delta: A critical examination of partnership initiatives. *Journal of Business Ethics, 90*(1).

Idemudia, U. (2011). Corporate social responsibility and developing countries: Moving the critical CSR research agenda in Africa forward. *Progress in Development Studies, 11*(1).

Julius Berger. (2022a). *Enabling progress expanding impact 2022 sustainability report*. Retrieved from jbn_sustainability_report_2022.pdf

Julius Berger. (2022b). *Consolidated financial statements for the year ended 31 December 2022.* Retrieved from https://africanfinancials.com/document/ng-jberge-2022-ar-00/
Lexis Nexis. (2023). *Round-up of recent major fines for alleged compliance failures – And how companies can mitigate this rising regulatory risk.* Retrieved from https://www.lexisnexis.com/blogs/gb/b/compliance-risk-due-diligence/posts/fines-compliance-failures
Meyer, M. (2015). Positive business: Doing good and doing well. *Business Ethics: A European Review, 24*, 175–197.
Nestle. (2022). *Annual report and accounts - The leading nutrition, health and wellness company.* Retrieved from https://www.nestle-cwa.com/sites/g/files/pydnoa346/files/2023-04/Nestle%20Annual%20Report%202022_0.pdf
Nestle. (2023). *Creating shared value and sustainability report 2023 advancing regenerative food systems at scale.* Retrieved from creating-shared-value-sustainability-report-2023-en.pdf (nestle.com)
Organization for Economic Co-operation and Development. (2023). *Guidelines for multinational enterprises on responsible business conduct.* OECD. https://doi.org/10.1787/81f92357-en
Ruggie, J. G. (2018). Multinationals as global institution: Power, authority and relative autonomy. *Regulation & Governance, 12*(3), 317.
Samat, N., & Ali, H. (2015). A legal perspective of shareholders' meeting in the globalised and interconnected business environment. *Procedia - Social and Behavioural Sciences, 172*, 762–769.
Schaltegger, S., & Burritt, R. (2018). Business cases and corporate engagement with sustainability. Exploring ethical motivations. *Journal of Business Ethics, 147*(2), 241–259.
Shell. (2022). *Shell sustainability report.* Available at Shell Sustainability Report 2022 - Shell plc Sustainability Report 2022.pdf
Shell. (2023). *Shell sustainability report.* Retrieved from https://reports.shell.com/sustainability-report/2023/powering-lives/contributing-to-communities/contributing-to-nigerias-economy.html
Total. (2020). *Staying resilient evolving towards a sustainable world. 2020 sustainability report.* Retrieved from https://services.totalenergies.ng/system/files/atoms/files/2020_totaenergies_marketing_nigeria_plcsustainability_report.pdf
Total. (2022a). *Total energies social development goal reporting 2021-2022.* Retrieved from TOTAL SDG_Report_2021-2022.pdf
Total. (2022b). *Sustainability & climate 2022 Progress report.* Retrieved from https://totalenergies.com/sites/g/files/nytnzq121/files/documents/2022-05/Sustainability_Climate_2022_Progress_Report_accessible_version_EN.pdf
Unilever. (2021). *Unilever sustainable living plan 2010 to 2020 summary of 10 years' progress.* Retrieved from the-unilever-compass.pdf
Unilever. (2022). *Business and purpose sustainability report.* Retrieved from https://www.unilever-ewa.com/files/92ui5egz/production/1fb2192ba2968cd01cdb550c0a4ef1916208d886.pdf
United Nations (UN) Global Compact. (2015). *Supply chain sustainability: Practical guide for continuous improvement* (2nd ed.). United Nations (UN) Global Compact.
Vallentin, S. (2015). Governmentalities of CSR: Danish government policy as a reflection of political difference. *Journal of Business Ethics, 127*(1), 33–47.
Wade, M. (2005). Good company citizenship. In U. Petschow, J. Rosenau, & E. von Weizsacker (Eds.), *Governance and sustainability: New challenges for states, companies and societies* (pp. 186–199). Greenleaf.
Wettstein, F. (2012). CSR and the debate on business and human rights: Bridging the great divide. *Business Ethics Quarterly, 22*(4), 739–770.
World Bank. (2022). *Nigeria poverty assessment 2022: A better future for all Nigerians.* Retrieved from https://documents1.worldbank.org/curated/en/099730003152232753/pdf/P17630107476630fa09c990da780535511c.pdf
World Bank. (2024). *The World Bank in Nigeria.* Retrieved from https://www.worldbank.org/en/country/nigeria/overview#3.
World Commission on Environment and Development. (1987). *Our common future.* WCED, Oxford University Press.

Adebimpe Adesua Lincoln holds an LLB and an LLM in Commercial Law from Cardiff University and an MBA in International Business and PhD in Entrepreneurship and Business Development from the University of South Wales. She holds a Postgraduate Diploma in Legal Practice and also holds various teaching qualifications in Higher Education. She is a Fellow of the Higher Education academy and an Associate Member of the Chartered Institute of Personnel Development. She has 20 years of experience in Higher Education. She has held positions in Cardiff University, University of South Wales and was a Senior Lecturer in Law at Cardiff Metropolitan University before taking up a position as an Assistant Professor in Saudi Arabia. Adebimpe Lincoln is a legal practitioner. She also works with University of Liverpool online Master's in Law Programs, Kaplan Online Learning and is an Adjunct Professor with Robert Kennedy College Switzerland. Adebimpe Lincoln has vast experience supervising research students. She has supervised a wide range of Master's level research including M.Sc., MBA, and LLM dissertations. She has also supervised students at Doctoral Level in the area of Entrepreneurship and Corporate Governance. Her research interests lie in the area of Female Entrepreneurship, Small and Medium Sized Enterprise (SME) Sector, Leadership Practices of Nigerian Entrepreneurs, Corporate Governance and Board Diversity, Sustainability and Corporate Social Responsibility among Nigerian SMEs. She has published and presented several articles on Leadership, CSR, Governance, and Entrepreneurship in Nigeria.

Brendhain Diamond is a graduate of the University of Western Kentucky (USA), where he holds a Bachelor of Science with double majors: Business Management and Philosophy. Brendhain holds a Master's Certificate in Problem Based Learning from Stenden University, Leeuwarden, The Netherlands. He is also a graduate of the University of Liverpool earning a Master's of Law (LLM) with Distinction. Brendhain received the graduation awards of Student of the Year (highest grade point average) and Dissertation of the Year. Brendhain Diamond lectures Economics and Marketing in Göteborg, Sweden. The main focus has been toward management programs within higher education for the Hospitality and Tourism Industries. Additionally, he has over 25 years of experience supervising student research projects within these programs. His research interests lie in the area of Economic and Social Development, Sustainability and Corporate Social Responsibility. He has published articles on Corporate Social Responsibility, Sustainability, Governance, and Entrepreneurship.

Conceptions of Sustainability and Their Impact on Practices in Oil and Gas Corporations Operating in the Niger Delta

Oluchukwu Jane Richard-Osu and Sarah Buckler

1 Introduction

Numerous efforts have been made by organisations, from the context of environmental, social and economic development, to ensure their strategic and operational activities are in alignment with notions of sustainability whilst maintaining profitability. Despite these efforts, issues surrounding sustainability, such as environmental degradation and social inequality due to the negative impacts of environmental degradation, persist. This can be keenly observed where efforts to facilitate the social acceptance of oil and gas organisations by host communities have been witnessed. For instance, oil and gas business operations have caused significant damage to the Niger Delta region in Nigeria over the last five decades despite there being a number of 'socially responsible' initiatives developed by multinational energy corporations active in the area. Indeed, the region which is one of the 10 most valuable wetlands in the world has been described as the world's most harshly affected ecosystem by oil (Okonkwo et al., 2015) due to the incessant oil spills in the region.

The persistence of such negative impacts has led to a call for a change and regulation of human actions on the environment. This will involve behavioural change to address these issues. We argue, however, that such behavioural change or action can only be taken within the organisation based on knowledge of that subject matter. That is to say, necessary and relevant actions to mitigate the negative effects of business practices can only be taken if there is awareness of the negative effects of current behaviours and an understanding of what actions can be taken to mitigate these, and this is not always as straightforward as we might expect.

O. J. Richard-Osu (✉) · S. Buckler
Robert Gordon University, Aberdeen, UK
e-mail: j.richard-osu1@rgu.ac.uk; e.s.buckler@rgu.ac.uk

S. O. Idowu, S. Vertigans (eds.), *Sustainability in Global Companies*, CSR, Sustainability, Ethics & Governance, https://doi.org/10.1007/978-3-031-77971-8_12

2 Sustainability in Oil and Gas Industry Setting: Perception

Many large organisations appear to accept that sustainable development is vital (Giddings et al., 2002). Such awareness has led to the creation of the World Business Council for Sustainable Development—which consists of more than 200 of the world's leading businesses in different sectors including mining, oil and gas, banking and finance who aim to work together to create a more sustainable world by targeting the realisation of the sustainable development goals (SDGs) (World Business Council for Sustainable Development (WBCSD), 2000). This level of commitment shows that more and more organisations are buying into the idea of sustainability either because of growing pressure by the society to 'do the right thing' or because they are beginning to realise the benefits of being sustainability conscious.

However, many academics and sustainability professionals have identified concerns about the sustainability practices of the oil and gas industry, questioning their adequacy with relation to how effectively or meaningfully the concept has been integrated into their business operations and activities (Schulz, 2015). Numerous critics claim that oil and gas companies just 'green wash' by exaggerating their commitment towards sustainability in order to improve their image in the public eye (Schulz, 2015).

When considering the stated desires of oil and gas multinationals to move towards a more sustainable future, and the accompanying observations of the limited progress towards achieving this aim, one thing we might want to explore is whether there is a shared vision of what that sustainable future refers to. In other words, when oil and gas corporations talk about sustainability, are they talking about the same thing as their critics? If they are not, we might wonder what role the SDGs may have in helping various players arrive at a shared understanding and vision. This chapter draws on research carried out amongst employees of oil and gas corporations operating in the Niger Delta to explore exactly these issues.

3 Background to the Study

The oil and gas industry has had a long history from the discovery of petroleum up to current day exploration and production of many types of goods from petrochemicals. Africa is richly gifted with a variety of natural resources, including the presence of abundant oil and gas in the West African region. This has meant that the region has continually dealt with a wide range of issues linked with the oil and gas business sector. The oil and gas industry remains a critical case in the discourse of sustainability, given the exceptionally high-level impact of their business operations and activities upon the ecosystem and communities. The environmental risks of corporate operations increase with oil and gas deposits that are located in areas of high ecological vulnerability such as in the Niger Delta (ND) region of Nigeria which has been the centre of oil exploration and production in Nigeria since 1956.

Nigeria is a sub-Saharan African country with proven vast oil reserves (World Bank, 2020), it is the eleventh largest oil producer and also the eighth largest exporter of oil in the world and the largest producer of oil in Africa. Moreover, 90–95 per cent of Nigeria's export revenues is accounted for by the country's oil and gas industry (Okotie, 2018). Nigeria earns billions of dollars through the exportation of oil whilst associate oil spills and gas flaring appear to be the two main harmful consequences of the oil and gas practices within the industry (Murphy et al., 2016).

Oil spill occurs either on land or water depending on the location of operations. In the case of the Niger Delta, it is mainly onshore. In addition to oil spill, a further cause for national and global concern is gas flaring, which is in no way environmentally, socially or economically friendly, yet it is quite widespread as organisations find it easier to choose the option of flaring due to very high costs of harnessing such gases for economic use (Malumfashi, 2007). Among the countries with the highest gas flare figures is Nigeria (Ishisone, 2004), and this has caused substantial climate and infrastructural damage as well as health concerns.

Together, gas flaring and oil pollution have been frequently cited as the major challenges facing the Niger Deltans, their environment and its sustainability (Idemudia & Ite, 2006). These activities impact negatively on the local communities as their livelihoods such as farming, fishing and hunting are destroyed. In fact, it has been argued that the 'oil-soaked' Niger Delta communities should be 'Ground zero' for human oil exposure (Murphy et al., 2016). In addition to environmental degradation and loss of livelihood, there has been incessant conflict in the Niger Delta regions because of the adverse environmental effects and it has become imperative that actions are taken by oil and gas organisations to mitigate the negative impacts of their business practices in the region. This fact was highlighted through Richard-Osu's (2023) research conducted within the context of the Niger Delta region of Nigeria to understand the peculiarities of implementing sustainable business practices by oil and gas organisations operating within the region.

4 CSR in the Oil and Gas Industry: Actions for Remediation

Research has shown the intent of the oil and gas organisations to implement sustainability measures to ensure environmental damage is mitigated. Such sustainability initiatives which have been implemented in an attempt to mitigate environmental degradation and its impact have been reported to be mainly ineffective. An example of such initiative is corporate social responsibility (CSR) initiatives in the Niger Delta.

CSR has been frequently linked to sustainability in businesses, it is seen not only as a means through which businesses can counteract the negative impacts brought about by their operations/business practices but also as a way to contribute to sustainable development (Idemudia, 2009). However, CSR practices in the Niger Delta have been perceived as actions taken to protect the organisation's reputations or sometimes described as cosmetic attempts to act in a socially responsible way (Ite, 2004). There appears to be no real sense of urgency with regard to CSR activities in the Niger Delta where concerns of communities are not dealt with urgently. An example can be seen in the continuous gas flaring within the region despite complaints of its negative effects by the host communities (Idemudia, 2009). Indeed, ground-breaking research carried out by Amnesty International found that the Ogoniland in the Niger Delta region is still being polluted through oil and gas practices (Amnesty International, 2018). To compensate for the consequences of their action, companies are said to be spending considerable amounts of money in the communities they operate in under the guise of CSR. The diversion of CSR intended funds might be linked to the fact that there are numerous cases of abandoned CSR projects. Such diversion of funds could be a result of corruption as suggested here by one of the participants in Richard-Osu's research:

> you will see the corruption that is taking place there… That is why I say that that effectiveness is what I can't say… It is a Nigerian factor (SBP008).

These factors have resulted in the success and usefulness of CSR initiatives in the oil and gas sector being questioned more and more (Idemudia, 2009). Similarly, an argument has been put forward that CSR approaches have failed to solve many of the problems they are linked with creating despite increased investment by companies (Vertigans, 2012). In fact, CSR policies are worsening existing problems while creating new ones such as corruption and lack of trust.

With the realisation that CSR initiatives are not achieving the desired impact on the environment, alternative approaches have been focused on and some scholars have proposed the concept of sustainable business practices (SBPs) (Høgevold et al., 2014). This focuses on the business practices themselves as against focusing on counteracting the impact of these practices. This has become a major area of focus in discussions about sustainability due to the increased pressure on organisations to be held accountable for the effect of their processes on the natural and social environments where they operate.

We can see that although organisations appear to be aware of their actions and the effects these have on both the environment and the societies in which they operate; problems such as environmental degradation, loss of livelihood of host communities as a result of the environmental degradation and the incessant conflict due to a struggle for basic human rights are still rife. One reason for this could be that there are uncertainties around the understanding and subsequent implementation of SBPs despite the increased awareness.

5 The Research

It was in this context that Richard-Osu carried out her research which involved interviews with personnel from 10 oil and gas organisations operating in the Niger Delta region. The research participants included both HR and sustainability personnel within the organisations and was designed to shed light on understandings and implementation of sustainable business practices in O&G corporations.

As noted above, multinational corporations publicly acknowledge the importance of developing and delivering on sustainability initiatives. Further, they have implemented a variety of projects and programmes under a 'Corporate Social Responsibility' agenda that are ostensibly intended to ameliorate the damage caused by the direct and indirect impacts of oil and gas extraction. This is an incredibly complex area with tense relations impacting upon the success or otherwise of numerous activities and the underlying reasons for the continued degradation and poor outcomes for the Niger Delta region are many and varied (see Richard-Osu, 2023). In this chapter we will focus on two questions that we have identified as being particularly relevant in influencing the activities and impacts of the multinational corporations:

> To what extent is there a shared vision and understanding of sustainability amongst oil and gas professionals?
>
> To what extent might the Sustainable Development Goals form a basis on which to arrive at a shared vision of a sustainable future?

6 Methods

The research adopted a qualitative, interpretive approach and was designed to explore how certain factors or principles within an organisation can either hinder or encourage implementation of SBPs and how HR can align with these. Following the completion of exploratory survey and informal discussions with sustainability and HR professionals in the oil and gas industry in the UK and Nigeria, semi-structured interviews were then conducted in Nigeria. The sample scope was narrowed from all oil and gas companies to only oil and gas producing and exploration companies in Nigeria. Purposive sampling technique was utilised for the study whereby the researcher selected professionals who were (or should be) directly involved in the implementation of the organisations' sustainability initiatives as well as HR personnel who were (or should be) familiar with the organisations' policies relevant to sustainability. It was intended that those selected would be involved in informal conversations, pilot survey and interviews. Although a purposive sampling technique was initially utilised, it became imperative to employ a snowballing technique due to the difficulties encountered in the course of recruiting respondents for the study.

A total of 24 interview questions were divided into eight sections covering: role and organisational structure, understanding sustainability/SBPs, Nigeria, SBPs implementation, challenges and HR involvement, HR, authority and power and finally, identity—behaviours and interests. The interview questions were set with the aim of exploring the opinions and experiences of the research participants with regard to sustainability awareness, SBPs implementation (as well as challenges encountered in SBPs implementation) and HR involvement—actual and potential.

Consent to audio record the interview was obtained from all respondents to encourage a focus on the respondents, their responses to the questions asked and non-verbal reactions (Bryman, 2006). This also enabled the use of follow-on questions to get more information out of what had been said. These recorded interviews varied in length, ranging from 15 min to 1 h 45 min with an average of 1 h. The transcription of data was done verbatim and all the pauses, voice inflection, laughter as well as hesitations were all captured. These were used to emphasise some points and findings, added to the rich data, and provided a more robust and comprehensive data analysis.

The research findings will now be discussed, and arguments presented on the need for understanding of sustainability to facilitate required actions for implementation of sustainable business practices in organisations.

7 Understanding Sustainability

For those who work in transnational and multinational corporations, terms such as sustainability and sustainable development are commonly used and can provide a useful starting point for discussions around social and environmental impacts of organisational activity.

At present, a universally agreed characterisation of sustainability is non-existent (Opon & Henry, 2019) rather with the discourse on sustainability being more widespread and socially complex, different social forces are beginning to define the meaning which best reflects their values and interests (Lima, 2003). The definition of the term sustainability based on interests and values placed on it may imply that different actors may have varying understanding and, ultimately act differently with regard to sustainable practices based on their understanding. Therefore, we argue that it is important that understanding and knowledge of sustainability are increased and controlled within the organisation to enable desired sustainable behaviours and actions.

The sections below demonstrate the differences in understanding of the concept of sustainability.

8 Sustainability Knowledge and Awareness

Sustainability and SBPs as concepts are viewed differently by different individuals therefore making the understanding of sustainability to be subjective. We can further demonstrate this through the responses given by respondents when talking about their understanding of SBPs. Respondents expressly highlight that the concepts of SBPs and sustainability can be subjective, i.e., understood differently:

> What I understand that to be, when you say sustainable business practices it depends on your own interpretation of it... (SBP009[1]).

> Again, it is contextual the issue of sustainability. Even the eerrmm... it will mean different things to different people. (HR007).

It was not surprising then that the respondents had varying interpretations of what they believe sustainability and SBPs are. Some respondents explained that they consider SBPs are ways to ensure the business sustainability:

> ...In the sense that whatever it is you are doing, what will give you longevity and have something to continue tomorrow. What would you do in your business today that definitely will make you last long beyond your own time in the company? (SBP012).

> So it is essentially how does eerrm... what should we be doing to make sure this business doesn't die tomorrow? (HR010).

> Sustainability is all about how you carry out your business activities in such a way that the business itself is run for a very long time (HR011).

> So there are 2 levels of it. If you ask me, the first thing that comes to my mind is business continuity. Essentially talking about that how to ensure that your documents are safe, how to ensure that knowledge is passed from one generation to another et cetera (HR002).

> business practice that outlives the business. So business practice that ensures that your stakeholders trust you. That it helps your brand because you are synonymous with eerrmm the best way to work in every situation both ethically and morally and eerrmmm imbibing the best standards actually that help you have a sound, all round capacity to do your business in a safe and orderly way both with your people, environment, government, unions and regulatory bodies generally (HR006).

Responses range from SBPs being viewed as a way of ensuring business sustainability and continuity through processes and practices within the organisation to SBPs being seen as practices that help build the brand of the business resulting in business longevity. Noteworthy is the fact that within the views of SBPs as way to sustain the business, there are varying degrees of understanding—from internal processes to profitability to ethical and moral practices that contribute to business sustainability.

Also, within these varying thoughts of SBPs as a means of achieving business sustainability is the idea that SBPs are proper controls as mentioned by the

[1] Respondents are categorised as Sustainable Business Professionals or Human Resources and the letters SBP or HR before the number allow the reader to identify which profession they belong to.

respondent below. That is, effective controls which when implemented within the organisation will help guarantee business sustainability and continuity:

> What comes to my mind when they talk about sustainable business practices is something like controls, having proper internal controls within your company… (HR005).

> … eeerrmm care or caution will be laid in place to ensure that the right practices are cultivated, and it becomes a culture too. (HR007).

> Well… first of all, to me it connotes ethical business practices eerrrm it equally suggests practices that would support businesses thriving and as well as in ensuring that the environment is left in a way that future generations will continue to benefit from the resources therein. (SBP003).

> What I understand that to be, when you say sustainable business practices it depends on your own interpretation of it… we work with the regulatory bodies to ensure that what we do is not contravening the acceptable standards and what we also do is in conformity to acceptable global standards. So, it is not just about us, it is about every interested party (SBP009).

These organisations are portrayed as engaging in sustainable and ethical business practices in order to ensure the environment and resources are preserved for future generations. Adding to the mix of varying understanding, HR005 portrayed SBPs as being the controls within the organisation while HR007 though still in line with SBPs as controls has a different view. For this respondent SBPs are perceived as practices that *create* controls within the organisation. These controls then ensure that necessary actions are taken, or relevant activities engaged in within the organisation to achieve set organisational goals of sustainability. It is noteworthy that in a number of cases respondents with varying understandings of sustainability and SBPs are employees in the same organisation. This raises the question: if members of the same organisation have varying understanding of sustainability, how then can the goal of sustainability be achieved? Although both HR and SBP personnel believed that SBPs are controls within the organisation, HR viewed these controls as needed for business continuity while the SBP personnel viewed them as necessary for prevention of environmental degradation. This variation further highlights differences in understanding possibly influenced by area of work or expertise.

Different perspectives of sustainability as portrayed by the respondents include SBPs seen as ethical business practices, as a way of having controls in the business, widely acceptable business practices and sustainable business. The interpretation of the concept of sustainability can be subjective which we argue can have further impact on the actions of these actors based on divergent understandings. Furthermore, the knowledge and awareness of sustainability/SBPs as important in the generation of the desired behaviours and actions aimed at creating a culture of sustainability was highlighted. Therefore, it is important for all employees to have not only the knowledge and awareness of SBPs but also the same understanding of sustainability and SBPs to ensure that actions are all aligned to the same goal.

There is an expectation that knowledge and awareness of the issue of sustainability are needed to generate interest on the subject matter, and this will then result in the development of desired behaviours. These behaviours will lead to desired actions or activities being carried out in order to achieve sustainability as well as build a culture of sustainability. However, in order for this to happen it is important that individuals within the organisation have a shared understanding of sustainability and how it can be achieved; for example, how a business can make its practices more sustainable and the need for it to do so. It can be argued that with such shared knowledge, employees will then develop the desire or interest to implement (or seek ways to implement) sustainable business practices. Such behaviours then become the norm within the organisation where sustainability is interwoven in every decision, task, and action within the organisation. However, if there is no shared knowledge and understanding, it is very difficult to see how this could be achieved.

9 Knowledge vs Culture Creation

In addition to demonstrating how different understandings of sustainability can impede implementation of SBPs the research also explored how an organisation's culture can contribute to or hamper implementation of SBPs.

A frequently repeated theme amongst participants in the research was that consistent participation in sustainable actions and behaviours can lead to a creation of a culture of sustainability. This culture of sustainability when created should be imbibed by all employees within the organisation:

> You need to build the culture of sustainability as part of your business and is gonna cut across whether if you are from the boardroom... from the gateman to the boardroom ... it's a culture that we've built. So, when there's a culture on sustainability it affects all of us (SBP001).

> So there is a collective effort to ensure that things are done right, in the right way... Now of course when you are coming into the company, it is the culture of the company so it is easy for you to follow (HR005).

The hope is that creation of this culture of sustainability and employee engagement supports the achievement of organisations' sustainability agenda. Achieving a culture of sustainability would ensure that every employee function and behaviour within the organisation would be aligned to sustainability and the implementation or adoption of sustainable business practices based on shared understanding.

It is important that sustainability is embedded within the ethos of the organisation to achieve this, it is important that there is a clear communication of sustainability plans. Having knowledge and awareness of a subject matter results in the generation of interests which would in turn lead to the development of desired values or behaviours in employees within the organisation. Increased knowledge and awareness can come from communication of relevant information. The research responses demonstrated that in order to change the organisational culture or build desired behaviours and values within the organisation, clear and regular communication of top management plans and strategy to employees at all levels is important.

> I think one of the things we do is continuous education. So... awareness of staff (HR006).

> So how we pass message across in terms of the change in culture is about communication. You can never over communicate. You communicate, communicate and communicate. So with that helps the employees to have a better understanding of what you are talking about, your expectations and why the change in culture. So, it is surrounded by communication (HR017).

Returning to the matter of harmonised understanding of sustainability, we argue that alongside clear and regular communication, there needs to be common understanding right from the top management. This understanding should then be cascaded to all employees within the organisation. This brings to the fore the importance of effective communication of shared, common knowledge and understanding to all employees within the organisation. It is crucial for all employees to have not only the knowledge and awareness of sustainability but also the same understanding of sustainability and SBPs to ensure that actions are all aligned to the same goal.

10 Knowledge vs Action

It was considered important for the research to discover the level of knowledge about sustainability and SBPs among HR personnel in oil and gas companies to understand what the impact of this is on the actions taken with regard to sustainable business practices.

Having identified that sustainability and SBPs can mean different things to different people, we argue that actions toward sustainability will differ based on understanding and perception. In addition to this, findings revealed that not only shared, but increased engagement with and awareness of sustainability and SBPs are needed to enable adoption of or participation in sustainable practices. This can be observed through belief that relevant actions or activities with regard to a particular issue can only occur if there is interest and passion towards said issue:

> you can only get involved in what you are interested in. if you are not passionate about an issue, the chances are that you won't give in your best (SBP003).

To generate the required interest needed to create necessary actions towards sustainability, continuous education in the subject matter is key.

> "interest will create awareness…" "with awareness, more and more people are beginning to think about the issue of sustainability". "I think one of the things we do is continuous education. So… awareness of staff" (SBP006).
>
> "we need to have a very strong educational campaign about the impact of some of these things and generally about sustainable behaviours and sustainable business operations." "If you can have a way of getting everybody to be aware, that will help people to better connect to some of these initiatives and of course at the individual levels and the company levels" (HR011).

Awareness creates engagement in the sustainability discourse leading to development of relevant behaviours and desired actions.

As previously mentioned, although some actions have been carried out by the oil and gas organisations in the Niger Delta region, these have not been successful, nor have they led to any meaningful mitigation of the harmful impacts of O&G activity. Building on this observation Richard-Osu's research highlighted the fact that the SDGs have not been adopted by O&G corporations. Rather, there has been some degree of blame shifting by the oil and gas organisation in a bid to avoid dealing with the consequences of their business operations. This blame shifting could in part be a result of the lack of understanding and common knowledge of sustainability and required actions as previously demonstrated. This is demonstrated below using the research data to show how the oil and gas organisations may be, to some degree, transferring some blame to the communities and not taking full responsibility of the effects of their practices on the environment.

> In situation where for instance there are usually occasional sabotages even from the host communities probably for reasons… either to attract attention, to make some claims and possibly to attract resources to themselves. So, there are sabotage of pipelines and those are some of the isolated issues and incidences that could also jeopardise laid systems and processes for ensuring that the right thing is done (SBP003).

> The oil industry are probably not the major polluters of the environment because there are other activities that pollute even more than the oil industry (HR006).

> For example, spillage, no organisation really wants to have oil spillage but things happen of course. In Nigeria for example, spillage is a complex issue… in Nigeria you have to add the sabotage to most spillages that happen in Nigeria. They happen because somebody went and tampered with some of the infrastructure… So, there is not much the companies could do in terms of…. It is not that there is not much they could do. There is a lot they can do but the effort does not really guarantee whether they are going to be a sabotage or not. Sometimes they do everything they can to engage, collaborate but what will happen, still happens (HR011).

Participants appear to suggest that the organisations work to prevent oil spills, but they still happen and, in an attempt, to mitigate their sense of responsibility they transfer the blame on to the host communities for the spills and subsequent environmental degradation. Respondents seem to believe that the organisations are taking the necessary actions and that it is others that are the source of the problems.

> "there's been a whole lot of agreements and frameworks that are being put in place to eliminate gas flaring. Some of them requires a number of commitments on the part of the different stakeholders…" "So, some of the communities obviously they haven't worked according to plan…" (HR011).

In addition, there appears to be a transference of blame of environmental degradation to other actors both within and outside the industry by research participants:

> Who are the big violators? The Chinese. I don't think Chinese companies are going to respect your green rules because they can get money from China… if you check the biggest violators it's probably going to be the Chinese. They are looking for the shortest, the fastest way and somewhere that can give them the fastest return (SBP001).

There is the additional burden of handling the responsibilities which the respondents believe should be handled by other actors. For example, SBP004 claims that:

> As international oil companies, the government of Nigeria at different times and in different ways has requested international oil companies to get involved in what is important to them. It has been important to this country for a long time the problem of power… for us is not only being sustainable in the way we do the work, our main business in that country but is also making sure we have sustainable relationships with important stakeholders. Stakeholders being people who can have an impact on what we do and who what we do can also have an impact on

There is a reference to the shortcomings of the government in providing basic infrastructure in the country therefore requiring assistance from the international oil companies (IOCs) operating in the country. The respondent opines that there is government reliance on oil and gas organisations to address these gaps within the country for example, the issue of limited power supply. This has been mentioned by another respondent that the government places a lot of burden on the oil and gas organisations by asking them to perform duties that the government should all in the name of CSR.

In addition to transference of blame of environmental degradation to other actors within the oil and gas industry, there is a perception that oil and gas organisations are burdened with the responsibility of catering for the communities and ensuring that basic needs are met. "…but you now see that they (the government) are now transferring it as a form of corporate social responsibility to companies, especially to oil and gas companies" (SBP008). Based on all of these facts above, it is not surprising that respondents view the organisations as the only actors within the industry doing the right thing. This can be called into question if it cannot be demonstrated that there is common knowledge effectively communicated, desired behaviours developed, and relevant actions taken to effectively implement sustainable business practices. In fact, the persistence of the problems faced despite efforts have led to one of the main negative consequences of oil and gas operations to be seen as normal:

> Well… what I would say talking about oil spillage is that it is part of the business. (HR006).

> "For example, spillage, no organisation really wants to have oil spillage but things happen of course." "We understand that you cannot eliminate the impact completely…" (HR011).

In addition to the fact that not only is oil spill considered to be a part of the business, despite the fact that it results in environmental degradation and loss of livelihood and other negative impacts for the host communities,

> So there is not much the companies could do in terms of…. It is not that there is not much they could do. There is a lot they can do but the effort does not really guarantee whether they are going to be a sabotage or not. Sometimes they do everything they can to engage, collaborate but what will happen, still happens (HR011).

We demonstrate here that there is indeed a lack of clear understanding of what the responsibilities of different actors are within the oil and gas industry with regard to sustainability. This, we argue, has resulted in blame shifting as discussed above. Therefore, numerous questions can be asked with regard to the issue of blame shifting: how can the responsibilities of the oil and gas organisations be defined? How can common knowledge of sustainability and required actions be effectively communicated to all employees within the organisation?

It is interesting to note that though awareness of sustainability exists, responses from this study suggest that the organisations would rather focus on other actors' (in)activities rather than on what actions can be taken by the organisations to prevent these negative impacts. It is important to point out here that it is not enough to be aware of the concept of sustainability, it is also crucial to have clear knowledge of what actions the organisation can take and defined responsibilities for ensuring those actions are taken. This is absolutely necessary in order to spur on the action of employees and build the desired behaviours and interest in carrying out actions that are essential to enable implementation of SBPs.

11 SDGS: The Solution?

The United Nations hoped that the SDGs (see Fig. 1) could help address such issues by creating a coherent framework and accompanying narrative around sustainability, however at present in Nigeria they do not seem to be working. Is there any way they could? The preceding discussions demonstrate a need to reassess the approach towards understanding and implementing sustainability initiatives within the oil and gas industry in Nigeria. That industry is characterised by a complex political, social and economic environment that affects the implementation of SBPs. Within this complexity a major concern is highlighted—the issue of whether a shared understanding exists and arising from that, the extent to which collaboration among stakeholders within the industry is possible. We put forward that all stakeholders working with a common framework and towards a common goal will serve as a good beginning for the achievement of effective sustainable business practices implementation and it is possible the SDGs could provide this.

When the United Nations launched the sustainable development goals (SDGs) there was awareness that a change of approach was needed in order to ensure sustainable development is achieved. These SDGs were launched at the end of 2015 by the United Nations (UN). The twofold part of the human element as both the originator and the beneficiary of the implementation of SDGs can be recognised (Chams & Garcia-Blandon, 2018). SDGs aim to instigate sustainability integration into firms' operations, focusing on present and future stakeholders' needs thereby leading to the attainment of SD for society (Rosati & Faria, 2019).

Clearly then, the intention is to build a framework that can serve as the foundation for a shared narrative encompassing a shared understanding of sustainability. Nevertheless, even though many IOCs operating in the Nigerian oil and gas industry are members of global sustainability bodies/associations who are aligned to the achievement of the SDGs, it is noteworthy that the research respondents did not mention SDGs while discussing their organisations' strategies and effort in implementing sustainable business practices. For instance, many IOCs are members of oil and gas climate initiative (OGCI), a voluntary CEO-led initiative taking

Fig. 1 Sustainable development goals (SDGs). Source: UN (2015) https://sustainabledevelopment.un.org/

'focused on accelerating action to a net zero future consistent with the Paris Agreement' (OGCI, 2024) which comprises 12 members including Total, Shell, Eni, ExxonMobil, Chevron all of whom operate in Nigeria. The OGCI objectives of reducing energy value chain footprint, accelerating low-carbon solutions and enabling a circular carbon model are aligned with Nigerian targets which have been set to achieve zero routine flaring by 2030 and to reduce methane and recycle carbon as well as reduce carbon emissions. These objectives are aligned with SDGs 7 and 13 and could serve as a platform to set SDGs as a framework for the implementation of SBPs in the oil and gas industry in Nigeria. None of this was mentioned by any of the respondents to the research suggesting a real breakdown in understanding and/or communication.

So, whilst sustainability in oil and gas is being debated at a global level through organisations such as OGCI, and this debate does align to the SDGs, it does not become effectively cascaded to the level of national operations and thus fails to deliver on its promise. This is evidenced by the fact that, as previously mentioned, despite the IOCs operating in Nigeria being members of international sustainability organisations such as OGCI, the employees (interviewees) in these organisations do not mention SDGs. This is an indication that there is a lack of or inadequate 'trickle down' of information which can have an impact on actions at a national and local level.

In seeking a solution, we ask what could be put in place to make the SDGs more than just a 'label'? Could SDGs form a meaningful template or model from which corporations can build a shared understanding of sustainability, or not? SDGs' adoption is quite weak especially with regard to oil and gas organisations where there is a limited shared understanding of sustainability. We propose that the international organisations such as OCGI should develop a shared understanding as to what the SDGs mean, and this should be signed up to by the participating organisations who should then incorporate these into their sustainability reports. Richard-Osu's research identified that HR has the potential to cascade information throughout organisations if there is buy-in from top management. International fora should facilitate this buy-in.

12 Conclusion

A (possibly unconscious) desire of the oil and gas organisations to shift sustainability responsibilities to other stakeholders within the industry and a lack of shared understanding among these stakeholders suggests that at present multinational oil and gas corporations have no real desire to adopt the SDGs. It is evident that current organisational cultures do not encourage or enable the adoption of SDGs. Therefore, for SDGs to be effectively taken up and SBPs implemented, there needs to be increased buy-in from top management and acceptance of their role in creating the necessary organisational culture. This would in turn enable regular communication of these goals thereby building the required knowledge and culture, with that shared understanding of sustainability which we posit is required for the effective implementation of sustainable business practices.

An increased awareness and commitment from top management can similarly help address the issue of the creation of monitoring and evaluation mechanisms that ensure relevant SDGs are signed up to as part of sustainability reporting. This research has shown that blame shifting among stakeholders within the oil and gas industry has resulted in inaction of the oil and gas organisations when it comes to implementing sustainable business practices. We contend that SDGs could and should be used to define the responsibilities of the oil and gas organisations with regard to sustainability and also which sustainable business practices should be adopted by the organisations. Furthermore, without some sort of accountability, the SDGs are toothless; therefore, there should be some kind of agreement or standard, adopted at an international level which sets out the responsibilities of various stakeholders including IOCs. If such an agreement was monitored and reported upon through organisations such as OGCI and also communicated as part of the reporting cycle of businesses, this could persuade the IOCs to pass on the relevant understanding and awareness to local/national actors and could influence the actions of those corporations at a more local level. It could help to ensure that knowledge is cascaded down and a meaningful approach to sustainable business practices adopted throughout the organisation.

In short, if SDGs are going to help achieve a meaningful move towards greater sustainability in the oil and gas sector then there needs to be a shared understanding and narrative amongst all stakeholders regarding what they actually mean and whose responsibility delivery is. Corporations, governments, communities and multinational organisations such as the UN and OGCI all have a part to play in this—the blame shifting has to stop.

References

Amnesty International. (2018, March 16). *Nigeria: Negligence in the Niger Delta: Decoding Shell and Eni's poor record on oil spills*. Accessed March 22, 2022, from https://www.amnesty.org/en/documents/afr44/7970/2018/en/, Index Number: AFR 44/7970/2018.

Bryman, A. (2006). Integrating quantitative and qualitative research: How is it done? *Qualitative Research, 1*, 97–113.

Chams, N., & Garcia-Blandon, J. (2018). On the importance of sustainable human resource management for the adoption of sustainable development goals. *Resources, Conservation and Recycling, 141*, 109–122.

Giddings, B., et al. (2002). Environment, economy and society: Fitting them together into sustainable development. *Sustainable Development, 10*, 187–196.

Høgevold, N. M., Svensson, G., Wagner, B., Petzer, D. J., Klopper, H. B., Varela, J. C. S., Padin, C., & Ferro, C. (2014). Sustainable business models: Corporate reasons, economic effects, social boundaries, environmental actions and organizational challenges in sustainable business practices. *Baltic Journal of Management, 9*(3), 357–380.

Idemudia, U. (2009). Oil extraction and poverty reduction in the Niger Delta: A critical examination of partnership initiatives. *Journal of Business Ethics, 90*, 91–116.

Idemudia, U., & Ite, U. E. (2006). Corporate–community relations in Nigeria's oil industry: Challenges and imperatives. *Corporate Social Responsibility and Environmental Management, 13*, 194–206.

Ishisone, M. (2004). *Gas flaring in the Niger Delta: The potential benefits of its reduction on the local economy and environment*. Accessed December 10, 13.

Ite, U. E. (2004). Multinationals and corporate social responsibility in developing countries: A case study of Nigeria. *Corporate Social Responsibility and Environmental Management, 11*(1), 1–11.

Lima, G. C. O. (2003). The sustainability discourse and its implications for education. *Ambiente & Sociedade, 6*(2), 99–119.

Malumfashi, G. I. (2007). Phase-out of gas flaring in Nigeria by 2008: The prospects of a multi-win project (Review of the regulatory, environmental and socio-economic issues). *Nigeria Gas Flaring Petroleum Training Journal, 4*(2), 1–39.

Murphy, D., et al. (2016). An in-depth survey of the oil spill literature since 1968: Long term trends and changes since deepwater horizon. *Marine Pollution Bulletin, 113*, 371–379.

Oil and Gas Climate Initiative (OGCI). (2024). Accessed July 9, 2024, from https://www.ogci.com/

Okonkwo, C. N. P., et al. (2015). The Niger Delta wetland ecosystem: What threatens it and why we should protect it? *African Journal of Environmental Science and Technology, 9*(5), 451–463.

Okotie, S. (2018). The Nigerian economy before the discovery of crude oil (chapter 5). In P. E. Ndimele (Ed.), *The political ecology of oil and gas activities in the Nigerian aquatic ecosystem* (pp. 71–81). Academic Press.

Opon, J., & Henry, M. (2019). An indicator framework for quantifying the sustainability of concrete materials from the perspectives of global sustainable development. *Journal of Cleaner Production, 218*, 718–737.

Richard-Osu, J. (2023). *A critical evaluation of the role of human resources in the transition towards sustainable business practices in the oil and gas industry in Nigeria*. PhD Thesis. Robert Gordon University.

Rosati, F., & Faria, L. G. D. (2019). Business contribution to the sustainable development agenda: Organizational factors related to early adoption of SDG reporting. *Corporate Social Responsibility and Environmental Management, 26*(3), 588–597.

Schulz, N. (2015). *Exploring the development of corporate social and environmental reporting in an emerging oil and gas producing country: A case study of Tanzania*.

United Nations (2015). The 17 goals. Available from: "https://sdgs.un.org/goals" THE 17 GOALS | Sustainable Development

Vertigans, S. (2012). Paying the price for corporate social responsibility: Social costs and dividends of oil and gas company approaches in Nigeria. *Social Responsibility Review, 2012*(1), 35–48. Accessed May 5, 2018, from http://www.socialresponsibility.biz/2012-1.pdf

World Bank. (2020). *Global gas flaring data*. Accessed September 17, 2020, from https://www.ggfrdata.org/

World Business Council for Sustainable Development (WBCSD). (2000). *Corporate social responsibility: Making good business sense*. WBCSD.

Dr Oluchukwu Jane Richard-Osu is a HR specialist particularly interested in how HR personnel within the organisation can support in the attainment of set organisational goals. This includes exploring what factors can encourage or hinder effective strategy implementation and how HR can ensure that interactions within the organisation can be effectively aligned to support the organisation in achieving its goals. Further insights on how this can be achieved was gained through her PhD research which was focused on how HR can support the business in implementing sustainable business practices effectively in the oil and gas industry in Nigeria. This resulted in the design of a conceptual framework to support HR professionals within organisations to support in effective strategy implementation.

Dr Sarah Buckler is a social anthropologist with an interest in the relationships between people and organisations, organisational cultures, human creativity and development. An interest in narrative and rhetoric and the ways these are used by individuals and organisations in order to achieve particular ends began with her PhD research with Gypsy Travellers and continued into her subsequent employment carrying out research for local and national government departments. More recently this interest has expanded to focus on the ways different CSR and other development projects are delivered, especially in West Africa where she spent some time working on evaluation of a variety of CSR projects and exploring the tensions that can arise when organisations and groups have different expectations around development projects.

Sustainability Reporting Practices in Nigeria: A Study of Firms Quoted on the Nigerian Exchange Group

Gloria O. Okafor, Chinedu U. Asogwa, Amaka E. Agbata, and Sunday C. Okaro

1 Introduction

Sustainability reporting refers to the process an organization accounts for the impact—positive and/or negative, of its operations on the economy, environment, and society in monetary, qualitative, or quantitative terms to its stakeholders (Burhan & Rahmanti, 2012; Christensen et al., 2019; Songi & Dias, 2019). According to Boston College Center for Corporate Citizenship (2022), sustainability reporting is the disclosure and reporting of environmental, social, and governance (ESG) goals in addition to firms' advancement in the direction of such goals. It includes environmental performance reporting as well as other wider facets of a company's activities such as social, economic, and governance performance (Croner-i, 2022). Simply put, it is a practice of measuring, analyzing, and reporting on social and environmental impacts of companies, as well as on their economical sustainability (Gacser & Szoka, 2021).

The need for sustainability reporting sprung from the recognition that the information contained in the financial reporting was not sufficient for proper decision-making (Beattie, 2000); the acknowledgement of the role of business organizations toward the achievement of the sustainable development goals (SDGs) set out by the United Nations to achieve a prosperous world and the stakeholders̓ need for greater transparency by business organizations (Chipalkatti et al., 2021). Sustainability reports are usually prepared in such a way to enable easy comparison and measure growth from one period to another (Saji in Eneh & Amakor, 2019). Globally, stakeholders call for more transparent, verifiable, and comparable sustainability-related reporting for better assessment of a company's value.

G. O. Okafor (✉) · C. U. Asogwa · A. E. Agbata · S. C. Okaro
Department of Accountancy, Nnamdi Azikiwe University, Awka, Anambra State, Nigeria
e-mail: go.okafor@unizik.edu.ng

S. O. Idowu, S. Vertigans (eds.), *Sustainability in Global Companies*, CSR, Sustainability, Ethics & Governance, https://doi.org/10.1007/978-3-031-77971-8_13

Interestingly, the use of sustainability reporting has been credited with more effective risk management by an organization (Corporate Citizenship, 2012; Saridewi & Koesrindartoto, 2014; Uwuigbe, 2011); easy identification of opportunities for innovation (Buniamin et al., 2011); better stakeholder engagement which results in trust and long-term growth of the business (Burhan & Rahmanti, 2012; Corporate Citizenship, 2012).

Early publications of sustainability reporting can be traced to environmental reports published by companies whose activities were considered very detrimental to the environment. These reports were mainly for public relations—to inform the stakeholders on their efforts to mitigate the harmful effects of their activities on the environment; and, to manage their reputation. These reports later metamorphosed into corporate social responsibility reporting that included organizational labor policies and practices. Sustainability reports are published either as standalone documents or as part of annual reports of companies.

2 Sustainability Reporting

The concept of corporate sustainability is rapidly growing into a crucial concern in many territories across the globe (Okafor et al., 2022). It is a development that encourages management of organizations to manage their resources, both internal and external resources, in such a way that the needs of the present generation would be met without compromising the ability of future generations to meet their own needs. This development seeks to meet human development goals while also enabling natural systems to provide necessary natural resources and ecosystem services to humans. Corporate sustainability is described as the triple bottom line of an organization which explains management approach to social, economic, and environmental activities of an organization; in other words, it entails management of issues relating to social equality, economic prosperity, and environmental protection (Okafor et al., 2022; Gupta & Gupta, 2020; Okaro & Okafor, 2017). In this study, the components of sustainability reporting are as contained in the Global Reporting Initiative (GRI) version 2021. They include: general disclosures, economic performance disclosures, environmental disclosures, and the social disclosures.

The general disclosures contain information about organizations reporting practices; activities and workers; governance; strategy, policies, and practices; and stakeholder engagement. These disclosures help for proper understanding of the magnitude and background of the organizations and their impacts. The economic performance disclosures present information on the creation and distribution of economic value. The disclosures indicate how the organizations have created wealth for stakeholders, the economic value generated and distributed, and provide economic profiles of organizations, which can be useful for regulating other performance figures. This component of sustainability disclosures has different elements with several disclosure requirements. The elements include—the economic performance, market presence, indirect economic impacts, proportion of spending on local suppliers, anti-corruption, anti-competitive behavior and, tax (GRI 200, 2016).

The environmental dimension of sustainability discloses the impacts organizations have on living and non-living natural systems, including land, air, water, and ecosystems. This dimension presents information about the organization's contribution to resource conservation, indicating clearly their approaches to recycling, reusing, and reclaiming materials, products, and packaging. It provides information about an organization's impacts related to energy, and how it manages them; access to fresh, clean, safe, affordable, and good quality water; impacts related to biodiversity and how they are being protected and managed; and the generation, treatment, and disposal of waste, spills of chemicals, oils, fuels, and other substances. Clear disclosures are also required about an organization's compliance with environmental laws and regulations (GRI 300, 2016).

The social aspect of sustainability reporting discloses the impact an organization has on the social system where it operates. It discloses an organization's approach to employment or job creation; training and upgrading of employees skills; consultative practices with employees and their representatives and how they communicate significant operational changes; healthy and safe work conditions; and promotion of equality and diversity at work, with clear avoidance of discrimination against workers. It also addresses the organizations' conduct with respect to human rights, and other human rights issues like child labor; freedom of association and right of workers to bargain the terms and conditions of work as a group; and forced or compulsory labor. Other key issues that are disclosed in this section include: the security practices of the organization; the rights of indigenous peoples; local communities participation in the development of public policy; systematic efforts to address health and safety of customers across the life cycle of a product or service; social evaluation of suppliers; proper marketing communication; breach of customers privacy; and so on (GRI 400, 2016).

3 Sustainability Reporting in Nigeria

Nigeria has a robust regulatory framework for sustainability reporting in different sectors of the economy, even though the Companies and Allied Matters Act of 2020, which contains regulation on the conduct and publication of annual reports, does not contain any provision for sustainability reporting (Udo Udoma and Belo-Osagie Consulting, 2023). The Financial Reporting Council of Nigeria (FRCN) which was established by the Financial Reporting Council of Nigeria Act No 6 of 2011, was saddled with the responsibility to regulate the standards to be used for financial reporting and other related reports. The FRCN published the Nigeria Code of Corporate Governance in 2018; the aim was to promote best business practices among Nigerian companies and enhance investors' trust. In 2023, the Financial Reporting Council of Nigeria adopted the IFRS sustainability standards. This makes the country the first in Africa to fully adopt the IFRS sustainability standards. The effective date of use by companies is January 2024. In addition to the above, the Climate Change Act which was enacted in 2021, requires business organizations that have more than 50 employees to publish sustainability reports, the Nigeria Council on Climate Change was established to implement the law and the target is that Nigeria would achieve net zero emission by 2060 (KPMG, 2023).

Furthermore, companies that are quoted on the Nigerian stock exchange are expected to publish sustainability disclosures according to the Nigerian Exchange Group Sustainability Disclosure Guidelines (NGX SD Guidelines). The Securities and Exchange Commission also released the Securities and Exchange Commission Nigeria Guidelines on Sustainable Financial Principles (SEC SFPs). The SEC SFPs require companies who are players in the capital market to make Environmental, Social, and Governance (ESG) disclosures that concern them. Finally, for Nigerian banks, the Nigerian Sustainable Banking Principles (NSBPs) make it compulsory to publish sustainability reports that are reliable (Agbaje et al., 2023).

The NGX SD Guidelines are made up of nine principles which cover transparency and accountability in governance, responsible business practices, sustainable products and services, employee welfare, protection of the interests of the disadvantaged, human rights, inclusive growth and development and environmental preservation (The Nigerian Exchange Group Sustainability Disclosure Guidelines, 2018). The SEC SFPs are made up of five broad principles which include taking environmental, social, and governance (ESG) issues into account in business operations and decision-making, to collaborate with stakeholders for the sustainability of the Nigerian financial sector, financing of priority sectors to optimize ESG impacts, company's policy on human rights, women's economic empowerment, financial inclusion and job creation, and mandatory report on the progress in achieving the ESG goals of the companies (The Securities and Exchange Commission Nigeria Guidelines on Sustainable Financial Principles, 2021). Finally, the NSBP is made up of nine principles which include environmental and social considerations in risk management and business operations, human rights, women empowerment, financial inclusion, capacity building, collaborative partnership, and ESG reporting (Nigeria Sustainable Banking Principles, 2012).

Empirically, studies on sustainability reporting in Nigeria have shown that sustainability reporting has a positive significant effect on economic value addition of Nigeria quoted companies (Iliemena et al., 2023); return on assets, return on equity, Tobin Q, net profit margin and earnings per share (Alhassan et al., 2021; Ofoegbu & Asogwa, 2020; Umar et al., 2021; Agbata et al., 2021; Mautalib et al., 2020; Ezeokafor & Amahalu, 2019; Asuquo et al., 2018) and cash value added (Nzekwe et al., 2021), market value (Emeka-Nwokeji & Osisioma, 2019; Syder et al., 2020). Conversely, other studies in Nigeria have shown that sustainability reporting has a non-significant positive effect on return on assets and a non-significant negative effect on return on equity (Agbo & Joel, 2023; Amadiegwu, 2021); non-significant effect on corporate survival (Mbu-Ogar et al., 2023) and non-significant effect on asset growth (Agbata et al., 2021).

Furthermore, studies that examined the effects of firm characteristics on sustainability reporting found that the ownership structure and profitability had a negative significant effect on Nigerian companies while those of South Africa had a positive significant effect (Onoja et al., 2021); corporate strategic posture and organizational culture have a positive significant effect on sustainability reporting practices; while, institutional pressure has a non-significant negative effect on sustainability reporting practices (Mustapha, 2023); finance, risk, and audit committees have a significant positive association with sustainability reporting, while remuneration committees have an significant association with sustainability reporting (Abdulwahab et al., 2023). Board gender diversity has a non-significant effect on sustainability reporting (Adeniyi & Fadipe, 2018).

In summary, it can be seen that sustainability reporting affects the performance of business organizations in Nigeria and is in turn affected by other characteristics of the reporting organization. But, not much has been said about the current sustainability reporting practices of quoted companies. Studies like Tilt et al. (2021) found out that sustainability reporting in Africa was at the nascent stage; but the focus was on 48 African countries. Erin et al. (2022), however, examined the SDG reporting of the top 50 companies in Nigeria and found out that Nigerian business organizations perform poorly on sustainability reporting. The issue with the study is that the data were based on 2016–2018 annual reporting and on the assumption that sustainability reporting was voluntary. To have an up-to-date insight in sustainability reporting practices, there is need for a more recent and more detailed study in the area. This study provides an up-to-date assessment of the sustainability reporting practice of listed companies in Nigeria.

4 Methods

The study is a cross-sectional study of the sustainability reporting practices among listed companies in Nigeria. 86 out of 98 most largely capitalized listed companies as reported by Trading View 2022 were sampled and their annual reports reviewed. The sampled 86 companies include 24 companies in the manufacturing sector (Man) which consists of both the consumer goods and industrial goods sectors, 29 companies in the financial sector (fin) made up of banks and insurance companies, 5 companies in the agricultural sector (Agric), 6 in the oil and gas sector (O&G), and 22 companies from other sectors (Others). Other sectors are made up of the conglomerates, construction/real estate, healthcare, ICT, natural resources, and services. The 2022 sustainability reports of the companies were reviewed. Some companies have standalone sustainability reports, some have their sustainability reports embedded in the annual reports, and some have their sustainability practices presented in the directors', chairman's, and governance reports. Data on the sustainability reporting practices of companies were extracted from the reports using content analysis based on the features of the GRI 2021 version. Four components of sustainability reporting based on the GRI 2021 version were reviewed; they are the general disclosures, the economic disclosures, the environmental disclosures, and the social disclosures. The general disclosures are made up of 30 required disclosures; the economic disclosures are made up of 7 elements and 17 required disclosures; the environmental disclosures are made up of 7 elements and 31 required disclosures; and the social disclosures are made up of 17 elements and 36 disclosures. The extent of sustainability practices of the companies sampled was presented in this study using percentages.

5 Findings and Discussions

The extent of sustainability reporting practices among Nigerian companies is presented in Tables 1, 2, and 3. The first component of the sustainability disclosures which is the general disclosures has 30 required disclosures, only 8 out of the 30 have average disclosures of 50% and above, the other remaining 22 items have less than 50% disclosures. The prominent general disclosures based on the analysis are: organizational details; reporting period, frequency and contact point; governance structure and composition; chair of the highest governance body; activities, value chain, and other business relationship; policy commitments; mechanisms for seeking advice and raising concerns; and employees. The most prominent general disclosures are the organizational details and the reporting period, frequency and contact point with 100% disclosures each. The average general disclosure is 42.1%, which shows that on the average; about 57.9% of the required disclosures were not made. A greater proportion of the required general disclosures are yet to be made.

Table 1 Sustainability reporting practices according to GRI standards

	Sustainability reporting elements disclosed in the annual reports	Disclosures (%)	Average disclosures (%)
A	GRI 2: General disclosures		42.1
	2-1 Organizational details	100	
	2-2 Entities included in the organization's sustainability reporting	39.5	
	2-3 Reporting period, frequency and contact point	100	
	2-4 Restatements of information	12.8	
	2-5 External assurance	14	
	2-6 Activities, value chain and other business relationships	84.9	
	2-7 Employees	55.8	
	2-8 Workers who are not employees	19.8	
	2-9 Governance structure and composition	91.9	
	2-10 Nomination and selection of the highest governance body	37.2	
	2-11 Chair of the highest governance body	91.9	
	2-12 Role of the highest governance body in overseeing the management of impacts	45.3	
	2-13 Delegation of responsibility for managing impacts	24.4	

(continued)

Table 1 (continued)

	Sustainability reporting elements disclosed in the annual reports	Disclosures (%)	Average disclosures (%)
	2-14 Role of the highest governance body in sustainability reporting	32.6	
	2-15 Conflicts of interest	25.6	
	2-16 Communication of critical concerns	27.9	
	2-17 Collective knowledge of the highest governance body	19.8	
	2-18 Evaluation of the performance of the highest governance body	25.6	
	2-19 Remuneration policies	29.0	
	2-20 Process to determine remuneration	15.1	
	2-21 Annual total compensation ratio	10.5	
	2-22 Statement on sustainable development strategy	45.3	
	2-23 Policy commitments	62.8	
	2-24 Embedding policy commitments	45.3	
	2-25 Processes to remediate negative impacts	18.6	
	2-26 Mechanisms for seeking advice and raising concerns	60.5	
	2-27 Compliance with laws and regulations	47.7	
	2-28 Membership associations	26.7	
	2-29 Approach to stakeholder engagement	37.2	
	2-30 Collective bargaining agreements	14.0	
B	Economic disclosures		23.5
	GRI 201: Economic performance		37.8
	201-1 Direct economic value generated and distributed	96.5	
	201-2 Financial implications and other risks and opportunities due to climate change	14.0	
	201-3 Defined benefit plan obligations and other retirement plans	32.6	
	201-4 Financial assistance received from government	8.1	
	GRI 202: Market presence		9.3
	202-1 Ratios of standard entry level wage by gender compared to local minimum wage	9.3	
	202-2 Proportion of senior management hired from the local community	9.3	
	GRI 203: Indirect economic impact		45.4
	203-1 Infrastructure investments and services supported	55.8	
	203-2 Significant indirect economic impacts	34.9	
	GRI 204: Spending on local suppliers		16.3
	204-1 Proportion of spending on local suppliers	16.3	
	GRI 205: Anti-corruption		21.7
	205-1 Operations assessed for risks related to corruption	17.4	
	205-2 Communication and training about anti-corruption policies and procedures	30.2	
	205-3 Confirmed incidents of corruption and actions taken	17.4	
	GRI 206: Anti-competitive behavior		8.1

(continued)

Table 1 (continued)

	Sustainability reporting elements disclosed in the annual reports	Disclosures (%)	Average disclosures (%)
	206-1 Legal actions for anti-competitive behavior, anti-trust, and monopoly practices	8.1	
	GRI 207: Tax		12.2
	207-1 Approach to tax	17.4	
	207-2 Tax governance, control, and risk management	14.0	
	207-3 Stakeholder engagement and management of concerns related to tax	11.6	
	207-4 Country-by-country reporting	5.8	
C	Environmental disclosures		15.5
	GRI 301: Material Usage		8.5
	301-1 Materials used by weight or volume	5.8	
	301-2 Recycled input materials used	12.8	
	301-3 Reclaimed products and their packaging materials	7.0	
	GRI 302: Energy		21.4
	302-1 Energy consumption within the organization	26.7	
	302-2 Energy consumption outside of the organization	11.6	
	302-3 Energy intensity	17.4	
	302-4 Reduction of energy consumption	32.6	
	302-5 Reductions in energy requirements of products and services	18.6	
	GRI 303: Water and effluent		17.2
	303-1 Interactions with water as a shared resource	25.6	
	303-2 Management of water discharge-related impacts	17.4	
	303-3 Water withdrawal	11.6	
	303-4 Water discharge	12.8	
	303-5 Water consumption	18.6	
	GRI 304: Biodiversity		4.9
	304-1 Operational sites owned, leased, managed in, or adjacent to, protected areas and areas of high biodiversity value outside protected areas	4.7	
	304-2 Significant impacts of activities, products and services on biodiversity	5.8	
	304-3 Habitats protected or restored	9.3	
	304-4 IUCN Red List species and national conservation list species with habitats in areas affected by operations	4.7	
	GRI 305: Emission		11.8
	305-1 Direct (Scope 1) GHG emissions	15.1	
	305-2 Energy indirect (Scope 2) GHG emissions	11.6	
	305-3 Other indirect (Scope 3) GHG emissions	9.3	
	305-4 GHG emissions intensity	12.8	
	305-5 Reduction of GHG emissions	27.9	
	305-6 Emissions of ozone-depleting substances (ODS)	5.8	

(continued)

Table 1 (continued)

	Sustainability reporting elements disclosed in the annual reports	Disclosures (%)	Average disclosures (%)
	305-7 Nitrogen oxides (NOx), sulfur oxides (SOx), and other significant air emissions	9.3	
	GRI 306: Waste		23.3
	306-1 Waste generation and significant waste-related impacts	24.4	
	306-2 Management of significant waste-related impacts	32.6	
	306-3 Waste generated	20.9	
	306-4 Waste diverted from disposal	19.8	
	306-5 Waste directed to disposal	18.6	
	GRI 308: Supplier environmental assessment		14.5
	308-1 New suppliers that were screened using environmental criteria	17.4	
	308-2 Negative environmental impacts in the supply chain and actions taken	11.6	
D	Social disclosures		27.9
	GRI 401: Employment		15.5
	401-1 New employee hires and employee turnover	16.3	
	401-2 Benefits provided to full-time employees that are not provided to temporary or part-time employees	9.3	
	401-3 Parental leave	20.9	
	GRI 402: Labor/management		40.7
	402-1 Minimum notice periods regarding operational changes	40.7	
	GRI 403: Occupational health and safety		40.1
	403-1 Occupational health and safety management system	64.0	
	403-2 Hazard identification, risk assessment, and incident investigation	20.9	
	403-3 Occupational health services	50.0	
	403-4 Worker participation, consultation, and communication on occupational health and safety	44.2	
	403-5 Worker training on occupational health and safety	43.0	
	403-6 Promotion of worker health	60.5	
	403-7 Prevention and mitigation of occupational health and safety impacts directly linked by business relationships	34.9	
	403-8 Workers covered by an occupational health and safety management system	51.2	
	403-9 Work-related injuries	19.8	
	403-10 Work-related ill health	12.8	
	GRI 404: Training and Education		43.8
	404-1 Average hours of training per year per employee	16.3	
	404-2 Programs for upgrading employee skills and transition assistance programs	78.0	

(continued)

Table 1 (continued)

	Sustainability reporting elements disclosed in the annual reports	Disclosures (%)	Average disclosures (%)
	404-3 Percentage of employees receiving regular performance and career development reviews	37.2	
	GRI 405: Diversity and equal opportunity		37.2
	405-1 Diversity of governance bodies and employees	58.1	
	405-2 Ratio of basic salary and remuneration of women to men	16.3	
	GRI 406: Non-discrimination		39.5
	406-1 Incidents of discrimination and corrective actions taken	39.5	
	GRI 407: Freedom of association and collective bargaining		18.6
	407-1 Operations and suppliers in which the right to freedom of association and collective bargaining may be at risk	18.6	
	GRI 408: Child labor		18.6
	408-1 Operations and suppliers at significant risk for incidents of child labor	18.6	
	GRI 409: Forced or compulsory labor		14.0
	409-1 Operations and suppliers at significant risk for incidents of forced or compulsory labor	14.0	
	GRI 410: Security practices		10.5
	410-1 Security personnel trained in human rights policies or procedures	10.5	
	GRI 411: Rights of indigenous peoples		4.7
	411-1 Incidents of violations involving rights of indigenous peoples	4.7	
	GRI 413: Local communities		45.4
	413-1 Operations with local community engagement, impact assessments, and development programs	77.9	
	413-2 Operations with significant actual and potential negative impacts on local communities	12.8	
	GRI 414: Supplier social assessment		14.5
	414-1 New suppliers that were screened using social criteria	20.9	
	414-2 Negative social impacts in the supply chain and actions taken	8.1	
	GRI 415: Public policy		24.4
	415-1 Political contributions	24.4	
	GRI 416: Customer health and safety		12.8
	416-1 Assessment of the health and safety impacts of product and service categories	17.4	
	416-2 Incidents of non-compliance concerning the health and safety impacts of products and services	8.1	
	GRI 417: Marketing and labelling		7.4
	417-1 Requirements for product and service information and labeling	9.3	

(continued)

Table 1 (continued)

Sustainability reporting elements disclosed in the annual reports	Disclosures (%)	Average disclosures (%)
417-2 Incidents of non-compliance concerning product and service information and labelling	5.8	
417-3 Incidents of non-compliance concerning marketing communications	7.0	
GRI 418: Customer privacy		12.8
418-1 Substantiated complaints concerning breaches of customer privacy and losses of customer data	12.8	
Overall average disclosure		**27.6**

Source: 2022 Annual and sustainability reports of companies

Table 2 Commonly disclosed sustainability practices among Nigerian companies

	Agric (%)	Fin (%)	Man (%)	O&G (%)	Others (%)	Overall Ave (%)
GRI 2: General disclosures						
2-1 Organizational details	100	100	100	100	100	100
2-3 Reporting period, frequency, and contact point	100	100	100	100	100	100
2-6 Activities, value chain, and other business relationships	100	82.8	79.2	100	86.4	84.9
2-7 Employees	40	65.5	37.5	50	68.2	55.8
2-9 Governance structure and composition	100	93.1	91.7	100	86.4	91.9
2-11 Chair of the highest governance body	60	93.1	91.7	100	95.5	91.9
2-23 Policy commitments	60	72.1	54.2	33.3	68.2	62.8
2-26 Mechanisms for seeking advice and raising concerns	80	62.1	50	33.3	72.7	60.5
Economic disclosures						
GRI 201: Economic performance						
201-1 Direct economic value generated and distributed	100	93.1	100	100	95.5	96.5
GRI 203: Indirect economic impact						
203-1 Infrastructure investments and services supported	40	55.2	70.8	66.7	40.9	55.8
Environmental disclosures						
Social disclosures						
GRI 403: Occupational health and safety						
403-1 Occupational health and safety management system	60	65.3	58.3	66.7	68.2	64.0
403-3 Occupational health services	80	48.3	45.8	50	50	50.0
403-6 Promotion of worker health	100	58.6	54.2	50	63.6	60.5
403-8 Workers covered by an occupational health and safety management system	60	48.3	37.5	50	68.2	51.2
GRI 404: Training and education						

(continued)

Table 2 (continued)

	Agric (%)	Fin (%)	Man (%)	O&G (%)	Others (%)	Overall Ave (%)
404-2 Programs for upgrading employee skills and transition assistance programs	100	79.3	62.5	100	81.8	78.0
GRI 405: Diversity and equal opportunity						
405-1 Diversity of governance bodies and employees	60	62.1	50	66.7	59.1	58.1
GRI 413: Local communities						
413-1 Operations with local community engagement, impact assessments, and development programs	80	75.9	83.3	83.3	72.7	77.9

Source: 2022 Annual and sustainability reports of companies

Table 3 Average sustainability disclosures by sectors

	Agric (%)	Fin (%)	Man (%)	O&G (%)	Others (%)	Overall average (%)
General disclosures	38.7	43.3	38.3	43.3	43	42.1
Economic disclosures	21.2	20.6	27.6	30	18.2	23.5
Environmental disclosures	12.9	11.9	23.9	17.1	11.0	15.5
Social disclosures	27.2	26.9	31.1	26.9	26.4	27.9
% number of companies that disclosed more than the average general disclosures	60	41.4	37.5	33.3	54.5	44.2
% number of companies that disclosed more than the average economic disclosures	20	41.4	41.7	33.3	22.7	29.1
% number of companies that disclosed more than the average environmental disclosures	20	37.9	41.7	33.3	40.9	32.6
% number of companies that disclosed more than the average social disclosures	60	37.9	33.3	50	50	32.6

Source: 2022 Annual and sustainability reports of companies

The economic disclosures have an average disclosure of 23.5%, showing that about an average of 76.5% of the required disclosures are not reported. There are 7 elements of economic disclosures with 17 individual items required to be disclosed. The element of economic disclosures with the highest average disclosure is the indirect economic impact with an average disclosure of 45.4%, followed by the economic performance with an average disclosure of 37.8%, anti-corruption with an average disclosure of 21.7%, spending on local suppliers with an average disclosure of 16.3%, market presence with an average disclosure of 9.3%, and anti-competitive behavior with an average disclosure of 8.1%. Only two disclosures out of the entire 17 required disclosures have an average disclosure of 50% and above. The more prominent of the two is the direct economic value generated and distributed with 96.5% disclosures, followed by the infrastructure investment and services supported with 55.8% disclosures.

The environmental disclosures have an average disclosure of 15.5%, showing that about an average of 84.5% of the required disclosures are not reported. The environmental disclosures have 7 different elements with 31 required disclosures. But no disclosure has an average disclosure of up to 50%. The material usage element has an average disclosure of 8.5%, energy has an average of 21.4%, water and effluent has an average disclosure of 17.2, biodiversity has an average of 4.9%, emission—11.8%, waste—23.3%, and suppliers environmental assessment—14.5%. Waste management and energy were disclosed more than other elements, though the average disclosures are still very low. Among the required 31 disclosures, no disclosure has up to 50% disclosures, only 5 required disclosures attracted up to 25% disclosures. The 5 items that attracted up to 25% disclosures include: management of significant waste-related impacts with an average disclosure of 32.6%; reduction of energy consumption with average disclosure of 32.6%; reduction of GHG emissions with an average disclosure of 27.9%, energy consumption within the organization and interactions with water as shared resources with 26.7% and 25.6%, respectively. The environmental disclosures were the least presented. One will think that for companies to be seen as environmentally friendly companies that disclosing their environmental activities in their annual reports should be one of the key strategies to pursue.

The average social disclosure is 27.9%, showing that about an average of 72.1% of the required disclosures are not reported. Social disclosures have different elements but none of the 17 different elements has an average disclosure of up to 50%. Moreover, out of the 36 required disclosures, some disclosures attracted up to 50% disclosures. The elements that have up to 25% disclosures are: the local community—45.4%; training and education—43.8%; the labor/management—40.7%; occupational health and safety—40.1%; non-discrimination—39.5%, and diversity and equal opportunity—37.2%. Other elements have average disclosures of less than 25%. 7 out of the 36 required disclosures have an average disclosure of 50% and above; these are the most commonly reported social disclosures. They include occupational health and safety management system with 64% average disclosure; occupational health services—50% disclosure; promotion of worker health—60.5% average disclosure; workers covered by an occupational health and safety management system—51.2%; programs for upgrading employee skills and transition assistance programs—78%; diversity of governance bodies and employees—58.1%; and operations with local community engagement, impact assessments, and development programs with an average disclosure of 77.9%. Operations with legal community and programs for upgrading employee skills are the most prominent in this component.

Table 3 presented a sectoral comparison of all the average disclosures. In the general disclosures, the agricultural sector shows an average disclosure of 38.7%, the financial sector and oil &gas sector show an average disclosure of 43.3% each, the manufacturing sector shows an average disclosure of 38.3%, and the others show an average disclosure of 43%. The overall average disclosure of the entire required general disclosures is 42.1% and only 44.2% of all the sampled companies (38 out of 86 companies) disclosed up to the average general disclosure. In the economic disclosures, the agricultural sector has an average of 21.2%, the financial sector has an average 20.6%, manufacturing has an average of 27.6, oil and gas has an average of 30%, and the others have an average disclosure of 18.2%. Oil and gas and manufacturing have more disclosures than other sectors, though not so significant. The average economic disclosure is 23.5% and only 29.1% (25 out of 86) of the companies were able to disclose up to this average. The environmental disclosures with an average disclosure of 15.5% are the most uncommonly disclosed component. Manufacturing has the highest disclosures among the companies—23.9%, followed by oil and gas—17.1%, agricultural—12.9%, financial—11.9%, and others have 11%. Only 32.6% of the companies (28 out of 86 companies) disclosed up to 15.5% of the required environmental disclosures. In social disclosures, though the manufacturing sector has the highest average disclosure of 31.1%, yet there is no significant difference in the average disclosures among the sectors. Agricultural sector has 27.2% average, financial and oil & gas sectors have 26.9% average disclosures each, and others have 26.4% average disclosure. Only 32.6% of the companies (28 out of 86 companies) disclosed up to 15.5% of the required environmental disclosures.

On the overall, the average sustainability practices among Nigerian companies considering all components and aspects are about 27.6%, which points that in the sustainability reports prepared among the companies, about 72.4% of the required disclosures are not being disclosed. This is poor and calls for improvement in both internal sustainability practices and reporting.

6 Conclusion

Though many Nigerian companies have started preparing sustainability reports, the level of sustainability activities disclosures is still very low and could be classified as poor. This finding is in line with the findings of Okaro and Okafor (2017) that sustainability reporting is at low ebb in Nigeria. It also agrees with the findings of that of Erin et al. (2022) which revealed that business organizations in Nigeria perform poorly on sustainability reporting. These poor sustainability disclosures were prevalent among the four components of the sustainability practices used in this study. The general disclosures were reported more than others, followed by the social disclosures, then economic disclosures and environmental disclosures. The environmental disclosures were the least presented. The average disclosures among the sectors do not show significant differences; there seems to be similar behaviors among the sectors. The poor sustainability disclosures could be as a result of poor internal management of organizational resources or as a result of non-mandatory statute of sustainability report as at 2022 financial year, or could be as a result of a lack of interest.

There is a great need for improvement in sustainability reporting. As at 2022, the time at which the reports for the study were prepared, publication of sustainability reports were voluntary in Nigeria, though globally best practice. Now, with Nigeria being part of the adoption of the IFRS S1 and IFRS S2 launched on the 26th of June 2023 with effective date from January 1, 2024, there is a strong feeling that the majority of Nigerian companies will improve greatly in their sustainability disclosures. This adoption has both long-term and short-term benefits of improving the image of organizations and boosting sustainable business performances and will help to improve the transparency and comparability of sustainability information among companies in Nigeria and outside Nigeria.

References

Abdulwahab, A. I., Bala, H., Yahaya, O. A., & Abdullahi, M. (2023). Corporate governance committee and sustainability reporting of listed consumer goods firms in Nigeria. *International Journal of Research and Innovation in Social Science, 7*(7), 1761–1770. https://doi.org/10.47772/IJRISS.2023.7083

Adeniyi, S. I., & Fadipe, A. O. (2018). Effect of board diversity on sustainability reporting in Nigeria: A study of beverage manufacturing firms. Indonesian Journal of Corporate Social Responsibility and Environmental Management, 1(1), 43–50. Retrieved from https://www.researchgate.net/publication/328654263_Effect_of_Board_Diversity_on_Sustainability_Reporting_in_Nigeria_A_Study_of_Beverage_Manufacturing_Firms

Agbaje, A., Anjola, B., Alada, B., & Bella, S. (2023, January 6). *In review: Sustainable development requirements and taxonomies in Nigeria*. Banwo & Ighodalo. Retrieved from https://www.lexology.com/library/detail.aspx?g=d81b2beb-ba3e-40fd-9314-a7c23146d5d2

Agbata, A. E., Eze, M. N., & Uchegbu, C. U. (2021). Corporate sustainability reporting and corporate financial performance of brewery firms quoted on the Nigeria Stock Exchange. *Nigeria Academy of Management Journal, 16*(91), 1–15. Retrieved from https://namj.tamn-ng.org/index.php/home/article/view/125

Agbo, A., & Joel, I. G. (2023). Sustainability reporting and financial performance of listed environmentally sensitive companies in Nigeria. *Advance Journal of Financial Innovation and Reporting, 7*(05), 1–17. Retrieved from https://aspjournals.org/ajfir/index.php/ajfir/article/download/25/25/50

Alhassan, I., Islam, K. M. A., & Haque, S. (2021). Sustainability reporting and financial performance of listed industrial goods sectors in Nigeria. *International Journal of Accounting & Finance Review, 9*(1), 46–56. Retrieved from https://www.cribfb.com/journal/index.php/ijafr/article/view/1541

Amadiegwu, I. G. (2021). Effect of sustainability reporting on listed manufacturing firms in Nigeria. *Mouau.afribary.org, 7*(2). Retrieved from https://repsoitory.mouau.edun.ng/work/view/effect-pf-sustainability-reporty-on-listed-manufacturing-firms-in-nigeria-7-2

Asuquo, A., Dada, T. E., & Onyaogaziri, R. U. (2018). The effect of sustainability reporting on corporate performance of selected quoted brewery firms in Nigeria. *International Journal of Business and Law Research, 6*(3), 1–10. Retrieved from https://seahipaj.org/journals-ci/sept-2018/IJBLR/Full/IJBLR-S-1-2018.pdf

Beattie, V. (2000). The future of corporate reporting: A review article. *Irish Accounting Review, 7*(1), 1–3. Retrieved from http://eprints.sla.ac.uk/archive/000008291

Boston College Centre for Corporate Citizenship. (2022). *Sustainability reporting BC CCC*. Retrieved from https://ccc.bc.edu/content/ccc/research/corporate-citizenship-news-and-topics/sustainability-reporting.html

Buniamin, S., Alrazi, B., Johari, N. H., & Abd Rahman, N. R. (2011). Corporate governance practices and environmental reporting of companies in Malaysia: Finding possibilities of double thumbs up. *Journal Pengurusan, 32*, 55–71. Retrieved from http://journalarticle.ukm.my/2105/1/jurus_32-06-lock.pdf

Burhan, A. H. N., & Rahmanti, W. (2012). The impact of sustainability reporting on company performance. *Journal of Economics, Business, and Accountancy Ventura, 15*(2), 257–272. Retrieved from https://pdfs.semanticscholar.org/711a/6e1788e94667dbf489a219eb80172a83140f.pdf

Chipalkatti, N., Le, Q. V., & Rishi, M. (2021). Sustainability and society: Do environmental, social and governance factors matter for foreign direct investment? *Energies, 14*(19), 1–18. https://doi.org/10.3390/en14196039

Christensen, H. B., Hail, L., & Leuz, C. (2019). *Adoption of CSR and sustainability reporting standards: Economic analysis and review*. ECGI Working Paper Series in Finance Working Paper.

Corporate Citizenship. (2012). *Adding value through sustainability reporting*. Retrieved from https://corporate-citizenship.com/wp-content/uploads/Adding-Value-Report-Final.pdf

Croner-i. (2022). *Environmental reporting: In-depth*. Retrieved from app.croneri.co.uk/topics/environmental-reporting/indepth

Emeka-Nwokeji, N. A., & Osisioma, B. C. (2019). Sustainability disclosures and market value of firms in emerging economy: Evidence from Nigeria. *European Centre for Research, Training & Development, 7*(3), 1–19. Retrieved from https://www.researchgate.net/publication/340607905_SUSTAINABILITY_DISCLOSURES_AND_MARKET_VALUE_OF_FIRMS_IN_EMERGING_ECONOMY_EVIDENCE_FROM_NIGERIA

Eneh, O., & Amakor, I. C. (2019). Firm attributes and sustainability reporting in Nigeria. *International Journal of Academic Accounting, Finance & Management Research, 3*(6), 36–44. Retrieved from https://www.researchgate.net/publication/334251502

Erin, O. A., Bamigboye, O. A., & Oyewo, B. (2022). Sustainable development goals (SDG) reporting: An analysis of disclosure. *Journal of Accounting in Emerging Economics, 12*(15), 761–789. https://doi.org/10.1108/JAEE-02-2020-0037

Ezeokafor, F. C., & Amahalu, N. N. (2019). Effect of sustainability reporting on corporate performance of quoted oil and gas firms in Nigeria. *Journal of Global Accounting, 6*(2), 217–228. Retrieved from https://journals.unizik.edu.ng/joga/article/view/2339

Gacser, N., & Szoka, K. (2021). Sustainability accounting - Historical development and future perspectives of the discipline. *PressAcademia Procedia (PAP), 14*, 1–4.

GRI 200. (2016). *Economic*. Global Reporting Initiative. Retrieved from standards@globalreporting.org., www.globalreporting.org

GRI 300. (2016). *Environmental*. Global Reporting Initiative. Retrieved from standards@globalreporting.org., www.globalreporting.org

GRI 400. (2016). *Social*. Global Reporting Initiative. Retrieved from standards@globalreporting.org., www.globalreporting.org

Gupta, A. K., & Gupta, N. (2020). Effect of corporate environmental sustainability on dimensions of firm performance - Towards sustainable development: Evidence from India. *Journal of Cleaner Production, 253*, 119948. https://doi.org/10.1016/j.jclepro.2019.119948

Iliemena, R. O., Ijeoma, N. B., & Uagbale-Ekatah, R. (2023). Sustainability reporting and economic value added: Empirical evidence from listed manufacturing entities in Nigeria. *International Journal of Social Science and Educational Research Studies, 3*(4), 660–677. Retrieved from https://ijssers.org/wp-content/uploads/2023/04/18-2004-2023.pdf

KPMG. (2023). *Big shifts, small steps: 2022 survey of sustainability reporting in Nigeria*. Retrieved from https://assets.kpmg.com/content/dam/kpmg/ng/pdf/2022-survey-of-sustainability-reporting-in-nigeria-updated.pdf

Mautalib, Y. O., Iriabije, E. U., Okon, A. E., & Odumegwu, E. C. (2020). Impact of sustainability reporting on corporate performance: Evidence from Nigeria Stock Exchange (NSE). *Journal of Management Science and Entrepreneurship, 20*(7), 345–367. Retrieved from https://www.hummingbirdpubng.com/wp-content/uploads/2020/10/HUJMSE_VOL20_NO7_JUNE2020_-26.pdf

Mbu-Ogar, G. B., Kangkpang, K. A., Nkiri, J. E., & Amoke, C. V. (2023). Sustainability reporting and survival of selected oil and gas companies in Nigeria. *Nigerian Journal of Management Sciences, 24*(16), 173–182. Retrieved from https://nigerianjournalofmanagementsciences.com/

wp-content/uploads/2023/02/18.-SUSTAINABILITY-REPORTING-AND-SURVIVAL-OF-SELECTED-OIL-AND-GAS-COMPANIES-IN-NIGERIA.pdf
Mustapha, A. (2023). Determinants of sustainability reporting practices among listed industrial and domestic firms in Nigeria. *Journal of Family Business and Management Sciences, 15*(1), 155–174. Retrieved from https://www.fbmsjournal.com/wp-content/uploads/2023/03/09_Determinants-of-Sustainability-Reporting-Practices....pdf
Nigeria Sustainable Banking Principles. (2012) *Nigeria sustainable banking principles*. Retrieved from https://www.cbn.gov.ng/out/2012/ccd/circular-nsbp.pdf
Nzekwe, O. G., Okoye, P. V. C., & Amahalu, N. N. (2021). Effect of sustainability reporting on financial performance of quoted industrial goods companies in Nigeria. *International Journal of Mgt Studies and Social Science Research, 3*(5), 265–280. Retrieved from https://www.ijmsssr.org/paper/IJMSSSR00536.pdf
Ofoegbu, G. N., & Asogwa, C. U. (2020). The effect of sustainability reporting on profitability of quoted consumer goods companies in Nigeria. *The International Journal of Innovative Research and Development, 9*(4), 271–282. https://doi.org/10.24940/ijird/2020/v9/i4/APR20075
Okafor, U. I., Philip, E. O., Edet, T. E., & Okon, N. B. (2022). Corporate sustainability practices and corporate financial performance of selected breweries in Nigeria. *Finance & Economic Review, 4*(1), 25–40. https://doi.org/10.38157/fer.v4il.390
Okaro, S. C., & Okafor, G. O. (2017). Integrated reporting in Nigeria: The present and future. In S. O. Idowu & R. Schmidpeter (Eds.), *Responsible corporate governance, towards sustainable and effective governance structures* (CSR, sustainability, ethics and governance series). Springer.
Onoja, A. A., Okoye, E. I., & Nwoye, U. (2021). Global reporting initiative and sustainability reporting practices: A study of Nigeria and South Africa oil and gas firms. *Research Journal of Management Practice, 1*(12), 1–25. Retrieved from https://www.ijaar.org/articles/rjmp/v1n12/Global-Reporting-Initiative-and-Sustainability-Reporting-Practices-A-Study-of-Nigeria-and-South-Africa-Oil-and-Gas-Firms.pdf
Saridewi, P. N. & Koesrindartoto, D. P. (2014). The link between social, environmental and financial performances of companies in Indonesia. In *A paper presented at the international conference on trends in economics, humanities and management (ICTEHM'14) Pattaya (Thailand).*
Songi, O., & Dias, A. K. (2019). Sustainability reporting in Africa: A comparative study of Egypt, Equatorial Guinea, Kenya, Nigeria, Botswana and South Africa. In *Cambridge handbook of corporate law, corporate governance and sustainability* (pp. 536–550). Cambridge University Press. https://doi.org/10.1017/9781108658386.045
Syder, I. D., Ogbonna, G. N., & Akani, F. N. (2020). The effect of sustainability accounting report on shareholder value of quoted oil and gas companies in Nigeria. *International Journal of Management Science, 7*(5), 44–57. Retrieved from https://www.arcnjournals.org/images/ASPL-IJMS-7-4-4.pdf
The Nigeria Exchange Group. (2018). *The Nigeria exchange group sustainability disclosure guideline*. Retrieved from https://ngxgroup.com/ngx-download/sustainability-disclosure-guidelines/
The Securities and Exchange Commission, Nigeria. (2021). *Securities and Exchange Commission Nigeria guidelines on sustainable financial principles*. Retrieved from https://sec.gov.ng/wp-content/uploads/2021/12/SEC-Guidelines-on-Sustainable-Financial-Principles-for-the-Capital-Market_Final.pdf
Tilt, C. A., Qian, W., Sanjaya, K., & Dissanayake, D. (2021). The state of business sustainability reporting in sub-saharan Africa: An agenda for policy and practice. *Sustainability Accounting, Management and Policy Journal, 12*(2), 267–296. https://doi.org/10.1108/SAMPJ-06-2019-0248
Udo Udoma & Belo-Osagie Consulting. (2023, January 11). *The companies and allied matters act 2020-what you need to know-part 12-directors under the CAMA 2020*. Retrieved from https://www.mondaq.com/nigeria/shareholders/1024130/the-companies-and-allied-matters-act-2020%2D%2Dwhat-you-need-to-know%2D%2D-part-12%2D%2Ddirectors-under-the-cama-2020

Umar, M. M., Mustapha, L. O., & Yahaya, O. A. (2021). Sustainability reporting & financial performance of listed consumer goods firms in Nigeria. *Journal of Advance Research in Business Management and Accounting, 7*(3), 21–32. Retrieved from https://nnpub.org/index.php/BMA/article/view/939

Uwuigbe, U. (2011). An empirical investigation of the association between firms' characteristics and corporate social disclosures in the Nigerian financial sector. *Journal of Sustainable Development in Africa, 13*(1), 60–74. Retrieved from https://core.ac.uk/download/pdf/17045102.pdf

Gloria O. Okafor, PhD, FCA, is Professor in Accountancy Department at Nnamdi Azikiwe University, Awka, Anambra State Nigeria. She has 20 years' experience in accounting lecturing and research, holds a fellowship of the Institute of Chartered Accountants of Nigeria (ICAN), and is a member of the Academy of Management Nigeria. She teaches taxation, environmental management accounting, financial management, financial accounting, and accounting standards. She has special research interest in taxation, corporate sustainability practices and reporting, accounting ethics, and corporate governance. Gloria has published her research in peer-reviewed journals, chapters in books, and conference proceedings.

Chinedu U. Asogwa, PhD, is a lecturer at the Department of Accountancy, Nnamdi Azikiwe University, Awka. Her research interest is in sustainable development especially as it relates to corporate reporting. Her research works are published in reputable journals, and she is open to collaborations from colleagues.

Amaka E. Agbata, PhD, is a lecturer in the Department of Accountancy, Nnamdi Azikiwe University, Awka, Nigeria. She is an associate member of the Institute of Chartered Accountants of Nigeria. Her academic research interest includes financial and corporate reporting, public sector accounting, corporate social responsibility reporting, taxation, and auditing. She has a good number of article publications to her credit.

Sunday C. Okaro, PhD, holds a Bachelor's degree, a Master's degree, and a Doctorate degree in Accountancy. He also holds a Master's degree in Banking and Finance. He is Professional Accountant and holds the Fellowship of the Chartered Association of Certified Accountants of London (FCCA). He is also an associate member of the Institute of Chartered Accountants of Nigeria (ICAN). Professor Okaro also belongs to the Chartered Institute of Taxation of Nigeria as associate member. Professor Okaro taught for 40 years on both a part-time and full-time basis before retiring in 2021 from full-time teaching career. He, however, still doubles as Adjunct Professor at UNN Business School and Probono Faculty Professor at the prestigious NnamdiAzikiwe University, Awka, Nigeria. It is not all academics for Professor Okaro as he also had extensive experience in the public sector as a civil servant and a banker in the private sector. He undertakes consultancy assignments. He is a prolific writer and has over 70 publications to his credit. He is happily married with children and grandchildren.

Part III
Sustainability in Asian Companies

India

Inclusion of Sustainability into Business Education: Understanding the Student's Awareness, Knowledge, Attitude, and Beliefs—A Study Based on India

Sumona Ghosh

1 Introduction

Global research has highlighted on the urgency for businesses to become sustainable (Aras & Crowther, 2009). However, within this ever-expanding literature there is a conflict regarding what constitutes sustainability since this area is addressed by different disciplines in their own way (van Marrewijk & Werre, 2003). Despite the varied definitions and approaches to sustainability, most of the researchers tend to follow the definition provided by the "Brundtland Report". The "Brundtland Report" tends to emphasize on the three pillars of sustainability, that is, environmental, economic, and social (Aras & Crowther, 2009). From the point of view of business, sustainability has been defined by Galbreath (2009) as "sustainability [is] a business approach that seeks to create long-term value for stakeholders by embracing opportunities and managing risks associated with economic, environmental, and social developments".

A study conducted by Ernst and Young and cited in van Marrewijk (2003) had observed that, 94% of Global 1000 list of companies believed that sustainability could result in financial benefits but hardly 11% incorporated it in their business activities. Survey carried on by KPMG International (2008) found out that it was challenging for 80% of the companies to understand how to make their businesses more sustainable. One way to address these challenges is to promote ESD or sustainability education. Sustainability literate people are those who "understand the need for change to a sustainable way of doing things; have sufficient knowledge and skills to decide and act in a way that favours sustainable development; [and are] able

S. Ghosh (✉)
Post Graduate and Research Department of Commerce,
St. Xavier's College, (Autonomous) Kolkata, Kolkata, India

S. O. Idowu, S. Vertigans (eds.), *Sustainability in Global Companies*, CSR, Sustainability, Ethics & Governance, https://doi.org/10.1007/978-3-031-77971-8_14

to recognise and reward other people's decisions and actions that favour sustainable development" (Forum for the Future, 2004).

Thus, HEIs have an immense responsibility of promoting sustainability (Sibbel, 2009). Education for sustainability and issues related with it was given immense importance in the Rio de Janeiro UN Conference held in the year 1992 and "Agenda 21" was endorsed. More recently, the dire need to incorporate business sustainability in management education was articulated in the "United Nations' (2008) Principles for Responsible Management Education initiative (PRME)". The first principle of PRME is to "develop the capabilities of students to be future generators of sustainable value for business and society at large and to work for an inclusive and sustainable global economy".

We argue in our paper that if future managers need to enhance their knowledge and understanding of sustainability so that they can develop strategies that will make their organizations more sustainable which in turn will be beneficial for the organization, then business education both at the undergraduate and post graduate levels must incorporate sustainability education in their curriculum. To the extent sustainability education can be incorporated into the curriculum across universities is dependent on students' understanding, attitude, and belief in sustainability and its related issues. Thus, our paper will address two aspects. First to capture the students' perception regarding inclusion of sustainability into the business education. Second to capture and understand the students' awareness, knowledge, attitude, and beliefs regarding sustainability, SDGs, and business education.

The remaining paper is organized as follows. Section 2 gives an overview of the literature review. Section 3 deals with the methods. Section 4 discusses the results and discussions. Section 5 focuses on concluding remarks and future directions.

2 Literature Review

2.1 Sustainability

Both researchers and practitioners have started recognizing the importance of sustainability. Today we have a vast body of literature that speaks not only about sustainability in general but also the about the diverse facets of sustainability. According to Linton et al. (2007), the term sustainability has been defined from the perspective of environmental science, management, and social science. The most accepted definition of sustainability is: "meeting the needs of the present without compromising the ability of future generations to meet their own needs" (Brundtland, 1987). Hockerts (1999) felt that sustainability was a state which fulfilled the stakeholders' needs with compromising its ability also to meet their needs in future. From the above observations, the following point arises and that is, sustainability concept differs from the orthodox management theory in its realization that economic sustainability alone can lead to success only in the short run, however for a long-term success we need social and environmental sustainability as well (Elkington, 1997). Porter and Kramer (2011) proposed the Creating Shared Value (CSV) concept which focused on enhancing the competitiveness of a company while simultaneously improving the social, environmental, and economic conditions in the communities where it is operating. Thus, included within this broad spectrum of sustainability are issues like understanding how an economic activity affects the environment (Erlich & Erlich, 1991); how we can provide for worldwide food security (Lal et al., 2002); how we can meet the basic human needs (Savitz & Weber, 2006); and how we can protect our non-renewable resources (Whiteman & Cooper, 2000). Unfortunately, the macro-economic definition of sustainability provided by the Brundtland Commission's was difficult for organizations to apply and they failed to determine their individual roles within this broad definition (Starik & Rands, 1995; Shrivastava, 1995).

Hence, more microeconomic applications of sustainability in the fields of management, operations, engineering, social sciences started getting investigated. Within the management, operations, social sciences literature, the term sustainability has focused more on ecological (e.g. the natural environment) sustainability, as compared to social and economic sustainability (Shrivastava, 1995; Starik & Rands, 1995).

Starik and Rands (1995, p. 909), for example, define sustainability as:

> the ability of one or more entities, either individually or collectively, to exist and flourish (either unchanged or in evolved terms) for lengthy timeframes, in such a manner that the existence and flourishing of other collectivities of entities is permitted at related levels and in related systems.

Shrivastava (1995) describes sustainability as offering, "the potential for reducing long-term risks associated with resource depletion, fluctuations in energy costs, product liabilities, and pollution and waste management".

Interestingly, the engineering literature on sustainability has been more encompassing, and has been defined as, "a wise balance among economic development, environmental stewardship, and social equity" (Sikdar, 2003) and as including "... equal weightings for economic stability, ecological compatibility and social equilibrium" (Go'ncz et al., 2007).

2.2 Sustainable Development

More than 200 years ago, researchers and thinkers were plagued with the question as to what would be the impact on the environment and resources of our planet with the fast evolution of our civilization (Peter et al., 2008). Much later the results of computer simulations made by MIT technicians were published in the well-known book "*The Limits to Growth*" (Meadows, 1972). The results predicted depletion of non-renewable resources and increasing commodity prices. Then in 1974, we had Lester Brown setting up the "World Watch Institute and the Earth Policy Institute". Both presented facts about global use of natural resources and suggested alternate approaches (Brown, 2006). The above stated facts were a reminder to the society and the world at large that unsustainable consumption would put us all in trouble.

If we must trace the development of the concept of sustainable development, then we need to study some of the major conferences that were devoted to this issue from 1972 to 2002. We will observe that environment was the focus in the Stockholm Conference held in 1972 (Vogler, 2007). However, there was a shift in focus from environment to sustainability both social, economic, and environmental in the Rio de Janeiro Earth Summit held in 1992 (Prizzia, 2007), to emphasizing more on poverty alleviation at the Millennium Summit in 2000 (Neil & Alexandra, 2007) and at the Johannesburg World Summit in 2002 (Vogler, 2007). The Industrial Revolution with "limitless human and environmental exploitation" is the biggest challenge that sustainable development needs to tackle. Sustainable development simply means "development that can be continued either indefinitely or for the given time" (Stoddart et al., 2011). Even though there is a plethora of definitions defining sustainable development, the most cited definition is the one proposed by the "Brundtland Commission Report" stated above (Schaefer & Crane, 2005).

Mensah and Casadevall (2019) argued that the entire issue of sustainable development centres around "inter- and intragenerational equity anchored essentially on three-dimensional distinct but interconnected pillars, namely the environment, economy, and society. Decision-makers need to be constantly mindful of the relationships, complementarities, and trade-offs among these pillars and ensure responsible human behaviour and actions at the international, national, community and individual levels to uphold and promote the tenets of this paradigm in the interest of human development".

Sustainable development rests on the principle of conservation of the ecosystem (Kanie & Biermann, 2017), effective human resource management (Wang, 2016), principle of participation (Guo, 2017), and promoting progressive social and political culture (Tjarve & Zemīte, 2016).

The aim of sustainable development is to take care of economic, social, and environmental sustainability equally which can be achieved through increased public participation, enhanced commitment of the people, and integrated initiatives and efforts taken at different levels.

2.3 *Education for Sustainable Development (ESD)*

"Sustainability Education" or "Education for Sustainability" (ES) or "Education for Sustainable Development" (ESD) aims to empower individuals so that they can reflect on their own actions and its impact on the social, cultural, economic, and environmental levels, from a local as well as a global perspective (UNESCO, 2017). It helps learners to fight the major social, economic, and environmental challenges of the twenty-first century and at the same time helps to create transformational leaders who would foster sustainability in organizations. ESD can be defined as a "twenty-first century education" which has the potential to achieve sustainability globally (Bell, 2016), and Higher educational institutions have a major role to play in this regard (Kir'aly & G'ering, 2019). The importance of ESD is also recognized by SDGs themselves. SDG 4.7, states that: "By 2030, ensure that all learners acquire the knowledge and skills needed to promote sustainable development" (UNESCO, 2018). HEIs through research, teaching, and learning can play a pivotal role in helping the society to achieve SDGs (Ferguson & Roofe, 2020). In fact, according to SDSN (2020), without the support of academia it would be very difficult to achieve SDGs. ESD in higher education encourages young minds to be proactive in building more sustainable societies (Finnveden et al., 2020). Universities and HEIs have started to integrate ESD into their business curriculum to enable students to develop a mindset towards sustainability (Farrag & Obeidat, 2020; Winfield & Ndlovu, 2019), and failure to do so may cause the future leaders to become only profit-oriented (Hernandez-Lopez et al., 2020). It has become essential nowadays for graduates of higher education, especially business school graduates to have a sustainability mindset and the capability to implement it in practice (Haney et al., 2020; Winfield & Ndlovu, 2019). Winfield and Ndlovu (2019) in their survey found that students who studied "sustainability-focused modules" as a part of their curriculum got better employment opportunities than others. Therefore, Universities and HEIs need to rethink about integrating sustainable development in business education (Winfield & Ndlovu, 2019).

Implementation of ESD, effective performance assessment and its mark on society is a global challenge. ESD encompasses diverse areas like climate change, renewable energy, sustainable production and consumption, and poverty reduction (Venkatraman, 2009). Nevertheless, some researchers have tried to resolve this global challenge. Didham and Ofei-Manu (2012) propagated a policy framework called "ESD Learning Performance Framework (ESD-LPF)" that could address these global challenges. Rudsberg and Ohman (2010) introduced a policy framework called "Pluralism" to understanding the complexities of sustainable development. Santone et al. (2014) also introduced a new paradigm called "Education for sustainability" (EfS) to educate teachers who would then share their learnings and knowledge to the students. Reynolds and Cavanagh (2009) introduced "sustainability quotients" as a part of ESD. Pauw et al. (2015) suggested that "sustainability consciousness" must be there amongst the school teachers to get the desirable ESD learning outcomes. Branden (2012) had suggested four drivers "Intrinsic motivation, Instructional improvements, Team work and Allness" to foster ESD at all levels. Lotz-Sisitka et al. (2015) suggested that schools and HEIs need to transform and innovate learning styles to foster social learning in the organization.

2.4 Implementation of the PRME Framework in Business Education

Transformational leadership and innovative thinking are what is required to achieve Sustainable Development Goals, and, hence, PRME initiative was launched to develop responsible leaders of tomorrow. Principles for Responsible Management Education (PRME) was launched at the United Nations Global Compact Leaders'

Summit in Geneva, in 2007. According to Haertle (2012), "the Principles for Responsible Management Education initiative is the first organized relationship between business and management schools and the United Nations, with the PRME Secretariat housed in the United Nations Global Compact Office". The PRME comprising six principles encouraged universities and HEIs to integrate sustainability development in their curriculum (Storey et al., 2017; PRME, 2020). The six principles of PRME are shown in Table 1. PRME became the largest "sustainability-focused business school network" in 2015 (PRME, 2016) propagating the significance of achieving SDGs for businesses, in line with ESD (Parkes, 2017). It was an initiative to reform business education and move beyond just the financial parameters (Forray & Leigh, 2012). Universities and HEIs can enhance their impact on business by sharing research on CSR and sustainability (Carruthers et al., 2017).

2.5 *ESD in India*

Literature suggests that globally it has been accepted that there is a need to integrate ESD at all levels of education across disciplines (Badjanova et al., 2014). Studies showed that that the launch of "Decade of Education for Sustainable Development (DESD)" inspired ESD moment in India (Chhokar, 2010). However, this did not have an impactful effect with respect to integrating ESD in business curricula. Literature suggested that the efforts that were taken were "mostly environmentally dominated", missing out on the social, economic, and cultural dimensions of ESD

Table 1 Six principles of PRME

Principle 1	"We will develop the capabilities of students to be future generators of sustainable value for business and society at large and to work for an inclusive and sustainable global economy"
Principle 2	"We will incorporate into our academic activities, curricula, and organizational practices the values of global social responsibility as portrayed in international initiatives such as the United Nations global compact"
Principle 3	"We will create educational frameworks, materials, processes and environments that enable effective learning experiences for responsible leadership"
Principle 4	"We will engage in conceptual and empirical research that advances our understanding about the role, dynamics, and impact of corporations in the creation of sustainable social, environmental, and economic value"
Principle 5	"We will interact with managers of business corporations to extend our knowledge of their challenges in meeting social and environmental responsibilities and to explore jointly effective approaches to meeting these challenges"
Principle 6	"We will facilitate and support dialogue and debate among educators, students, business, government, consumers, media, civil society organizations and other interested groups and stakeholders on critical issues related to global social responsibility and sustainability"

Note. This table Adapted from https://www.unprme.org/what-we-do. Copyright 2020 by PRME presents the detailed definitions of the PRME

(Iyengar & Bajaj, 2011). Mishra (2002) also opined that understanding of the concept of sustainability was mostly environmental in nature. Sustainability is not a sought-after area when it comes to inclusion in any formal curricula and hence, we observe that there were very limited universities and HEIs in India that addressed sustainability issues at the graduate and post graduate levels (Gafoor & Mumthas, 2011). This highlights the need to reform our curricula especially in the domains of business, management, and marketing to achieve this twenty-first century educational objective of SD (Isa et al., 2020).

3 Methodology

3.1 Study Design

This study used an online questionnaire to study the students' perception on sustainability development and its inclusion in the business curricula. The first part of the questionnaire consisted of the basic demographic questions with respect to their age, gender, grade, and subject area specialization. The second part of the questionnaire focuses on (i) students' *awareness* of sustainability related terms and their meanings, (ii)students' *understanding* of sustainability and the 17 SDGs, (iii) students' *attitude* towards sustainability and the 17 SDGs, (iv) students' *commitment towards* sustainability, (v) students' perception of the *college/university's role in sustainability*, (vi) students' perception about *inclusion* of sustainability issues into curricula and research, (vii)students' knowledge about *university/college support* to sustainability and the SDGs within its Curriculum of Business Education, (viii) students' *engagement* with sustainability and SDGs, (ix) students' perception about the *barriers* to further adoption of sustainability and SDGs in business curriculum, (x) students' *involvement* for campus sustainability, (xi) *practices* adopted by students for furthering the cause of sustainability, (xii) students' *information sources* about sustainability and SDGs, and (xiii) students' perception about the *way* in which sustainability should be included as a part of the business curriculum.

It consisted of 121 five-point Likert-type items/statements divided into 13 groups specified above. The questionnaire was given to the authors' colleagues for their suggestions and a pilot survey was also conducted amongst 100 respondents so that irrelevant material, ambiguity, and discrepancies were eliminated. After the review and the pilot survey, the final questionnaire was administered to the respondents. Google Forms were distributed amongst the sample after the questionnaire was developed. According to Lefever et al. (2007), "collecting data by using online survey could enhance the accuracy of the data collection because respondents must complete all the questions before they can move on to the next page. This ensures response validity". This enables elimination of missing data and hence time and error towards data cleaning gets minimized.

3.2 Sample Size and Study Period

The convenient sampling method which was of non-probabilistic type was adopted by the author for this study. A cross-sectional study design was used to obtain the primary data from college and university students in India between June 2023 and January 2024. Using snowball sampling, the author reached out to respondents across various colleges and universities in India. The questionnaire was administered to only those students who had a background in business and management related areas. The survey was returned by 304 respondents and there were no incomplete questionnaires. According to Anderson and Gerbing (1984), "for practical research applications 150 or larger sample size is required for parameter estimates with small standard errors", thus a sample size larger than 150 is acceptable. The respondents in the study belonged to 19 undergraduate colleges (bachelor's degree programme) and 39 universities (post-graduation/master's degree programme).

3.3 Method

Cronbach's α (alpha) statistic was used to study the reliability of factors/variables used to measure latent constructs and the answers. "It is commonly used as a measure of the internal consistency reliability of a psychometric instrument". Cronbach's alpha increases when the correlations between the items increase. For this reason, it is also called the "internal consistency" or "the internal consistency reliability of the test". If there is low reliability of the instrument alpha value will be close to zero and if there is high reliability of the instrument alpha value will be close to infinity. The recommended cut-off value is 0.60 (Nunnally & Bernstein, 1994). Table 2 shows the Cronbach's alpha as a measure of reliability of survey questionnaire. The table suggests strong reliability of the overall questionnaire.

Next qualitative analysis was used to arrive at the results to present an overview of the sample experience regarding the students' perception on sustainability development and its inclusion in the business curricula. Considering the sample size, the results are indicative of trends and a useful preliminary assessment rather than generalizable to larger populations.

Table 2 Cronbach's alpha as a measure of reliability of survey questionnaire

Name of the construct	Number of questions	Value of Cronbach's alpha
Students' awareness of sustainability related terms and their meanings	9	0.936
Students' understanding of sustainability and the 17 SDGs	17	0.955
Students' attitude towards sustainability and the 17 SDGs	17	0.957
Students' commitment towards sustainability	10	0.923
Students' perception of the college/university's role in sustainability	5	0.884
Students' perception about inclusion of sustainability issues into curricula and research	5	0.894
Students' knowledge about university/college support to sustainability and the SDGs within its curriculum of business Education	6	0.947
Students' engagement with sustainability and SDGs	10	0.959
Students' perception about the barriers to further adoption of sustainability and SDGs in business curriculum	8	0.932
Students' involvement for campus sustainability	5	0.927
Practices adopted by students for furthering the cause of sustainability	14	0.935
Students' information sources about sustainability and SDGs	11	0.908
Students' perception about the way in which sustainability should be included as a part of the business curriculum	4	0.810
Total	121	0.980

Source: Authors own work

4 Results and Discussions

4.1 Demographic Data

The sample of respondents is balanced in terms of gender (47.4% female, 52% male, and 0.7% other). Most respondents are undergraduates (86.5%), 11.8% were post graduates and 1.6% said they were doing technical or professional course. The most represented age group was those between 19 and 25 years (82.6%) of age category. The respondents come from varied areas: accounting and finance (48%), economics (5%), international business (5%), management (20%), and others (22%).

4.2 Students' Awareness of Sustainability Related Terms and their Meanings

Students' knowledge on sustainability and its related terms can be enhanced through formal education. Hay and Eagle (2020) found a positive co-relationship existing between a curriculum, students' awareness and sustainability education. Lindgren et al. (2006) showed that lack of inclusion of sustainability education in the curriculum at the graduate level had a negative impact on them being responsive towards sustainability. Lukman and Glavic (2007) in their study showed the direct and indirect impact of higher education on graduates and their future decisions. A study by Qureshi (2020) showed that the students' awareness about sustainability enhances when they are made to adopt sustainable living practices. Researchers have argued that students' awareness through ESD can help students to acquire "adequate understanding of the concept of sustainability" because it is "an important first step toward initiating or participating in or advocating for intentional sustainability behaviors" (Emanuel & Adams, 2011). From our study, we observe that our respondents were very familiar with climate change (72.04%), environmental protection (66.78%), energy conservation (61.18%), environmental sustainability and conservation both (60.53%), SDGs (54.93%), and climate change adaptation (46.38%). However, they were partially familiar with economic sustainability (42.43%) and social sustainability (39.47%). Figure 1 illustrates the results observed above.

4.3 Students' Understanding of Sustainability and the 17 SDGs

University through research, teaching, commitment to society and governance can help in building a fairer and a resilient society and thus help in the construction of SDGs. SDGs can be addressed through collaborative work of several disciplines and different areas of the same discipline. Dlouhá and Pospíšilová (2018) through his

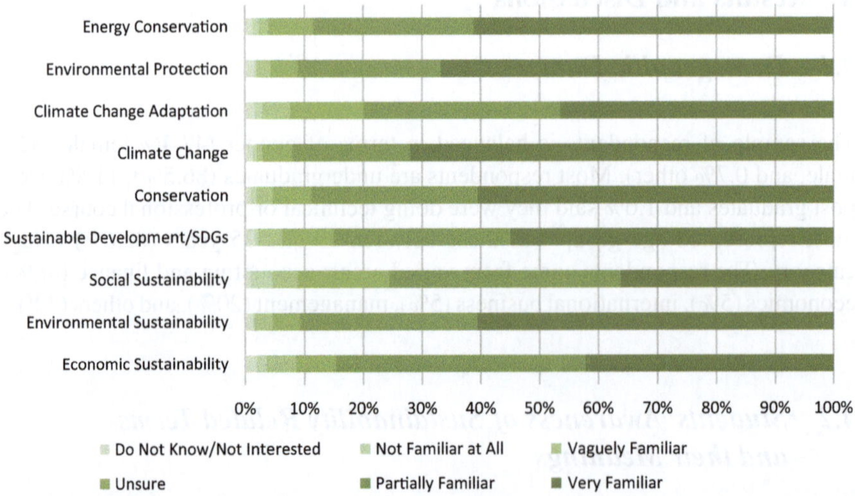

Fig. 1 Students' awareness of sustainability related terms and their meanings. Source: Authors own work

research proposed a framework to teach SDGs. Lazzarini et al. (2018) studied the level of integration of SDGs in the subjects taught by engineering professors and his research showed that the most developed themes were "climate change (SDG 13), conservation and sustainable use of ecosystems (SDG 14 and 15), and the use of water and sanitation (SDG 6). On the contrary, the least developed were the promotion of a global alliance (SDG 17), decent work and industrialization (8 and 9), and the promotion of peace and justice (SDG 16)". Successful implementation of SDGs requires progressive students who are committed to transform the world. Alvarez-Risco et al. (2021) administered a questionnaire amongst young university students to understand their knowledge of sustainability and SDGs. Ando et al. (2019) the perception of Kyoto University students about the knowledge, source, and understanding of SDGs. In 2018, the Vietnamese Centre for Sustainable Development Studies also conducted a study to understand the students' knowledge and understanding of the SDGs.

From our study, we observe that our respondents' understanding of sustainability and the 17 SDGs was high. Amongst the SDGs, the respondents' understanding about the consequences of poverty (50.3%) dealing with SDG1, consequences of malnutrition (48.4%) dealing with SDG2, and inequality being a major cause of societal problems (45.1%) dealing with SDG10 was noteworthy. Figure 2 illustrates the results observed above.

4.4 *Students' Attitude Towards Sustainability and the 17 SDGs*

There is very little research to show whether sustainability awareness has a positive influence on the attitude and behaviour of the students. According to some researchers, increasing awareness about sustainability may not bring about adequate changes

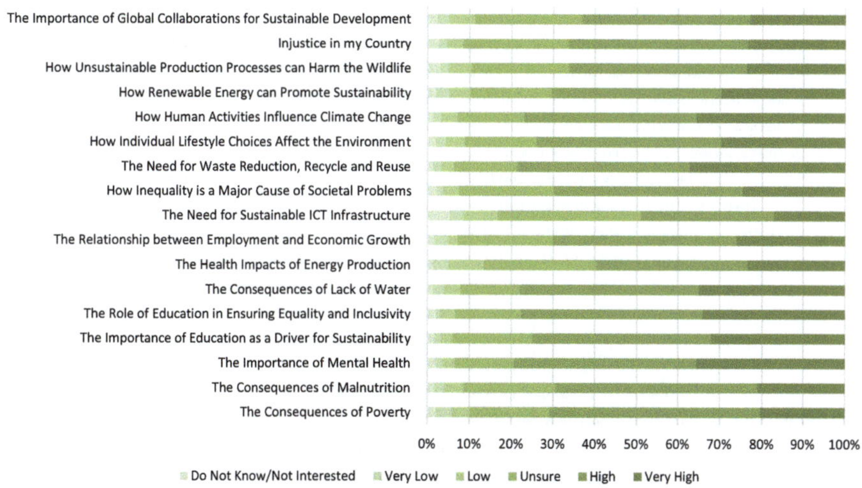

Fig. 2 Students' understanding of sustainability and the 17SDGs. Source: Authors own work

in the attitude and behaviour of students (Heeren et al., 2016). The question that arises therefore what can bring about this change? One possible factor may be engagement which enables the students to have a dialogue about the sustainability in a tangible way. According to Evans et al. (2015), the university acts like a "living laboratory". Thus, to understand sustainability, students must first become aware of new ideas and practices, engage themselves in sustainability related activities which in turn will bring about a radicle change in the attitude and behaviour of the students towards sustainability.

From our study, our respondents' attitude towards sustainability and the 17 SDGs was observed to be very positive. They could relate highly especially with SDG1 (56.3%) dealing with "feeling empathy for poor and vulnerable situations such as child labourers" followed by SDG4 (56.3%) dealing with "feeling the importance of education as a driver for sustainability", SDG14 (49%) dealing with "feeling that renewable energy can promote sustainability", and SDG 3 and SDG5 (48%) dealing with "feeling empathy for people suffering from various illness" and "feeling empathy for people who are suffering due to gender inequality and inclusivity". Figure 3 illustrates the results observed above.

4.5 *Students' Commitment Towards Sustainability*

Implementation of sustainable management in business life requires strong commitment towards sustainability especially amongst leaders who are responsible for shaping the social and economic environment. This is possible if within the higher education system, commitment and sustainability competencies are incorporated

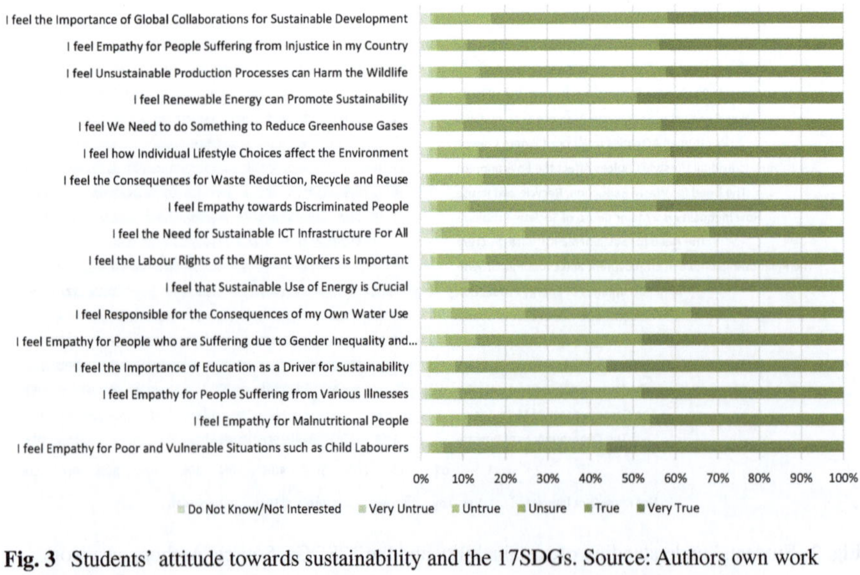

Fig. 3 Students' attitude towards sustainability and the 17SDGs. Source: Authors own work

for the students who would be the future leaders and problem solvers (Okręglicka et al., 2017). The business and management graduates need a responsible education which will help them to understand and realize how important it is to make a positive contribution towards societal well-being, economic stability, and ecological integrity—the three pillars of sustainable development (Baumann-Pauly et al., 2013). Responsible management education within the higher education which focuses on development of conceptual and managerial aptitude of students is now very pertinent (Rawal, 2013).

From our study, we observe that our respondents' belief on commitment towards sustainability is very high from their concerns and issues that they have agreed on. The respondents (57.6%) believe that economic development, societal development, and environmental protection are all necessary for sustainable development. 56.9% of our respondents have shown their concern about the fact that the overuse of natural resources is affecting the well-being of future generations. 53.3% of our respondents have agreed on the fact that there is wasteful consumption of natural resources and the destruction/pollution of the environment and that there is need to have stricter laws and regulations to protect the environment. Figure 4 illustrates the results observed above.

4.6 *Students' Perception of the College/University's Role in Sustainability*

Integrating sustainability into HEIs is not complete without the involvement of the students. Among all the university stakeholders, students are the most important. Thus, understanding their perception about sustainability and their involvement in sustainability can help the educational institutions to decide on what sustainability practices they would adopt (Emanuel & Adams, 2011).

Studies have been carried out to observe the students' perceptions and their involvement in campus sustainability. Examples of such studies could be experienced in several universities in the USA (Emanuel & Adams, 2011), Europe—such as in Germany (Barth & Timm, 2010), UK (Chaplin & Wyton, 2014)—and in Australia (Zeegers & Francis Clark, 2014), in China (Yuan & Zuo, 2013), in Malaysia (Nejati & Nejati, 2013), and in Turkey (Tuncer, 2008). Studies have shown that educational institutions have implemented campus sustainability by adopting measures such as energy efficiency and waste recycling, constructing green buildings, sustainable transport to name a few.

From our study, we observe that our respondents have strongly agreed on college/university's role in sustainability. 54.9% of our respondents have strongly agreed that everyone in HEIs should support environment sustainability. 53.9% of the respondents have strongly agreed on the fact that HEIs should make sustainability a "priority" while planning and developing a campus and day-to-day operations. They should contribute to societal well-being, tolerance, looking after the needs of the disabled, and social activities. Figure 5 illustrates the results observed above.

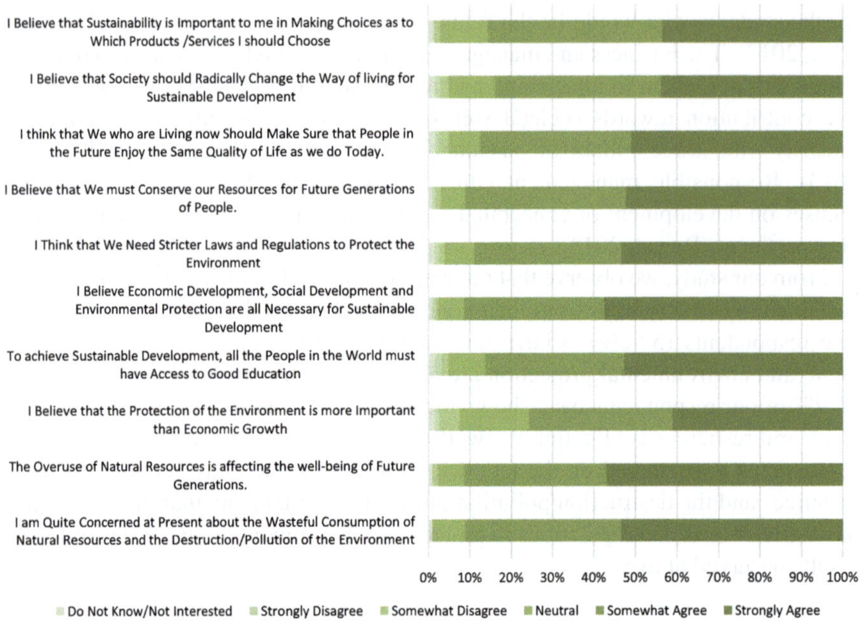

Fig. 4 Students' commitment towards sustainability. Source: Authors own work

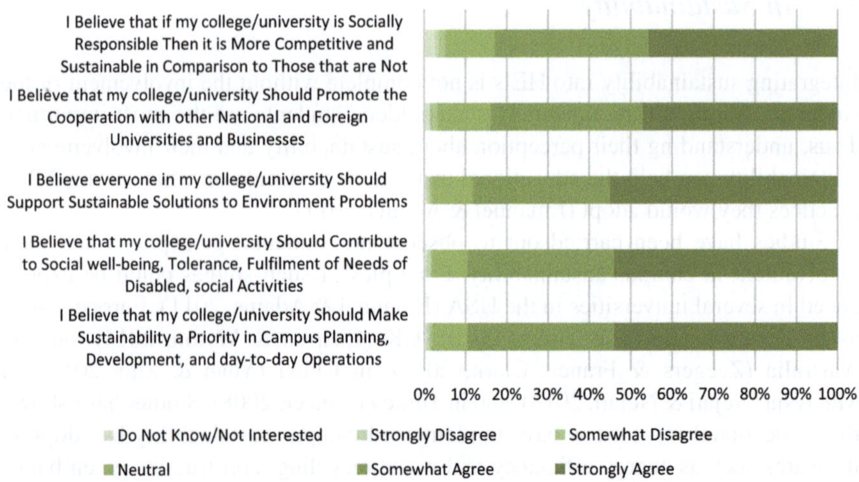

Fig. 5 Students' perception of the college/university's role in sustainability. Source: Authors own work

4.7 *Students' Perception About Inclusion of Sustainability Issues into Curricula and Research*

Inclusion of sustainability issues in curricula and research of educational institutes results in development of expertise, skills, awareness, principles, and attitudes amongst students that is required to build a society where our environment will be protected, where social justice shall thrive and people will experience economic stability (Cebrián & Junyent, 2015). It helps in creating a positive impact on the students' perception of sustainability. Lozano et al. (2013) observed that inclusion of sustainability issues in the curricula and research would help students to acquire the essential knowledge and skills to develop practical solution of the economic, environmental, and social problems. Amadei and Wallace (2009) felt that it is essential that such areas be included in the curricula and research since the students will be able to address the global sustainability issues better. Remington-Doucette et al. (2013) introduced the use of real-world problems in sustainability education where the students can solve real-world problems by working in teams. Students' conceptualization of sustainability issues and their participation in sustainability projects was studied by Rulifson and Bielefeldt (2017). These projects changed the students' perception of being responsible and resulted in better understanding of sustainability issues, safety, and ethics.

From our study, we observe that our respondents to a very large extent have agreed on the aspect of inclusion of sustainability in the curricula and research. 48.7% of the respondents agreed to the fact that educational institutions need to promote research and project related to environmental sustainability, whereas 45.7% believed educational institutions need to offer courses which address topics related to sustainability. 44.1% expressed their belief that HEIs need to incorporate issues in the curriculum that helps to limit the negative impact it has on the environment and society. Figure 6 illustrates the results observed above.

4.8 *Students' Knowledge About University/College Support to Sustainability and the SDGs Within Its Curriculum of Business Education*

Globally HEIs have been trying to incorporate sustainability concepts and knowledge in their teaching and research but if HEIs must be relevant to global sustainability and create futuristic responsible leaders, they will have to integrate their educational activities with those from private sector. Alvarez and Rogers (2006) had suggested three aspects that would form a basic framework for evaluating various approaches to teaching and learning of sustainable development. They were "(1) the definitions, history, and meanings, (2) implementation of sustainable concept, and (3) the discourse of sustainability". Besides Seatter and Ceulemans (2017) also observed in this regard that a good sustainable development curriculum needs to be

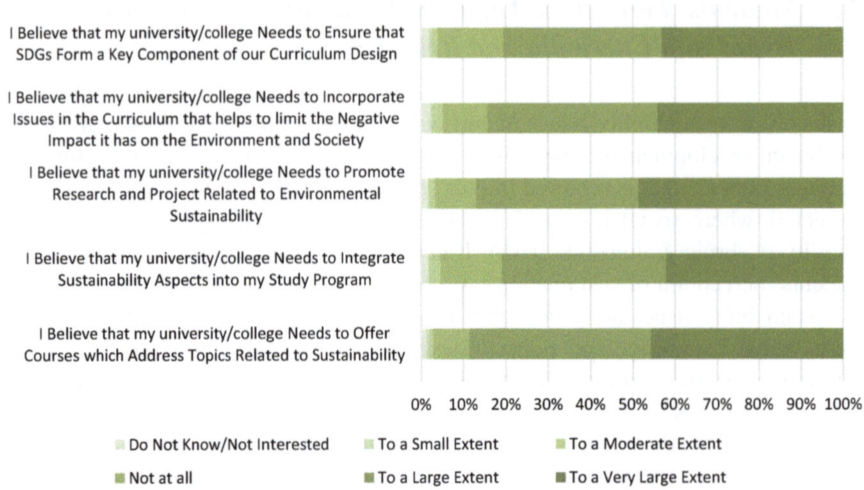

Fig. 6 Students' perception about inclusion of sustainability issues into curricula and research. Source: Authors own work

simple. According to Ramirez (2020), currently the curriculum of sustainable development across the globe comprises climate education and ESG.

Researchers have identified various ways in which sustainability issues can be incorporated into the curriculum that would be relevant and help students to develop the necessary skills. Ritchie (2013) suggested that experiential learning and engagement with the locals to understand the complex issues and problems associated with sustainability is necessary. Researchers have also identified that various approaches like "case studies, simulations and games, group projects, internships, study abroad or field trips, seminar, lectures, videos and group discussions" (Kurthakoti & Good, 2019) form a part of teaching and learning of sustainability.

From our study, we observe that the respondents have agreed to a large extent about the university/college support to sustainability and the SDGs within its curriculum of business education through internal training (40.8%), mentoring (37.8%), external training (36.2%) along with coaching, project-based learning, and collaborations. Figure 7 illustrates the results observed above.

4.9 Students' Engagement with Sustainability and SDGs

ESD of students requires "Normative competences", to understand the norms and values of "strategic thinking" or development of innovative actions; "self-awareness" to reflect on the role that each one needs to pay to make our planet a better planet to live in and, "integrated problem-solving" to find solutions to the complex problems raised by sustainability. It is difficult to teach and engage students in traditional instructional formats while dealing with an area as complex as

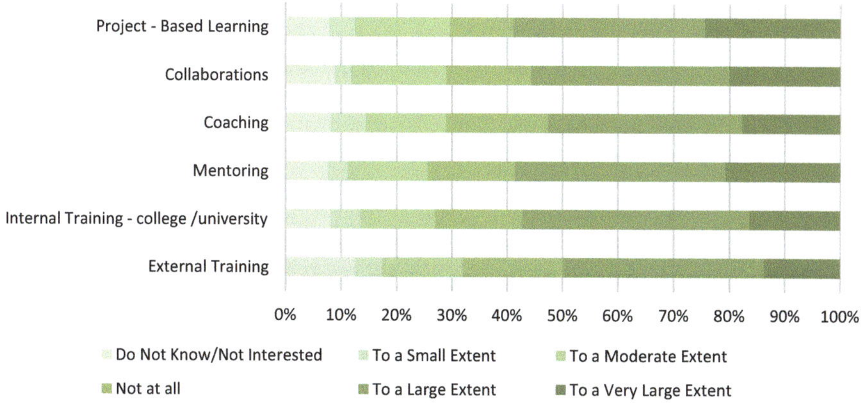

Fig. 7 Students' knowledge about university/college support to sustainability and the SDGs within its curriculum of Business Education. Source: Authors own work

sustainability unless the students are exposed to transformative learning. This creates more opportunities when there is collaboration with partners outside academia. For transformative learning, curriculum must adopt "more experiential, community-integrated, and practice-oriented approaches to teaching" (Wright et al., 2015). Researchers have identified various approaches like "case studies, simulations and games, group projects, internships, study abroad or field trips, seminar, lectures, videos and group discussions" (Kurthakoti & Good, 2019) are being adopted to engage students in this regard.

From our study, we observe that out of all the methods the three most popular ways of engaging students with sustainability and SDGs are student dissertations/theses (39.8%) followed by student project (39.5%) and Internships or placements (39.5%). Figure 8 illustrates the results observed above.

4.10 Students' Perception About the Barriers to Further Adoption of Sustainability and SDGs in Business Curriculum

HEIs if it wants to become sustainable it should be able to develop a "community oriented" towards sustainability where there will be involvement of staff, students, and extra-curricular activities (Green Office Model, 2019). However, research shows that there are barriers to further adoption of sustainability and SDGs in the HEIs and thereby the business curriculum. Aleixo et al. (2018) felt that sustainability concept because of its ambiguity is still believed to be an abstract and complex topic and hence looked upon as a barrier. Some researchers felt that lack of desire to change or teach sustainability as a theoretical concept without any practical application was a barrier by itself (Aleixo et al., 2018; Veiga et al., 2019). The

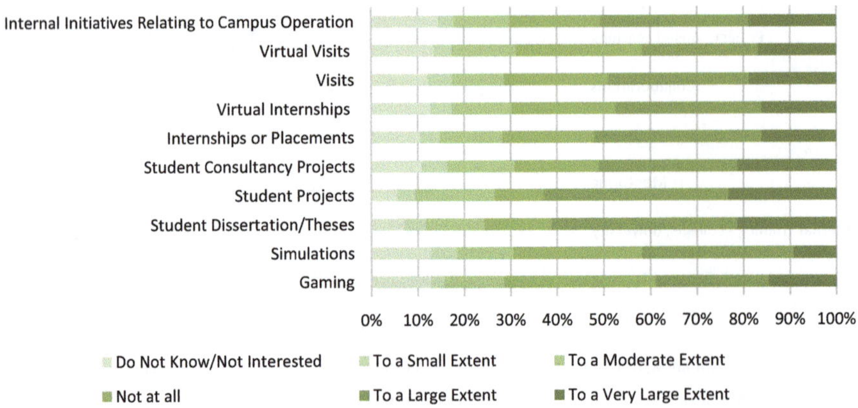

Fig. 8 Students' engagement with sustainability and SDGs. Source: Authors own work

organizational rigidity of the HEIs often prevents effective implementation of sustainability concepts within the organization. Lack of awareness of the sustainability issues amongst the staff, students, and authorities, lack of training, and specialization in this area, lack of proper working groups solely dedicated to sustainability related issues, lack of interest to change the institution behaviour/culture, lack of proper assessment and reporting methodology, lack of planning, lack of funds to support sustainability development within institutions, were some of the main barriers observed through previous research that has been undertaken (Aleixo et al., 2018; Veiga et al., 2019). From the perspective of students, the main barrier was observed to be their mentality and attitude (Novo-Corti et al., 2018). Respondents in our study to a large extent have felt that the main barriers were lack of student interest (38.8%), followed by lack of space in the curriculum (36.5%) and lack of interest amongst the faculty to teach sustainability and its related issues (33.9%). Figure 9 illustrates the results observed above.

4.11 Students' Involvement for Campus Sustainability

Globally HEIs have increased their awareness about sustainability and SDGs. Various events are being organized to generate awareness about sustainability regarding various areas like transportation, construction, energy, waste, food, water, and landscaping. Various conferences are being organized to spread the word about sustainable practices. But it is the knowledge of sustainability that results in successful implementation of sustainable practices especially with respect to campus sustainability. However, according to some researchers, knowledge and commitment to sustainable business practices have not reached at its desired level (Cosmann et al., 2006). The HEIs must talk more about sustainability and ensure sustainability practices on campus. The students, faculty, staff should all be encouraged to be a

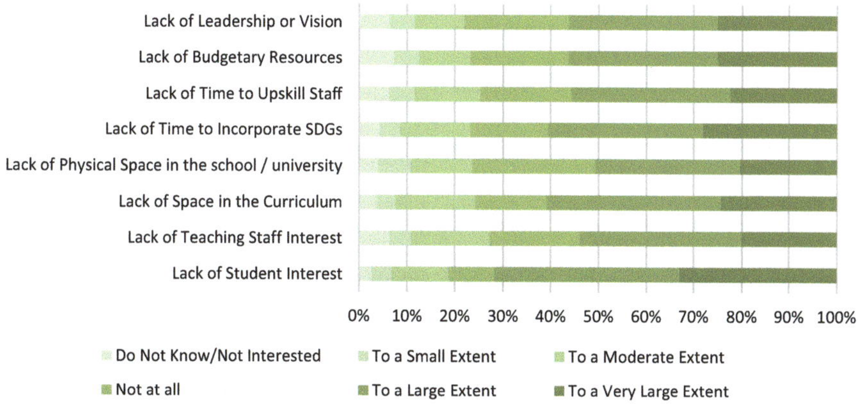

Fig. 9 Students' perception about the barriers to further adoption of sustainability and SDGs in business curriculum. Source: Authors own work

part of this because the students are tomorrows leaders. Through this the HEIs can become agents of change having far-reaching impact.

From our study, we find the respondents have strongly agreed on campus sustainability and the role they can play to improve and encourage it. 47.4% of the respondents strongly agreed to the fact that they want to create a sustainable campus. 46.7% wanted to participate and engage themselves into activities that would help in protecting the environment. 45.1% also agreed to participate in social activities organized by the educational institutions. Figure 10 illustrates the results observed above.

4.12 *Practices Adopted by Students for Furthering the Cause of Sustainability*

Undoubtedly the student community is aware of sustainability and its importance, is willing to engage in sustainable projects, hence HEIs must ensure that their awareness gets reflected in their daily activities.

Using environmentally friendly products (47%), switching off the lights when not needed (59.9%), and not wasting food (60.2%) are the three practices the respondents have strongly agreed on doing, however other activities in the form of recycling and reusing, adapting energy conservation practices, using "green transportation", reducing the current energy use, using environment friendly light bulbs, avoid buying from companies which are not environmentally conscious, avoid buying from companies who are violating human rights, taken up courses related to sustainability, purchasing organic products, buying sustainable new clothing, and helping in planting more trees are other important sustainable practices which our respondents have somewhat agreed to practice.

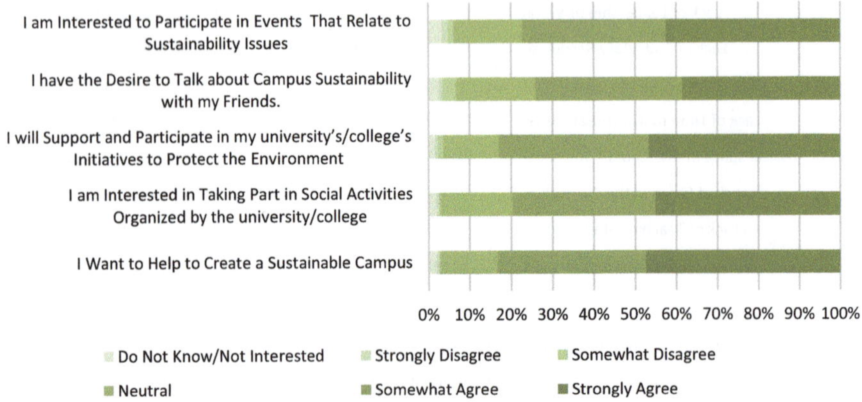

Fig. 10 Students' involvement for campus sustainability. Source: Authors own work

Therefore, a large proportion of respondents have expressed their concern for sustainability and have expressed their desire to participate in sustainable practices as observed through our earlier discussions. So, there seems to be scanty or no "knowledge gap" when it comes to sustainability practices in the campus but there does seem to be a "commitment gap". Figure 11 illustrates the results observed above.

4.13 Students' Information Sources About Sustainability and SDGs

Mass media plays a very important role in communicating information about sustainability to the public especially the student community who spends a lot of time on social media. Information disseminated through the mass media have been found to influence development of perceptions on the issues of sustainability (Boykoff & Boykoff, 2004). From the survey results, we observe that our respondents have strongly agreed that Internet (60.2%) and social media (57.9%) were the two main sources of information about sustainability and SDGs. Figure 12 illustrates the results observed above.

4.14 Students' Perception About the Way in Which Sustainability Should Be Included as a Part of the Business Curriculum

If the HEIs must achieve sustainability literacy, then it is essential for them to conduct either separate courses/programmes on sustainable development (SD) or integrate sustainability development throughout the study programme. SD can be

Fig. 11 Practices adopted by students for furthering the cause of sustainability. Source: Authors own work

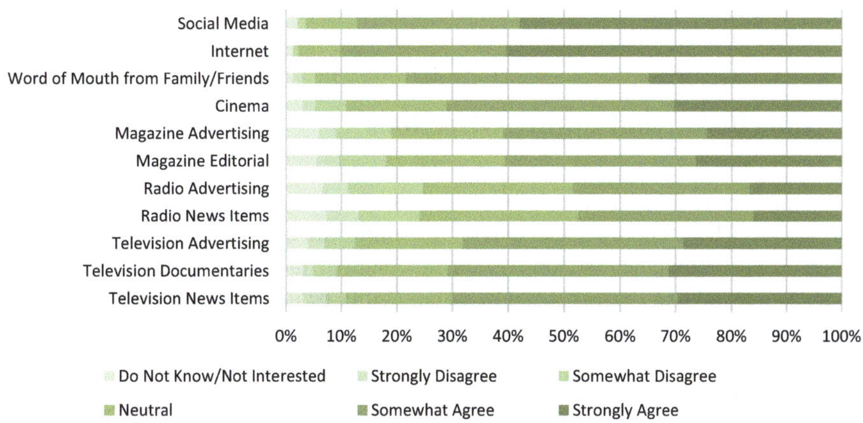

Fig. 12 Students' information sources about sustainability and SDGs. Source: Authors own work

included into selected courses such as governance and business ethics or introduce SD as a cross-disciplinary course or integrating SD into the core subjects of every programme. Teaching of SD can be either in the form of traditional teaching or by use of real-world case studies or through participatory or experiential learning or collaborative learning by inviting eminent resource persons to share their views and experiences. When our respondents were asked to give their opinion on the way in which sustainability should be included as a part of the business curriculum, 39.1% strongly agreed that it should be included as a Value-Addition Course. Figure 13 illustrates the results observed above.

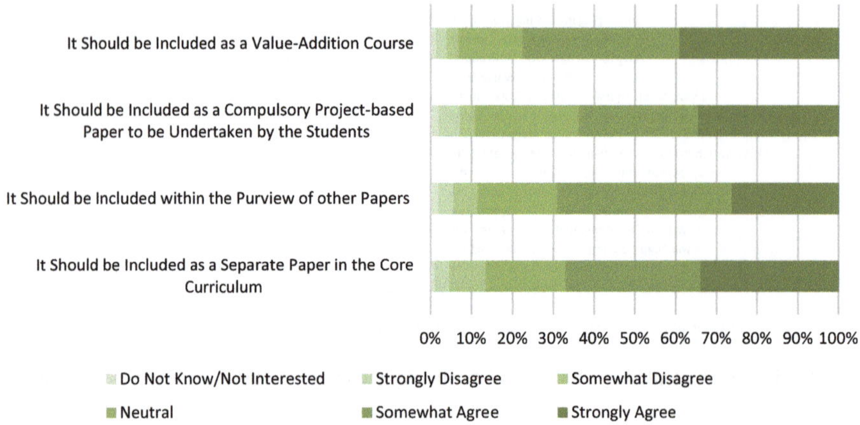

Fig. 13 Students' perception about the way in which sustainability should be included as a part of the business curriculum. Source: Authors own work

5 Conclusion and Future Directions

Sustainability is a key issue of the twenty-first century because business policies and practices have certain social and/or environmental consequences which make it essential for companies to adopt such practices in their strategies. Therefore, it becomes crucial that issues related to sustainability be included in the curriculum of business courses. The main aim of our paper was to know the students' viewpoints regarding this with reference to India. Further we wanted to have a clear understanding of the students' knowledge on various issues pertaining to sustainability.

The main challenges associated with the current research were the difficulty in obtaining field data by convincing students to participate in the survey. But despite the challenges the paper has succeeded in getting enough information to give us a holistic picture, some of the key findings have been highlighted below.

From our study, we observed that students were very familiar with sustainability related terms and their meanings particularly with climate change, environmental protection, energy conservation, environmental sustainability and conservation, SDGs, and climate change adaptation. Our respondents' understanding of sustainability and the 17 SDGs was high particularly with poverty, malnutrition, and inequality. The students' attitude towards sustainability and the 17 SDGs was observed to be very positive and they could relate themselves with critical situation's faced by the planet globally especially by the poor and vulnerable people, they could realize that education can bring in sustainability and they could relate to the environmental crisis faced by the world today.

Our respondents' belief on commitment towards sustainability was very high especially with the kind of concerns they have expressed regarding overuse of natural resources, wasteful consumption of natural resources, and the destruction/pollution of the environment. The students have strongly agreed on college/university's role in bringing about sustainability and they felt that HEIs should make this a priority. Our respondents to a very large extent have agreed on the aspect of inclusion of sustainability in the curricula and research and have expressed the desire that the HEIs should promote research and projects on various aspects of sustainability. The respondents have agreed largely about the university/college support to sustainability and the SDGs within its curriculum of business education through internal training, mentoring, external training along with coaching, project-based learning, and collaborations.

Our study has shown that out of all the methods the three most popular ways of engaging students with sustainability and SDGs were Student dissertations/theses followed by student project and Internships or placements. Research shows that there are barriers to further adoption of sustainability and SDGs in the HEIs and thereby the business curriculum. The main barriers that are preventing HEIs in India from incorporating sustainability in the business education according to the students were lack of student interest, followed by lack of space in the curriculum and lack of interest amongst the faculty to teach sustainability and its related issues. The respondents want and were very responsive towards HEIs developing "campus sustainability". A large proportion of respondents have expressed their concern for sustainability and have expressed their desire to participate in sustainable practices but when it came to sustainability practices in the campus there seemed to be a "commitment gap" as observed from our study.

Mass media plays a very important role in communicating information about sustainability especially the student community. From the survey results, we observed that our respondents had strongly agreed that Internet and social media were the two main sources which gave them information about sustainability and SDGs. Before inclusion of sustainability into business education, it is important we get to know how the students would like sustainability to be incorporated. Our study showed that students' strongly agreed that it should be included as a Value-Addition Course.

The Supreme Court of India in the year 2003 had directed all the States and educational institutions to incorporate environment as a compulsory subject at all levels. But sustainability is not only about environment it is rather a way of life. Mere theoretical teaching is not going to bring about the required change amongst the students that we are looking for, rather applicability and critical thinking are required. Making environmental studies compulsory therefore failed to bring about the desired behavioural change amongst the students and hence the Government of India introduced in 2020 the "New Educational Policy" which emphasized on the importance of "Education for Sustainable Development (ESD)" and recognized the need to integrate sustainability principles across all levels of education. NEP (2020) recognizes the importance of integrating environmental education across all levels of education, including sustainability as a part of faculty training, promotion of experiential learning, skill development and integrating technology in ESD.

There is ample scope of further research. Further studies may investigate into the sustainability behaviour and students' progression since they are the future leaders. More focus can be given on behavioural analysis of the students with respect to sustainability issues and the reasons for it.

References

Aleixo, A. M., Leal, S., & Azeiteiro, U. M. (2018). Conceptualization of sustainable higher education institutions, roles, barriers, and challenges for sustainability: An exploratory study in Portugal. *Journal of Cleaner Production, 172*, 1664–1673. https://doi.org/10.1016/j.jclepro.2016.11.010

Alvarez, A., & Rogers, J. (2006). Going out there: Learning about sustainability in place. *International Journal of Sustainability in Higher Education, 7*(2), 176–188.

Alvarez-Risco, A., Del-Aguila-Arcentales, S., Rosen, M. A., García-Ibarra, V., Maycotte-Felkel, S., & Martínez-Toro, G. M. (2021). Expectations, and interests of university students in COVID-19 times about sustainable development goals: Evidence from Colombia, Ecuador, Mexico, and Peru. *Sustainability, 13*, 3306.

Amadei, B., & Wallace, W. A. (2009). Engineering for humanitarian development. *IEEE Technology and Society Magazine, 28*, 6–15.

Anderson, J. C., & Gerbing, D. W. (1984). The effect of sampling error on convergence, improper solutions, and goodness-of-fit indices for maximum likelihood confirmatory factor analysis. *Psychometrika, 49*, 155–173. https://doi.org/10.1007/BF02294170

Ando, Y., Baars, R., & Asari, M. (2019). Questionnaire survey on consciousness and behavior of students to achieve SDGs in Kyoto University. *Journal of Environment and Safety, 10*, 21–25.

Aras, G., & Crowther, D. (2009). *The durable corporation*. Gower.

Badjanova, J., Iliško, D., & Drelinga, E. (2014). Holistic approach in reorienting teacher education towards the aim of sustainable education: The case study from the regional University in Latvia. *Procedia - Social and Behavioral Sciences, 116*, 2931–2935. https://doi.org/10.1016/j.sbspro.2014.01.682

Barth, M., & Timm, J. M. (2010). Higher education for sustainable development: Students' perspectives on an innovative approach to educational change. *Journal of Social Sciences, 7*, 13–23.

Baumann-Pauly, C., Wickert, S. L., & Scherer, A. G. (2013). Organizing corporate social responsibility in small and large firms: Size matters. *Journal of Business Ethics, 115*(4), 693–705.

Bell, D. V. (2016). Twenty-first century education: Transformative education for sustainability and responsible citizenship. *Journal of Teacher Education for Sustainability, 18*(1), 48–56. https://doi.org/10.1515/jtes-2016-0004

Boykoff, M. T., & Boykoff, J. M. (2004). Balance as bias: Global warming and the US prestige press. *Global Environmental Change, 14*, 125–136.

Branden, K. V. (2012). Sustainable education: Basic principle and strategic recommendations. *Journal School Effectiveness and School Improvement: An International Journal of Research, Policy and Practice, 23*(3), 285–304.

Brown, L. (2006). *Plan B 2.0: Rescuing a planet under stress and a civilization in trouble*. W.W. Norton & Company.

Brundtland, G. H. (1987). *Report of the World Commission on environment and development: "Our common future"*. United Nations.

Carruthers, S., Kakkar, M., Davidson, C., & Fox, L. (2017). Canadian academic uptake of the PRME. In P. M. Flynn, T. K. Tan, & Gudi'c, M. (Eds.), *Redefining success: Integrating sustainability into management education* (pp. 161–178). Routledge. https://doi.org/10.9774/gleaf.9781351268806

Cebrián, G., & Junyent, M. (2015). Competencies in education for sustainable development: Exploring the student teachers' views. *Sustainability, 7*(3), 2768–2786.

Chaplin, G., & Wyton, P. (2014). Student engagement with sustainability: Understanding the value–action gap. *International Journal of Sustainability in Higher Education, 15*, 404–417.

Chhokar, K. B. (2010). Higher education and curriculum innovation for sustainable development in India. *International Journal of Sustainability in Higher Education, 11*(2), 141–152. https://doi.org/10.1108/14676371011031865

Cosmann, N., Gray, M., Legge, T., Leous, J., Parry, N. and Valicenti, L. (2006, March 31). *Perceptions of campus environmental sustainability at Columbia University*. Retrieved from www.columbia.edu/cu/mpaenvironment/pages/CampussustainabilityReport.pdf

Didham, R. J., & Ofei-Manu, P. (2012). *Education for sustainable development country status reports: An evaluation of National Implementation during the UN decade of Education for sustainable development (2005-14) in east and Southeast Asia, Hayama, Japan*. Retrieved from http://pub.iges.or.jp/modules/envirolib/view.php?docid=4140

Dlouhá, J., & Pospíšilová, M. (2018). Education for sustainable development goals in public debate: The importance of participatory research in reflecting and supporting the consultation process in developing a vision for Czech education. *Journal of Cleaner Production, 172*, 4314–4327.

Elkington, J. (1997). *Cannibals with forks: The triple bottom line of 21st century business*. Capstone.

Emanuel, R., & Adams, J. N. (2011). College students' perceptions of campus sustainability. *International Journal of Sustainability in Higher Education, 12*, 79–92.

Erlich, P. R., & Erlich, A. H. (1991). *The population explosion*. Touchstone.

Evans, J., Jones, R., Karvonen, A., Millard, L., & Wendler, J. (2015). Living labs and co-production: University campuses as platforms for sustainability science. *Current Opinion in Environmental Sustainability, 16*, 1–6.

Farrag, D. A., & Obeidat, S. (2020). Integrating sustainability and CSR concepts in the college of business & economics (CBE) curriculum: An experiential learning approach. In N. El-Bassiouny, D. El-Bassiouny, E. K. Mohamed, & M. A. Basuony (Eds.), *Ethics, CSR and sustainability (ECSRS) Education in the Middle East and North Africa (MENA) region* (1st ed., pp. 127–144). Routledge.

Ferguson, T., & Roofe, C. G. (2020). SDG 4 in higher education: Challenges and opportunities. *International Journal of Sustainability in Higher Education, 21*, 959–975. https://doi.org/10.1108/IJSHE-12-2019-0353

Finnveden, G., Friman, E., Mogren, A., Palmer, H., Sund, P., Carstedt, G., et al. (2020). Evaluation of integration of sustainable development in higher education in Sweden. *International Journal of Sustainability in Higher Education, 21*, 685–698. https://doi.org/10.1108/IJSHE-09-2019-0287

Forray, J. M., & Leigh, J. S. (2012). A primer on the principles of responsible management education: Intellectual roots and waves of change. *Journal of Management Education, 36*(3), 295–309. https://doi.org/10.1177/105256291143303

Forum for the Future. (2004). *Sustainability literacy: Knowledge and skills for the future*. Report from Forum for the Future's Consultation Workshop.

Gafoor, K. A., & Mumthas, N. S. (2011). Greening teacher education: An analysis of teacher education curricula. In *Paper presented at UGC sponsored national seminar on Empowering teachers for sustainable development*.

Galbreath, J. (2009). Addressing sustainability: A strategy development framework. *International Journal of Sustainable Strategic Management, 1*(3), 303–319.

Go'ncz, E., Skirke, U., Kleizen, H., & Barber, M. (2007). Increasing the rate of sustainable change: A call for a redefinition of the concept and the model for its implementation. *Journal of Cleaner Production, 15*(6), 525–537.

Green Office Model. (2019). *What is a sustainable university? [online]*. Retrieved from https://www.greenofficemovement.org/sustainable-university/

Guo, F. (2017). The spirit and characteristic of the general provisions of civil law. *Law and Economics, 3*(5–16), 54.

Haertle, J. (2012). *Inspirational guide for the implementation of PRME: Placing sustainability at the heart of management education, 1-315*. GSE Research.

Haney, A. B., Pope, J., & Arden, Z. (2020). Making it personal: Developing sustainability leaders in business. *Organization and Environment, 33*(2), 155–174. https://doi.org/10.1177/1086026618806201

Hay, R., & Eagle, L. (2020). Impact of integrated sustainability content into undergraduate business Education. *International Journal of Sustainability in Higher Education, 1*, 131–143.

Heeren, A. J., Singh, A. S., Zwickle, A., Koontz, T. M., Slagle, K. M., & McCreery, A. C. (2016). Is sustainability knowledge half the battle? An examination of sustainability knowledge, attitudes, norms, and efficacy to understand sustainable behaviours. *International Journal of Sustainability in Higher Education, 17*(5), 613–632.

Hernandez-Lopez, L. E., Alamo-Vera, F. R., Ballesteros-Rodriguez, J. L., & De Saa-Perez, P. (2020). Socialization of business students in ethical issues: The role of individuals' attitude and institutional factors. *International Journal of Management Education, 18*(1), 100363. https://doi.org/10.1016/j.ijme.2020.100363

Hockerts, K. (1999). The SusTainAbility radar. *Greener Management International, 25*, 29–49.

Isa, F. M., Noor, S., Ahmdon, M. A. S., Setiawati, C. I., & Tantasuntisakul, W. (2020). Comparison of students' perception about curriculum design versus employability in Malaysia, Indonesia and Thailand. *International Journal of Management Education, 14*(4), 331–351. https://doi.org/10.1504/IJMIE.2020.10028165

Iyengar, R., & Bajaj, M. (2011). After the smoke clears: Toward education for sustainable development in Bhopal, India. *Comparative Education Review, 55*(3), 424ñ456. https://doi.org/10.1086/660680

Kanie, N., & Biermann, F. (2017). *Governing through goals; sustainable development goals as governance innovation*. MIT Press.

Kir'aly, G., & G'ering, Z. (2019). Introduction to 'futures of higher education' special issue. *Futures, 111*, 123–129. https://doi.org/10.1016/j.futures.2019.03.004

KPMG International. (2008). *A survey into the growth and sustainability issues driving consumer organizations worldwide*. KPMG/CIES Survey.

Kurthakoti, R., & Good, D. C. (2019). Evaluating outcomes of experiential learning: An overview of available approaches. In M. A. Gonzalez-Perez, V. Taras, & K. Lynden (Eds.), *The Palgrave handbook of learning and teaching international business and management*. Palgrave Macmillan.

Lal, R., Hansen, D. O., Uphoff, N., & Slack, S. A. (2002). *Food security and environmental quality in the developing world*. CRC Press.

Lazzarini, B., Pérez-Foguet, A., & Boni, A. (2018). Key characteristics of academics promoting sustainable human development within engineering studies. *Journal of Cleaner Production, 188*, 237–252.

Lefever, S., Dal, M., & Matthiasdottir, A. (2007). Online data collection in academic research: Advantages and limitations. *British Journal of Educational Technology, 38*(4), 574–582. https://doi.org/10.1111/j.1467-8535.2006.00638.x

Lindgren, A., Rodhe, H., & Huisingh, D. A. (2006). Systematic approach to incorporate sustainability into university courses and curricula. *Journal of Cleaner Production, 14*, 797–809.

Linton, J., Klassen, R., & Jayaraman, V. (2007). Sustainable supply chains: An introduction. *Journal of Operations Management, 25*(6), 1075–1082.

Lotz-Sisitka, H., Wals, A. E., Kronlid, D., & McGarry, D. (2015). Transformative, transgressive social learning: Rethinking higher education pedagogy in times of systemic global dysfunction. *Current Opinion in Environmental Sustainability, 16*, 73–80.

Lozano, R., Lozano, F. J., Mulder, K., Huisingh, D., & Waas, T. (2013). Advancing higher education for sustainable development: International insights and critical reflections. *Journal of Cleaner Production, 48*, 3–9.

Lukman, R., & Glavic, P. (2007). What are the key elements of a sustainable university? *Clean Technologies and Environmental Policy, 9*, 103–114.

Meadows, D. (1972). *Limits to growth*. Signet.

Mensah, J., & Casadevall, S. R. (Reviewing editor). (2019). Sustainable development: Meaning, history, principles, pillars, and implications for human action: Literature review. *Cogent Social Sciences, 5*, 1. https://doi.org/10.1080/23311886.2019.1653531

Mishra, B. P. (2002). *Analysis of the knowledge, attitude and perception regarding sustainable development among pre service and in service teachers at secondary level.* Doctoral dissertation, Utkal University.

National Education Policy. (2020). Retrieved from https://www.education.gov.in/sites/upload_files/mhrd/files/NEP_Final_English_0.pdf

Neil, A., & Alexandra, W. (2007). Vulnerability, poverty, and sustaining wellbeing. In *Handbook of sustainable development.* Edward Elgar.

Nejati, M., & Nejati, M. (2013). Assessment of sustainable university factors from the perspective of university students. *Journal of Cleaner Production, 48*, 101–107.

Novo-Corti, I., Badea, L., Tirca, D. M., & Aceleanu, M. I. (2018). A pilot study on education for sustainable development in the Romanian economic higher education. *International Journal of Sustainability in Higher Education, 19*(4), 817–838.

Nunnally, J. C., & Bernstein, I. H. (1994). *Psychometric theory.* McGraw Hill.

Okręglicka, M., Haviernikova, K., Mynarzova, M., & Lemańska-Majdzik, A. (2017). Entrepreneurial intention creation of students in Poland, Slovakia and Czechia. *Polish Journal of Management Studies, 15*(2), 162–172.

Parkes, C. (2017). The principles for responsible management education (PRME): The first decade–what has been achieved? The next decade–responsible management education's challenge for the sustainable development goals (SDGs). *International Journal of Management Education, 15*(2), 61–65. https://doi.org/10.1016/j.ijme.2017.05.003

Pauw, B. J., Gericke, N., Olsson, D., & Berglund, T. (2015). The effectiveness of education for sustainable development. *Sustainability, 7*(11), 15693–15717.

Peter, R., Jalal, K., & Boyd, J. (2008). *An introduction to sustainable development.* Earthscan.

Porter, M. E., & Kramer, M. R. (2011, January–February). *Creating shared value.* Harvard Business Review.

Prizzia, R. (2007). Sustainable development in an international perspective. In *Handbook of globalization and the environment.* CRC Press.

PRME. (2016). *The UN Sustainable Development goals realized through responsible management education: Strengthening PRME's network and aligning with UN priorities.* PRME. Retrieved February 20, 2024, from https://d1ngk2wj7yt6d4.cloudfront.net/public/uploads/PDFs/160517PRMEStrategicReviewFINAL.pdf

PRME. (2020). *What is PRME?* PRME. Retrieved February 20, 2024, from https://www.unprme.org/about

Qureshi, S. (2020). Learning by sustainable living to improve sustainability literacy. *International Journal of Sustainability in Higher Education, 1*, 161–178.

Ramirez, L. (2020). *The importance of climate education in a COVID-19 world.* World Economic Forum. Retrieved from https://www.weforum.org/agenda/2020/05/the-importance-of-climate-education-in-a-covid-19-world/

Rawal, D. M. (2013). Reorienting management education for sustainable prosperity. *Review of Management, 3*(1/2), 21–28.

Remington-Doucette, S. M., Connell, K. Y. H., Armstrong, C. M., & Musgrove, S. L. (2013). Assessing sustainability education in a transdisciplinary undergraduate course focused on real-world problem solving. *International Journal of Sustainability in Higher Education, 14*, 404–433.

Reynolds, P., & Cavanagh, R. (2009). Sustainable education: Principles and practices. In *Paper presented at the annual conference of the Australian Association for Research in Education, 29 November–3 December, Canberra, Australia.* Retrieved from https://www.aare.edu.au/data/publications/2009/rey091135.pdf

Ritchie, J. (2013). Sustainability and relationality within early childhood care and education settings in Aotearoa New Zealand. *International Journal of Early Childhood, 45*(3), 307–326. https://doi.org/10.1007/s13158-013-0079-0

Rudsberg, K., & Ohman, J. (2010). Pluralism in practice: Experiences from Swedish evaluation, school development and research. *Environment Education Research, 16*(1), 95–111.

Rulifson, G., & Bielefeldt, A. (2017). Motivations to leave engineering: Through a lens of social responsibility. *Engineering Studies, 9*, 222–248.
Santone, S., Saunders, S., & Seguin, C. (2014). Essential elements of sustainability in teacher education. *Journal of Sustainability Education, 6*(5), 1–15.
Savitz, A. W., & Weber, K. (2006). *The triple bottom line*. Jossey-Bass.
Schaefer, A., & Crane, A. (2005). Addressing sustainability and consumption. *Journal of Micromarketing, 25*(1), 76–92.
SDSN. (2020). *Accelerating education for the SDGs in universities: A guide for universities, colleges, and tertiary and higher education institutions*. Sustainable Development Solutions Network (SDSN).
Seatter, C. S., & Ceulemans, K. (2017). Teaching sustainability in higher education: Pedagogical styles that make a difference. *The Canadian Journal of Higher Education, 47*(2), 47–70. https://doi.org/10.47678/cjhe.v47i2.186284
Shrivastava. (1995). The role of corporations in achieving ecological sustainability. *Academy of Management Review, 20*(4), 936–960.
Sibbel, A. (2009). Pathways towards sustainability through higher education. *International Journal of Sustainability in Higher Education, 10*(1), 68–82.
Sikdar, S. K. (2003). Sustainable development and sustainability metrics. *AICHE Journal, 49*(8), 1928–1932.
Starik, M., & Rands, G. P. (1995). Weaving an integrated web: Multilevel and multisystem perspectives of ecologically sustainable organizations. *Academy of Management Review, 20*(4), 908–935.
Stoddart, H., Schneeberger, K., Dodds, F., Shaw, A., Bottero, M., Cornforth, J., & White, R. (2011). *A pocket guide to sustainable development governance*. Stakeholder Forum.
Storey, M., Killian, S., & O'Regan, P. (2017). Responsible management education: Mapping the field in the context of the SDGs. *International Journal of Management Education, 15*(2), 93–103. https://doi.org/10.1016/j.ijme.2017.02.009
Tjarve, B., & Zemīte, I. (2016). The role of cultural activities in community development. *Acta Universitatis Agriculturae et Silviculturae Mendelianae Brunensis, 64*(6), 2151–2160. https://doi.org/10.11118/actaun201664062151
Tuncer, G. (2008). University students' perception on sustainable development: A case study from Turkey. *International Research in Geographical and Environmental Education, 17*, 212–226.
UNESCO. (2017). *Education for sustainable development goals learning objectives*. Retrieved from United Nations Educational, Scientific and Cultural Organization http://unesdoc.unesco.org/images/0024/002474/247444e.pdf
UNESCO. (2018). *Issues and trends in education for sustainable development*. Retrieved from https://unesdoc.unesco.org/ark:/48223/pf0000261445
van Marrewijk, M. (2003). Concepts and definitions of CSR and corporate sustainability: Between agency and communion. *Journal of Business Ethics, 44*(2–3), 95–105.
van Marrewijk, M., & Werre, M. (2003). Multiple levels of corporate sustainability. *Journal of Business Ethics, 44*(2–3), 107–119.
Veiga, Á. L., Beuron, T., Brandli, L., Damke, L., Pereira, R., & Klein, L. (2019). Barriers to innovation and sustainability in universities: An international comparison. *International Journal of Sustainability in Higher Education, 20*(5), 805–821.
Venkatraman, B. (2009). Education for sustainable development. *Environment Science Policy Sustainable Development, 51*(2), 8–12.
Vogler, J. (2007). The international politics of sustainable development. In *Handbook of sustainable development*. Edward Elgar.
Wang, X. G. (2016). Civil law expression of environmental rights and interests—Reflections on the greening of civil code. *People Rule Law, 3*, 25–27.
Whiteman, G., & Cooper, W. H. (2000). Ecological embeddedness. *Academy of Management Journal, 43*(6), 1265–1282.

Winfield, F., & Ndlovu, T. (2019). "Future-proof your degree": Embedding sustainability and employability at Nottingham business school (NBS). *International Journal of Sustainability in Higher Education, 20*(8), 1329–1324. https://doi.org/10.1108/IJSHE-10-2018-0196

Wright, M. F., Cain, K. D., & Monsour, F. A. (2015). Beyond sustainability: A context for transformative curriculum development. *Transformative Dialogues: Teaching & Learning Journal, 8*, 1–19.

Yuan, X., & Zuo, J. (2013). A critical assessment of the higher Education for sustainable development from students' perspectives—A Chinese study. *Journal of Cleaner Production, 48*, 108–115.

Zeegers, Y., & Francis Clark, I. (2014). Students' perceptions of education for sustainable development. *International Journal of Sustainability in Higher Education, 15*, 242–253.

Sumona Ghosh has been associated with St. Xavier's College Kolkata since 2002. She was the head of the Department of Law from 2003 to 2018. Presently, she is the Joint Coordinator of the Foundation Course. After completing her postgraduation in Commerce with rare distinction, Prof. Ghosh has been conferred with the Degree of Philosophy in Business Management by the University of Calcutta on July 31, 2014. Her area of research was corporate social responsibility (CSR). The title of her doctoral dissertation was "Pattern of participation of Public and Private sector companies in Corporate Social Responsibility Activities." She has published in journals of national and international repute. Dr Ghosh has been highly acclaimed for her guest lectures on CSR in premier institutes of higher learning including the Indian Institute of Management (Calcutta) and Indian Institute of Management (Shillong). She has taken sessions in Management Development Programs conducted by premier institutes on CSR. She has presented papers on CSR at various national and international conferences. Her research interest lies in corporate social responsibility, sustainable development, integrated reporting, and philosophies of management. Dr Ghosh is also Certified Assessor for Sustainable Organizations (CASO), certification conferred upon her by UBB GmBH Germany. Dr Ghosh has been appointed as the Council Member of the Sustainable Businesses Council of the very first independent National Business Chamber for Women that has been established in India—Women's Chamber of Commerce and Industry (WICCI) (www.wicci.in).

Part IV
Sustainability in South American Companies

Bolivia

Enhancing Sustainable Supply Chains in Bolivia: Aligning CSR Practices with the SDGs

Boris Christian Herbas-Torrico, Carlos Alejandro Arandia-Tavera, Pamela Mirtha Zurita-Lara, and Pedro Alejandro Leoni-Peinado

Bolivia is undergoing significant social and economic changes, driven by strong economic growth and political transformations. As consumer expectations worldwide shift toward environmentally and socially responsible practices, the importance of managing sustainable supply chains has become evident, and Bolivia is no stranger to this trend. This chapter explores Bolivian firms' challenges and opportunities in achieving sustainable supply chains by providing valuable insights for CSR experts. In particular, it highlights the need for proactive measures to overcome the complexities posed by social and political instability that influence current supply chains. Additionally, we propose that Bolivian firms can enhance contributions to sustainable development by aligning CSR practices with the Sustainable Development Goals (SDGs). Notably, the study focuses on a mix of firms, analyzing the challenges and opportunities firms face in achieving SDGs. Our findings reveal that achieving SDGs in Bolivia is in the early stages, mainly due to high political instability and a significant informal economy. Specifically, our results indicate that Bolivian firms primarily focus on SDGs related to health, education, poverty reduction, economic growth, innovation, and environmental preservation. However, limitations exist in their ability to achieve a broader range of SDGs through their current CSR practices. By integrating these perspectives, this chapter

---

B. C. Herbas-Torrico (✉)
Tecnologico de Monterrey, Guadalajara, Mexico
e-mail: boris.herbas@tec.mx

C. A. Arandia-Tavera
Alicorp Bolivia, Santa Cruz de la Sierra, Bolivia

P. M. Zurita-Lara
YPFB Refinación S.A., Santa Cruz de la Sierra, Bolivia

P. A. Leoni-Peinado
Universidad de Cadiz, Cadiz, Spain

S. O. Idowu, S. Vertigans (eds.), *Sustainability in Global Companies*, CSR, Sustainability, Ethics & Governance, https://doi.org/10.1007/978-3-031-77971-8_15

provides CSR experts with comprehensive insights into the challenges and opportunities in achieving global sustainable supply chains in developing countries like Bolivia. It highlights the need for proactive measures to overcome the complexities posed by social and political instability. Additionally, aligning CSR practices with the SDGs can enhance contributions to sustainable development. The findings underscore the importance of continuous improvement and strategic actions for Bolivian firms to foster sustainable practices in their supply chains and contribute to broader development goals.

1 Introduction

Globalization has had a significant impact on Bolivia, both socially and economically. The country has experienced stability and growth in commodity prices in the past decade, contributing to positive changes (Kennemore & Weeks, 2011). However, Bolivia has recently encountered economic problems leading to political instability (Herbas-Torrico et al., 2021). In addition, the shift in consumer expectations toward environmentally and socially responsible practices has made the management of sustainable supply chains increasingly important (Taghikhah et al., 2019). This change affects developing countries like Bolivia, where production methods are often not environmentally responsible. Due to political instability and a large informal economy, Bolivia faces hurdles in implementing the SDGs. While Bolivian businesses prioritize SDGs linked to health, education, and poverty reduction, focusing on Corporate Social Responsibility (CSR) as a marketing tool limits their impact on other areas. This chapter explores these challenges and opportunities, emphasizing how Bolivian companies can refine their strategies to create sustainable supply chains and contribute more broadly to the SDGs.

2 Persistent Political Instability and Economic Inequalities

Bolivia's history, particularly its modern era, is marked by persistent political instability and stark economic inequality (Herbas-Torrico et al., 2021). This volatile political landscape significantly impacts economic policies and social structures. Next, we present an overview of the economic history of Bolivia and the challenges it faced in the last three decades:

1. *The Legacy of Economic Liberalism:* Bolivia embraced economic liberalism during the 1990s under President Gonzalo Sánchez de Lozada (1993–1997). Free-market policies and partial privatization aimed to stimulate growth. However, these reforms had unintended consequences. Widespread unemployment, poverty, and inequality fueled social discontent, culminating in political instability. In 2005, President Sánchez de Lozada resigned amidst growing tensions and protests.

2. *Evo Morales and the Shift Leftward:* Morales capitalized on the public's dissatisfaction as a leftist leader. His presidency ushered in significant economic reforms. Nationalizing privatized enterprises and increased state control over critical sectors starkly depart from the previous administration's free-market approach. This period has coincided with a rise in global commodity prices, leading to economic growth, reduced unemployment, and poverty alleviation through subsidies.
3. *Resurgence of Instability and the Pandemic:* Political stability proved elusive. Allegations of electoral fraud in 2019 led to Morales's resignation and President Jeanine Añez's rise. Her tenure coincided with the COVID-19 pandemic, requiring strict quarantines. These measures further strained the economy and fueled political unrest. The December 2020 elections, with Luis Arce's victory, marked a return to left-wing leadership.
4. *Shifting Left Again, Challenges Amidst Political Turbulence.* The ascendancy of Luis Arce, an economist personally chosen by Evo Morales, sparked optimism for economic revitalization and political stability. However, despite initial hopes, inadequate monetary measures, a substantial fiscal deficit, diminished energy exports (a primary revenue stream), and escalating subsidies precipitated a scarcity of dollars, fostering political instability again.

This persistent political instability significantly impacts business operations, particularly supply chain management. Uncertainty arising from political volatility requires higher safety stock inventories. Material and input shortages or distribution constraints can lead to operational inefficiencies. Furthermore, underemployment and a substantial informal economy pose additional challenges. Labor abuses, lack of transparency, and limited CSR, especially in smaller or informal enterprises, further complicate supply chain dynamics.

Therefore, based on the information presented above, this chapter delves deeper into the sustainability of Bolivian supply chains within this context. It aims to identify the challenges and opportunities for Bolivian companies to contribute to the SDGs through their CSR practices despite the ongoing political and economic uncertainties.

3 A Portrait of the Bolivian Economy

Bolivia's economic landscape is unique, characterized by a prevalence of small enterprises and limited large-scale industry (Morales, 2020). These small businesses are the nation's backbone, generating 80% of employment and 83% of total jobs (Opinión, 2017). However, this dominance comes with drawbacks. Bolivian employment falls within the micro and small enterprise sectors, often involving self-employment (entrepreneurship). Unfortunately, formalizing such businesses can be complex, hindering access to social benefits like healthcare, retirement plans, and unemployment protections. Morales (2020) highlights the cumbersome legal requirements and high taxes that make formalization a hurdle, pushing many toward the informal market. This translates to many of the workforce lacking essential social safety nets.

Furthermore, the pervasiveness of informality creates economic vulnerabilities. The Bolivian economy becomes susceptible to political instability and experiences erratic growth patterns due to the informal sector's size and the unprotected nature of its workers (Hussain, 2014). Innovation and entrepreneurship are stifled as well. The lengthy process of establishing a formal business takes at least 45 days (Del Castillo, 2020), and excessive regulations and administrative procedures discourage formalization (Strobel, 2010). This, in turn, fuels informality, hindering productivity and increasing economic costs. Consequently, Bolivia ranks among the world's least competitive economies (World Economic Forum, 2019).

4 Food and Beverage Industry Focus

Despite the challenges, Bolivia boasts a vibrant food and beverage sector. This industry stands out for its ability to add significant value to raw materials (Ministry of Productive Development and Plural Economy, 2019). It employs roughly 9% of the workforce (INE, 2020). Morales (2020) suggests a close relationship between this industry and the informal economy, which often relies on informal suppliers.

Given the predominance of small businesses and the unique characteristics of the food and beverage industry, this sector presents a valuable opportunity to research sustainable supply chain management practices in Bolivia. By focusing on CSR initiatives within this industry, we can explore how companies contribute to achieving the SDGs.

5 The Daunting Climb: Implementing SDGs in Developing Countries: The Bolivian Case

Sustainable development has emerged as a central focus for tackling pressing issues like climate change, inequality, and responsible consumption (Dasgupta et al., 2015). Governments recognize its importance and are developing methods to track progress. However, a crucial gap exists, i.e. a lack of straightforward assessment methods to determine if implemented programs are truly sustainable (Endo & Ikeda, 2022). This raises concerns that the SDGs might be more of an aspirational slogan than a practical roadmap.

A significant barrier for developing countries is financing. The UNCTAD estimates that developing nations like Bolivia require a staggering $2.5 trillion annually for SDG-related infrastructure alone (Akenroye et al., 2018). Coupled with limited economic resources, this creates immense financial pressure (United Nations Sustainable Development Report, 2022; Shayan et al., 2022). The result is a difficulty in achieving progress on the SDGs, as evidenced by Bolivia's struggles.

Similar challenges plague companies in developing countries. Economic factors often take priority, hindering their ability to demonstrate contributions to the SDGs (Bali-Swain & Yang-Wallentin, 2020). Beyond financial constraints, corruption, social conflicts, and energy access issues further complicate the picture (Aust et al., 2020; Kaygusuz, 2012; The Borgen Project, 2022). These obstacles are deeply entrenched in Bolivian society, highlighting the immense complexity for both government and businesses in pursuing the SDGs.

Bolivia is not alone. Studies reveal similar challenges in other developing nations. Lauwo et al. (2022) identify dysfunctional national governance structures hindering Tanzania's progress. In India, lack of data, fragmented coordination, and funding limitations are key hurdles (Khalid et al., 2021). Rahman (2021) points to vague government mandates in Bangladesh, leading to confusion and hampering stakeholder analysis, data availability, and accountability.

Zooming in on Bolivia, Agramont et al. (2019) reveal structural problems with rural water access. Hope (2021) highlights difficulties in tackling environmental issues, while Andersen et al. (2019) expose a lack of government action on educational problems and risks in vulnerable areas. This disconnect between operational realities and governmental efforts leaves crucial aspects of sustainable development unaddressed.

The COVID-19 pandemic has further exacerbated these issues. Liendo (2021) reports a decline in formal employment, a collapsing healthcare system, educational disruption, and market closures. This has pushed thousands back to rural areas, worsening poverty. As a result, achieving the SDGs in developing countries like Bolivia seems increasingly out of reach, overshadowed by more immediate needs.

6 Examining CSR Practices for SDG Alignment

Given these challenges, this study will explore CSR practices from an operational perspective. We will examine how CSR can be aligned with the SDGs to contribute to their fulfillment and propose practices that hold promise for the future.

7 Corporate Social Responsibility: A Balancing Act for Sustainable Development

CSR has evolved beyond simply doing good. Today, it is a strategic approach integrating social, economic, and environmental considerations into a company's operations (Montiel & Delgado-Ceballos, 2014). It is a commitment to sustainable development, extending beyond the company's walls to encompass its impact on stakeholders and the environment (Martinuzzi & Krumay, 2013; Seuring & Gold, 2013; Searcy, 2016). CSR can be implemented in stages, progressively improving a company's social, environmental, and financial performance (Keijzers, 2004). These stages typically involve:

- *Optimizing Operations:* Minimizing negative impacts through due diligence and compliance.
- *Quality and Sustainability Management:* Implementing systems to manage and control performance.
- *Business Model Integration:* Embedding environmental and social considerations into core business strategies (Keijzers, 2004; Martinuzzi & Krumay, 2013).

Consumers worldwide, including those in developing countries, increasingly expect companies to demonstrate how CSR practices contribute to sustainable development (Giannarakis & Theotokas, 2011; Jamali & Karam, 2016). However, research on CSR in developing countries remains scarce (Damoah et al., 2019).

8 The Bolivian Context: Challenges and Opportunities

Like other developing nations (Damoah et al., 2019), Bolivia has limited research on CSR practices. Herbas Torrico et al. (2018) suggest that CSR in Bolivia is often associated with philanthropy and environmental initiatives rather than core business strategies. Hence, a crucial shift is needed. Companies must align CSR practices to effectively engage stakeholders and contribute to the SDGs (ElAlfy et al., 2020).

Another challenge specific to Bolivia is its massive informal economy, accounting for over 62% of GDP (Medina & Schneider, 2018). This makes achieving complete supply chain transparency particularly difficult for robust CSR practices (Herbas-Torrico et al., 2021). The pandemic has further exacerbated this issue by expanding the informal market and hindering SDG progress (Morales, 2020).

9 Study Objective: Unveiling CSR Practices in Bolivia

This study analyzes CSR practices implemented in a sample of Bolivian industries and their connection to the SDGs. We aim to identify areas of progress and recommend additional practices that could be adapted to developing countries.

10 Methodology

10.1 Challenges and Data Acquisition

The limited availability of information posed a significant challenge for this research. Being a developing country, Bolivia has very few studies on the relationship between SDGs and CSR practices, which makes it difficult to conduct a meta-analysis. As a result, we relied on two previous studies on CSR challenges and

SDGs complementation, where companies requested privacy regarding sensitive information such as names or personal details. However, most of the companies are from the beverage and food industry, with only two from the goods manufacturing sector. Moreover, the companies that participated in these studies are mostly small and medium enterprises with 10 to 100 employees. The interviews were conducted with managers, supply chiefs, or those responsible for CSR.

10.2 Data Sources

Our research employed a two-step approach to explore the relationship between CSR practices and contributions to the SDGs within the Bolivian context.

Study 1: Mapping Sustainable Practices in Food and Beverage Supply Chains (Herbas-Torrico et al., 2021)

This study examined the present condition of sustainable supply chain management in Bolivia's food and beverage sector. Our main focus was to identify the primary challenges that companies encounter when they attempt to implement sustainable practices. To achieve this, we conducted interviews and collected data to scrutinize different elements of supply chain management, such as:

- Supplier complexity
- Supplier ethical considerations
- Environmental and social requirements
- Third-party integration
- Supply chain transparency
- Long-term supplier relationships
- Operational and social performance management
- Environmental performance management
- Supply chain strategic planning practices

By analyzing these characteristics, we aimed to understand where companies prioritize their CSR efforts within their supply chains.

Study 2: Assessing SDG Implementation and Complementary CSR Practices (Herbas-Torrico et al., 2023)

The second study delved deeper into the specific SDGs companies in Bolivia are addressing. We utilized the SDG Action Manager tool developed by B Lab Global (2022) to analyze the work of four Bolivian companies. In-depth interviews were conducted to explore how these companies integrate CSR practices into their efforts toward achieving the SDGs.

11 Data Analysis

We leveraged the information from these studies by exploring the relationship between CSR practices and SDG contributions. We analyze each SDG according to the following format:

- *SDG Number and Brief Description:* A concise explanation of the SDG based on the United Nations (2022) definition.
- *Current Practices:* A summary of CSR practices identified in the studies that relate to the specific SDG.
- *Improvement Opportunities:* Recommendations and suggestions for strengthening CSR practices to better address the SDG, considering the challenges of the informal economy.

12 Results and Discussion

This section presents a combined analysis of the findings from both studies to explore how CSR practices in Bolivia contribute to achieving the SDGs.

1. *No Poverty:* One of the SDGs is to eliminate poverty in all its forms. In Bolivia, poverty is a significant challenge due to the high incidence of informal work, which makes it difficult for companies to implement CSR practices (Benería & Floro, 2006). One of the biggest obstacles is the lack of documentation that establishes the formal establishment of supplier companies. Therefore, hiring policies for vulnerable individuals, training programs, or initiatives established by Samuel Doria Medina, who teaches sewing skills to women to provide them with access to job opportunities (Fundación Samuel Doria Medina, n.d.), should be implemented to address this SDG. Bolivia is a developing country, so finding ways to work on this SDG is essential, considering that accessing financing processes will be quite challenging.
2. *Zero Hunger:* Ensuring food security, improving nutrition, and promoting sustainable agriculture are crucial SDGs in the fight against hunger. However, Bolivian companies have not fully addressed these SDGs. Nevertheless, there are initiatives in Bolivia that aim to help achieve these goals. One example is the Bolivian Food Bank, which collects clothing and groceries through campaigns to support vulnerable populations or communities (Banco De Alimentos De Bolivia, n.d.). The critical aspect of this organization is its win–win relationship with other food-producing companies to establish strategic alliances. However, achieving these SDGs can be pretty challenging for established companies, and the only applicable practices may be implementing nutrition programs and health controls for employees. Moreover, the Bolivian government provides subsidies for food items like soy and corn and fuels like diesel, gasoline, and domestic gas for formal and informal workers. This policy helps

maintain stable economic indicators, but the sustainability of this policy is questionable due to the high fiscal deficit in the country (Los Tiempos, 2023).

3. *Good Health and Well-being:* The goal of ensuring healthy lives and promoting well-being for people of all ages, focusing on increasing life expectancy and reducing infant and maternal mortality rates, is one of the country's most actively pursued SDGs. This is mainly because all legally established companies must provide health benefits through private or government plans to care for their employees' health (Shapiro & Field, 1993). Additionally, some companies have made specific agreements to have medical buses visit their premises periodically, thus facilitating staff access to medical consultations (Nueva Economia, 2021). Some governmental policies allow the informal sector to provide basic health insurance to informal workers (Universal Health Insurance: SUS). However, its implementation is deficient because this insurance covers beds, if space exists, and medications, but not the most expensive ones (El Deber, 2021).
4. *Quality Education:* Ensuring quality, inclusive, and equitable education is essential to promote learning at all ages. This is a common practice in legally established companies, as they strive to train their employees on topics relevant to their businesses, such as good manufacturing practices or updates in emergency management. Moreover, many companies have scholarship programs encouraging their workers and immediate family members to pursue secondary, technical, or university degrees. However, the government must still do more work to improve education in rural areas, as the quality of education is still deficient in these regions, with many students unable to read, write, or perform simple mathematical operations (Perlman Robinson et al., 2016). Additionally, there are missed opportunities for students to apply and compete for scholarships in various universities that support the formation of new professionals (Heyneman, 1983). Moreover, none of the universities in Bolivia rank among the top 50 in Latin America, making their workforce among the least educated and productive in the region. (Times Higher Education, 2023).
5. *Gender Equality:* The SDG of achieving gender equality and empowering women and girls is still not being applied in the CSR practices of many companies. This is one of the most significant gaps. Many companies have significant gender gaps in hiring, as women are often offered benefits such as maternity leave or shorter working hours when they become pregnant (Zahidi & Ibarra, 2010). This makes it easier for companies to hire men, who can change roles, schedules, and availability more easily. There are no government training programs for women to learn skills for survival (Stromquist, 1995). However, entrepreneur Samuel Doria Medina established the CITES (Centers for Technological Innovation) program, which offers specialized training services and technical assistance (CITE, n.d.). The goal of these CITES, financed by the entrepreneur's foundation, is to provide training in sewing or carpentry to individuals with low or scarce skills to increase their chances of success in the job market. Despite being forgotten mainly within CSR, the CITES program is a

significant effort to address the gender gap in employment and empower women. According to Sumando Voces (2024), 34% of Bolivian adolescents believe it is acceptable for a husband to hit his wife. This alarming statistic highlights the long road that Bolivian society has ahead of it to achieve gender equality.

6. *Clean Water and Sanitation:* It is crucial to ensure water availability for sustainable management and to make it available for consumption, hygiene, and sanitation. However, the companies participating in the studies have not directly addressed this SDG. Even though they have mentioned treating their effluents before disposing of them in rivers, they have not presented any programs or practices related to reducing or optimizing their water consumption. Additionally, there are rural areas in the country that do not have access to resources for sanitation and hygiene. To add insult to injury, global warming is worsening access to water, causing draughts and emptying dams (Mongabay, 2022). NGOs like Start Americas Together have a program called "Water Is Gold," which aims to provide regular access to water for consumption, hygiene, and cleaning purposes to areas without access to it (Start Americas Together, 2023). However, effective government policies are needed to address this problem.
7. *Affordable and Clean Energy:* The goal is to provide accessible, reliable, and sustainable energy to all households. Companies can adopt CSR practices to support this goal. For example, they can seek transportation equipment that uses alternative energy sources like electric forklifts. However, there is still much room for improvement in achieving this goal. For instance, a beverage bottler implemented a photovoltaic system to supply energy to its plant in Cochabamba, which could contribute to achieving this goal. Even though some progress has been made, there are still issues with access to energy, especially in rural areas that do not have electricity (Andersen et al., 2023). Companies in nearby areas can replicate the initiative of the bottler and consider supplying households in the community. This could help achieve the goal of providing sustainable energy for all.
8. *Decent Work and Economic Growth:* One of the SDGs is to promote sustainable and inclusive economic growth, productive employment, and decent work. However, this goal has recently been heavily impacted in Bolivia due to political instability and the pandemic. As a result, factories and legally established companies have been forced to close down, leading to an increase in the informal economy (Benton & Benton, 1990). Although companies apply the conditions of the general labor law in terms of CSR practices, there is no progress beyond that. Improving compliance with this SDG requires policies and conditions that make it attractive for investors and entrepreneurs to establish legally recognized businesses. In addition, the significant increase in smuggling activity is impeding the growth of the legal economy. Smuggling accounts for 7.96% of the Bolivian Gross Domestic Product (GDP), affecting 600,000 formal jobs (EJV, 2023). The government should focus on developing policies that favor the formal economy and halt the expansion of the smuggling industry.

9. *Industry, Innovation, and Infrastructure:* The SDG we refer to is building resilient infrastructure, promoting inclusive and sustainable industrialization, and fostering innovation. Although no specific CSR practices promote this goal, many companies have mentioned their research and development programs to design new products. Companies could improve their research and development practices to significantly impact responsible consumption or reduce hazardous agents that may endanger workers. According to the 2023 Global Innovation Index (WIPO, 2023), Bolivia ranks in the third quartile, occupying 97th among 132 countries surveyed. Therefore, the Bolivian government and companies must improve the regulatory environment by strengthening the rule of law. Additionally, they should work on enhancing the business environment and avoid supporting radical political doctrines that can harm private companies. Moreover, companies and the government should create innovation linkages between each other and universities. Furthermore, the government should implement policies that boost foreign direct investment and promote online companies.
10. *Reduced Inequalities:* This is an SDG to reduce inequalities within and between countries. In our sample, local companies have rarely discussed how to reduce gaps or implement CSR practices to influence SDGs. However, multinational companies have commented that to balance resource equality, they manage similar operational budgets based on their production size. This means that the budget for a business unit in Argentina with an annual volume of several products will be the same for another business unit in Bolivia with the same number of products (Bruno, 1988). Furthermore, to better address this SDG, more practices and public policies are needed to optimize resource generation.
11. *Sustainable Cities and Communities:* The goal is to create safer, more resilient, and sustainable cities and communities by increasing access to public spaces, improving public transportation, and reducing marginal neighborhoods. Unfortunately, the companies involved in the studies have not mentioned any relevant practices. However, there is an opportunity to work on this aspect to make cities more modern and secure. One of the significant issues is the lack of access to energy, water, and quality education in rural areas. To address this, companies can propose alternative energy plants or generators in public spaces. There is progress in constructing green projects, such as buildings with regulated water consumption and space for green areas (Tian et al., 2023), but more work is needed.
12. *Responsible Production and Consumption:* The SDG of ensuring sustainable consumption and production is crucial to mitigate the adverse impact of climate change on the environment. Some companies have started addressing this through their CSR initiatives by seeking cleaner and renewable energy sources for their processes. This approach results in more responsible production. However, responsible consumption also needs to be improved, as production processes often use excessive water or gas that cannot be recovered in the long run (Lorek & Spangenberg, 2014). Therefore, we believe that public policies

should be implemented to address the excessive consumption of non-renewable resources and their impact on stakeholders.

13. *Climate Action*: The issue of combating climate change and its impacts, such as global warming, is a significant concern worldwide. In Bolivia, while CSR is mainly perceived as environmental activities, the focus is primarily on recycling rather than addressing climate change. The CSR practices of companies in Bolivia are oriented toward waste classification, reuse of waste, and waste recovery whenever possible. However, other initiatives are still pending to effectively address this, such as reducing greenhouse gas emissions in the atmosphere. Although these points are regulated by public policies with set limits to adhere to, there is no established frequency of control to ensure permissible limits are respected. Other aspects, such as soil erosion and contamination, should also be considered as they can impact this SDG.
14. *Life Below Water*: As a landlocked country, Bolivia does not have access to the sea, making it challenging to fulfill the SDG of conserving and sustainably using oceans and their biodiversity. However, this does not mean Bolivia does not face challenges maintaining life in aquatic ecosystems such as lakes or rivers. Unfortunately, companies and industries have not yet implemented practices to protect the lives of these ecosystems. To address this issue, it is suggested that they must consider aspects such as responsible water use and regulation, constant evaluation to ensure compliance with allowed parameters for river discharges and preventing water pollution from bodies with improper waste or debris. Moreover, the Bolivian government's political agenda has given free will to mining unions illegally mining gold. A study found that 180 illegal rafts have left 259 kilograms of mercury in the air and water. This affects people's health and the ecosystem (Mongabay, 2022). Therefore, it is crucial to enforce the rule of law on all companies without any political bias. Additionally, we require private initiatives to tackle this issue. For instance, a group of volunteers launched a project to clean up the banks of the Rocha River in Cochabamba to diminish pollution levels in the water body. This is just one of the many ways companies can contribute toward achieving this SDG.
15. *Life on Land:* This SDG is related to protecting, restoring, and promoting terrestrial ecosystems and biodiversity. Companies in our sample have not established actions or practices to work on this SDG. Furthermore, according to what has been mentioned, their efforts are focused on applying optimal routes or avoiding passages through forests or outdoor areas with terrestrial ecosystems. However, due to constant political blockades, it is observed that more routes are being opened up, affecting these ecosystems. Therefore, it is necessary to propose public policies to ensure that these ecosystems are not further affected.
16. *Peace, Justice, and Strong Institutions:* To achieve the goal of reducing conflicts, insecurity, and weak institutions, companies should follow ethical codes, implement internal audit controls, and adhere to social and environmental laws. This will help to maintain peace amongst institutions and strengthen ties.

However, there is still room for improvement in terms of SDG compliance. Companies can achieve this by including ethical conduct codes for their suppliers, which will help to assess any ethical risks that could lead to potential scandals.

17. *Partnership for the Goals:* This SDG aims to achieve strategic solid partnerships to work with SDGs collaboratively. Companies have stated that they lack formal agreements with the government or other companies to achieve the SDGs. However, they have contacts to establish service exchange relationships that are not currently tied to achieving the SDGs. This presents an opportunity for establishing formal relationships that can positively impact the SDGs, leading to greater compliance and replicating best practices.

This analysis reveals progress beyond identifying areas for action on SDGs 3 (Good Health and Well-being) and 4 (Quality Education). Companies are improving water and energy conservation, though these efforts remain in their early stages. Significant advancements require substantial investment to guide and accelerate impactful initiatives. Moreover, Bolivia's income levels are challenged by the need to secure the necessary funding. However, this also presents an opportunity to attract foreign investors who can co-finance these crucial changes. It is important to note that most documented CSR practices involve legally establishing companies. Hence, the impact of SDGs within the informal sector, a significant part of the Bolivian economy, remains unclear.

Additionally, simplifying business creation procedures through public policies could incentivize formalization and expand the reach of CSR practices. This would allow for a more comprehensive assessment of Bolivia's overall progress toward the SDGs. In conclusion, achieving significant progress on the SDGs requires a multifaceted approach:

- *Maintain a Stable and Predictable Legal Environment:* It emphasizes the importance of clear legal frameworks for attracting investors and encouraging formalization.
- *Strengthen Corporate Governance:* This promotes company transparency and accountability, which aligns with responsible business practices.
- *Promote Stakeholder Engagement:* This encourages companies to consider the needs and interests of all stakeholders, not just shareholders, which is a crucial aspect of CSR.
- *Develop SDG-aligned Metrics and Reporting:* This focuses on tracking and communicating a company's impact on the SDGs, fostering transparency and accountability.

By implementing these strategies, Bolivia can make substantial improvements toward achieving all the SDGs.

13 Conclusions

This analysis highlights the challenges and opportunities for developing countries, particularly Bolivia, to achieve the SDGs through CSR practices.

Investment and Collaboration:

- *Capital Injection:* Developing countries and legally established companies within them often lack the resources to implement CSR practices fully. Attracting foreign investors with a focus on sustainability can provide the necessary funding.
- *Informal Sector Integration:* There is a significant information gap regarding the informal sector's impact on SDGs. Establishing communication channels between formal and informal companies can facilitate sharing best practices and expand the reach of CSR initiatives.

Policy and Sustainable Practices:

- *Public Policy Development:* Government policies promoting responsible resource use (water, energy), cleaner production, and alternative energy generation are crucial for large-scale progress.

Addressing Limitations:

- *Transparency and Data Sharing:* Limited company openness regarding CSR practices hinders comprehensive analysis. Encouraging transparency and sharing success stories can improve data availability.
- *Research Expansion:* Further academic research on CSR in developing countries is needed to create a more robust knowledge base.

Future Research Opportunities:

- *Government Data Analysis:* Investigating data on companies utilizing the Ministry of Plural Economy and Productivity's CSR guide can provide valuable insights.
- *Strategic Partnerships:* Exploring collaborations with organizations like B Lab, which offer sustainability measurement and improvement tools, can benefit companies and researchers alike.

A multi-pronged approach is necessary for developing countries like Bolivia to achieve the SDGs. This includes increased investment, fostering collaboration across sectors, implementing supportive policies, and promoting transparency in CSR practices. Developing countries, such as Bolivia, can unlock a future of opportunity and prosperity for all by investing in the SDGs.

References

Agramont, A., Craps, M., Balderrama, M., & Huysmans, M. (2019). Transdisciplinary learning communities to involve vulnerable social groups in solving complex water-related problems in Bolivia. *Water, 11*(2), 385. https://doi.org/10.3390/w11020385

Akenroye, T. O., Nygård, H. M., & Eyo, A. (2018). Towards implementation of sustainable development goals (SDG) in developing nations: A useful funding framework. *International Area Studies Review, 21*(1), 3–8. https://doi.org/10.1177/2233865917743357

Andersen, L., Medinaceli, A., Maldonado, C., & Hernani-Limarino, W. (2019). *A country at risk of being left behind: Bolivia's quest for quality education*. Retrieved from http://southernvoice.org/wp-content/uploads/2019/07/190710_Summary-Bolivia_Final.pdf

Andersen, L. E., Branisa, B., & Guzmán, G. (2023). *Midiendo la pobreza y la desigualdad energética a nivel municipal en Bolivia*. Sdsn Bolivia. Retrieved from https://sdsnbolivia.org/midiendo-la-pobreza-y-la-desigualdad-energetica-a-nivel-municipal-en-bolivia/

Aust, V., Morais, A. I., & Pinto, I. (2020). How does foreign direct investment contribute to sustainable development goals? Evidence from African countries. *Journal of Cleaner Production, 245*. https://doi.org/10.1016/j.jclepro.2019.118823

B Lab Global. (2022). *SDG action Manager: Helping all businesses take action for the sustainable development goals*. Retrieved from https://www.bcorporation.net/en-us/programs-and-tools/sdg-action-manager

Bali-Swain, R., & Yang-Wallentin, F. (2020). Achieving sustainable development goals: Predicaments and strategies. *International Journal of Sustainable Development and World Ecology, 27*(2), 96–106. https://doi.org/10.1080/13504509.2019.1692316

Banco De Alimentos De Bolivia, B. (n.d.). *Banco de alimentos de Bolivia*. Retrieved from https://bab.org.bo/

Benería, L., & Floro, M. S. (2006). Labour market informalization, gender and social protection: Reflections on poor urban households in Bolivia and Ecuador. In *Gender and social policy in a global context: Uncovering the gendered structure of 'the social'* (pp. 193–216). Palgrave Macmillan UK.

Benton, L. A., & Benton, L. (1990). *Invisible factories: The informal economy and industrial development in Spain*. SUNY Press.

Bruno, M. (Ed.). (1988). *Inflation stabilization: The experience of Israel, Argentina, Brazil, Bolivia, and Mexico*. MIT Press.

CITE. (n.d.). *Que es un CITE?* Retrieved from http://www.cite.com.bo/pages/cite.php

Damoah, K. O., et al. (2019). The state of corporate social responsibility research in Ghana: A synthesis of literature. *Business Strategy & Development*.

Dasgupta, P., Duraiappah, A., Managi, S., Barbier, E., Collins, R., Fraumeni, B., Gundimeda, H., Liu, G., & Mumford, K. J. (2015). How to measure sustainable progress. *Science, 350*(6262), 748–748. https://doi.org/10.1126/science.350.6262.748

Del Castillo, J. I. (2020). Doing business in Bolivia: A case study in the Andean regulatory framework (01). In *Development research working paper series, 01/2020*. Institute for Advanced Development Studies. Retrieved from https://ideas.repec.org/p/adv/wpaper/202001.html

EJV. (2023). *Contrabando genera un riesgo para 38.000 industrias y genera una evasión tributaria de $us 930 millones, según CNI*. Retrieved from https://eju.tv/2023/07/contrabando-genera-un-riesgo-para-38-000-industrias-y-genera-una-evasion-tributaria-de-us-930-millones-segun-cni/

El Deber. (2021). *La salud en Bolivia, con un seguro universal que nunca es gratuito*. Retrieved from https://eldeber.com.bo/pais/la-salud-en-bolivia-con-un-seguro-universal-que-nunca-es-gratuito_261360

ElAlfy, A., Palaschuk, N., El-Bassiouny, D., Wilson, J., & Weber, O. (2020). Scoping the evolution of corporate social responsibility (CSR) research in the sustainable development goals (SDGs) era. *Sustainability, 12*(14), 5544. https://doi.org/10.3390/su12145544

Endo, K., & Ikeda, S. (2022). How can developing countries achieve sustainable development: Implications from the inclusive wealth index of ASEAN countries. *International Journal of*

Sustainable Development and World Ecology, 29(1), 50–59. https://doi.org/10.1080/13504509.2021.1910591

Giannarakis, G., & Theotokas, I. (2011). The effect of financial crisis on corporate social responsibility performance. *International Journal of Marketing Studies, 3*(1), 2–10.

Herbas Torrico, B., Frank, B., & Arandia Tavera, C. (2018). Corporate social responsibility in Bolivia: Meanings and consequences. *International Journal of Corporate Social Responsibility, 3*, 1–13.

Herbas-Torrico, B. C., Frank, B., & Arandia-Tavera, C. A. (2021). Corporate social responsibility in Bolivia: Context, policy, and reality. In *Current global practices of corporate social responsibility: In the era of sustainable development goals* (pp. 597–620). Elsevier.

Herbas-Torrico, B., Aranda-Tavera, C., & Leoni-Peinado, P. (2023). Sustainable development goals in Bolivia: assumptions and realities. In S. Idowu & L. Zu (Eds.), The Elgar companion to corporate social responsibility and the sustainable development goals (1st ed., pp. 332–354). Edward Elgar Publishing. https://doi.org/9781803927367.00028

Heyneman, S. (1983). Improving the quality of education in developing countries. *Finance and Development, 20*(1), 18.

Hope, J. (2021). The anti-politics of sustainable development: Environmental critique from assemblage thinking in Bolivia. *Transactions of the Institute of British Geographers, 46*(1), 208–222. https://doi.org/10.1111/tran.12409

Hussain, Z. (2014). *Can political stability hurt economic growth?* World Bank. Retrieved from https://blogs.worldbank.org/en/endpovertyinsouthasia/can-political-stability-hurt-economic-growth

INE. (2020). *Boletín estadístico 2020*. Retrieved from https://www.ine.gob.bo/index.php/comunicacion/boletines-estadisticos/#399-boletinestadistico-20-21

Jamali, D., & Karam, C. (2016). Consumers worldwide, including those in developing countries, increasingly expect companies to demonstrate how CSR practices contribute to sustainable development. *Business Strategy & Development*.

Kaygusuz, K. (2012). Energy for sustainable development: A case of developing countries. *Renewable and Sustainable Energy Reviews, 16*(2), 1116–1126. https://doi.org/10.1016/j.rser.2011.11.013

Keijzers, G. (2004). *Business, government and sustainable development*. Routledge.. https://doi.org/10.4324/9780203449394

Kennemore, A., & Weeks, G. (2011). Twenty-first century socialism? The elusive search for a post-neoliberal development model in Bolivia and Ecuador. *Bulletin of Latin American Research, 30*(3), 267–281.

Khalid, A. M., Sharma, S., & Dubey, A. K. (2021). Concerns of developing countries and the sustainable development goals: Case for India. *International Journal of Sustainable Development and World Ecology, 28*(4), 303–315. https://doi.org/10.1080/13504509.2020.1795744

Lauwo, S. G., Azure, J. D.-C., & Hopper, T. (2022). Accountability and governance in implementing the sustainable development goals in a developing country context: Evidence from Tanzania. *Accounting, Auditing & Accountability Journal, 35*(6), 1431–1461. https://doi.org/10.1108/AAAJ-10-2019-4220

Liendo, R. (2021). Desafío boliviano: El cumplimiento de los objetivos de desarrollo sostenible desde el sistema agroalimentario campesino indígena. *Revista Latinoamericana de Desarrollo Económico, 2021*, 13–34.

Lorek, S., & Spangenberg, J. H. (2014). Sustainable consumption within a sustainable economy–beyond green growth and green economies. *Journal of Cleaner Production, 63*, 33–44.

Los Tiempos. (2023, June 16). *Bolivia mantendrá política de subvención a alimentos y combustibles*. Los Tiempos. Retrieved from https://www.lostiempos.com/actualidad/pais/20230616/bolivia-mantendra-politica-subvencion-alimentos-combustibles-energia

Martinuzzi, A., & Krumay, B. (2013). The good, the bad, and the successful—How corporate social responsibility leads to competitive advantage and organizational transformation. *Journal of Change Management, 13*(4), 424–443. https://doi.org/10.1080/14697017.2013.851953

Medina, L., & Schneider, F. (2018). *Shadow economies around the world: What did we learn over the last 20 years? IMF working paper no. 18/17*. International Monetary Fund. https://ssrn.com/abstract=3124402

Ministerio de Desarrollo Productivo y Economía Plural. (2019). *Informe estadístico industrial de Bolivia*. La Paz, Bolivia. Retrieved from https://siip.produccion.gob.bo/noticias/files/BI_050820209ce83_bolivia.pdf

Mongabay. (2022). *Estudio revela que balsas mineras liberan mercurio en el río Madre de Dios en Bolivia*. Retrieved from https://es.mongabay.com/2022/10/estudio-revela-que-balsas-mineras-liberan-mercurio-en-el-rio-madre-de-dios-en-bolivia

Montiel, I., & Delgado-Ceballos, J. (2014). Defining and measuring corporate sustainability: Are we there yet? *Organization and Environment, 27*(2), 113–139. https://doi.org/10.1177/1086026614526413

Morales, R. (2020). The resilience of informal workers to COVID19 and to the difficulties of trade. *Sociology International Journal, 4*(4), 92–95. Retrieved from https://www.medcrave.org/index.php/SIJ/article/view/18878

Nueva Economia. (2021). *Empresas responsables durante la pandemia*. Retrieved from https://nuevaeconomia.com.bo/nota.php?url=EMPRESAS-RESPONSABLES-DURANTE-LA-PANDEMIA-15-04-2021

Opinión. (2017, February 19). *Ocho de cada 10 empleos, la mayor parte precarios, se crean en las mypes*. Informe Especial. Retrieved from https://www.opinion.com.bo/articulo/informe-especial/cada-10-empleos-mayor-parteprecarios-crean-mypes/20170219235300674422.html

Perlman Robinson, J., Winthrop, R., & McGivney, E. (2016). *Millions learning: Scaling up quality education in developing countries*. Available at SSRN 3956210.

Rahman, M. M. (2021). Achieving sustainable development goals of agenda 2030 in Bangladesh: The crossroad of the governance and performance. *Public Administration and Policy, 24*(2), 195–211. https://doi.org/10.1108/PAP-12-2020-0056

Searcy, C. (2016). Measuring enterprise sustainability. *Business Strategy and the Environment, 25*(2), 120–133. https://doi.org/10.1002/bse.1861

Seuring, S., & Gold, S. (2013). Sustainability management beyond corporate boundaries: From stakeholders to performance. *Journal of Cleaner Production, 56*, 1–6. https://doi.org/10.1016/j.jclepro.2012.11.033

Shapiro, H. T., & Field, M. J. (Eds.). (1993). *Employment and health benefits: A connection at risk*. National Academies Press.

Shayan, N., Mohabbati-Kalejahi, N., Alavi, S., & Zahed, M. A. (2022). Sustainable development goals (SDGs) as a framework for corporate social responsibility (CSR). *Sustainability, 14*(3), 1222. https://doi.org/10.3390/su14031222

Start Americas Together. (2023). *El agua es oro*. Retrieved from https://startamericastogether.org/el-agua-es-oro/

Strobel, T. L. (2010). Entry and exit regulations: The World Bank's doing business indicators. *CESifo DICE - Report Journal for Institutional Comparisons, 8*(1), 42–53. Retrieved from https://www.ifo.de/en/node/29072

Stromquist, N. P. (1995). Women's literacy and empowerment in Latin America. In *Education and social change in Latin America* (pp. 47–61). Sage.

Sumando Voces. (2024). *Con 34%, Bolivia tiene el porcentaje más alto de adolescentes que consideran justificado que el marido*. Retrieved from https://sumandovoces.com.bo/con-34-bolivia-tiene-el-porcentaje-mas-alto-de-adolescentes-que-consideran-justificado-que-el-marido-golpee-a-su-esposa/

Taghikhah, F., Voinov, A., & Shukla, N. (2019). Extending the supply chain to address sustainability. *Journal of Cleaner Production, 229*, 652–666.

The Borgen Project. (2022). *SDGs in developing countries*. Retrieved from https://borgenproject.org/sdgs-in-developing-countries/

Tian, Y., Cui, H., Feng, Y., & Zhao, Z. (2023). A geometric registration method for the 3D building models generated from oblique photogrammetry and building information modeling. In

Proceedings of the Second International Conference on Geographic Information and Remote Sensing Technology (GIRST 2023). https://doi.org/10.1117/12.3007522
Times Higher Education. (2023). *Latin America University rankings 2023*. Retrieved from https://www.timeshighereducation.com/student/best-universities/best-universities-latin-america
UN. (2022). *The sustainable development goals report 2021*. Retrieved from https://unstats.un.org/sdgs/report/2021/The-Sustainable-Development-Goals-Report-2021.pdf
WIPO. (2023). *Global innovation index 2023: Sixteen edition*. Retrieved from https://www.wipo.int/edocs/pubdocs/en/wipo-pub-2000-2023-en-main-report-global-innovation-index-2023-16th-edition.pdf
World Economic Forum. (2019). *The global competitiveness report 2019*. Switzerland. Retrieved from https://es.weforum.org/reports/global-competitivenessreport-2019
Zahidi, S., & Ibarra, H. (2010). *The corporate gender gap report 2010*. World Economic Forum.

Dr Boris Christian Herbas-Torrico is an industrial engineer with an international background and a passion for research and education. He received his PhD in Industrial Engineering from the Tokyo Institute of Technology in Japan and has an extensive background in the field. His research interests cover various topics, such as market research, corporate social responsibility, production, and applied statistics. He has published more than 60 papers in international refereed journals, conferences, and books and has received numerous awards for his contributions to the field, including the Outstanding Professor Award 2021 from IEOM Society International (USA) or the 2013 Nikkei QC Literature Prize awarded by the Japanese Union of Scientists and Engineers. Currently, Dr Herbas-Torrico is a full professor at the prestigious Tecnologico de Monterrey in the Department of Industrial Engineering at the Guadalajara Campus.

Carlos Alejandro Arandia-Tavera earned a bachelor's degree in industrial engineering from the Bolivian Catholic University and a master's degree from INCAE Business School. He currently works in production management at Alicorp Bolivia.

Pamela Mirtha Zurita-Lara earned a bachelor's degree in industrial engineering from Bolivian Catholic University. She works in human resources at YPFB Refinación S.A. in Bolivia.

Pedro Alejandro Leoni-Peinado has a bachelor's degree in Industrial Engineering from Bolivian Catholic University. He also holds a master's degree from OBS Business School and is currently pursuing another master's degree in Industrial Engineering at the Universidad de Cadiz in Spain.

Index

S. O. Idowu, S. Vertigans (eds.), *Sustainability in Global Companies*, CSR, Sustainability, Ethics & Governance, https://doi.org/10.1007/978-3-031-77971-8

MIX
Papier aus verantwortungsvollen Quellen
Paper from responsible sources
FSC® C105338

If you have any concerns about our products,
you can contact us on
ProductSafety@springernature.com

In case Publisher is established outside the EU,
the EU authorized representative is:
Springer Nature Customer Service Center GmbH
Europaplatz 3, 69115 Heidelberg, Germany

Printed by Libri Plureos GmbH
in Hamburg, Germany